GEOTECHNICAL SPECIAL PUBLICATION NO. 81

SOIL IMPROVEMENT FOR BIG DIGS

PROCEEDINGS OF SESSIONS OF GEO-CONGRESS 98

SPONSORED BY
The Geo-Institute of the American Society of Civil Engineers

October 18–21, 1998
Boston, Massachusetts

EDITED BY
Ali Maher
David S. Yang

WITHDRAWN

1801 ALEXANDER BELL DRIVE
RESTON, VIRGINIA 20191–4400

Abstract: This proceedings, *Soil Improvement for Big Digs*, contains papers presented at sessions sponsored by the Geo-Institute of ASCE in conjunction with the ASCE Annual Convention held in Boston, Massachusetts, October 18-21, 1998. These papers mainly discuss the use of deep soil mixing, ground freezing, and minipiles and soil anchors for in-situ soil improvement.

Library of Congress Cataloging-in-Publication Data

Soil improvement for big digs: proceedings of sessions sponsored by the Geo-Institute of the American Society of Civil Engineers in conjunction with the ASCE National Convention in Boston, Massachusetts, October 18-21, 1998 / edited by Ali Maher and David S. Yang.
      p.   cm.   —(Geotechnical special publication; no. 81)
Includes bibliographical references and index.
ISBN 0-7844-0388-0
1. Soil stabilization–Congresses. 2. Piling (Civil engineering)–Congresses. 3. Guy anchors–Congresses. 4. Frozen ground–Congresses. 5. Soil consolidation–Congresses. I. Maher, Ali. II. Yang, David S. III. American Society of Civil Engineers. Geo-Institute. IV. ASCE National Convention (1998: Boston, Mass.) V. Series.
TA710.A1S5227   1998   98-39144
624.1'5–dc21   CIP

Any statements expressed in these materials are those of the individual authors and do not necessarily represent the views of ASCE, which takes no responsibility for any statement made herein. No reference made in this publication to any specific method, product, process or service constitutes or implies an endorsement, recommendation, or warranty thereof by ASCE. The materials are for general information only and do not represent a standard of ASCE, nor are they intended as a reference in purchase specifications, contracts, regulations, statutes, or any other legal document.
ASCE makes no representation or warranty of any kind, whether express or implied, concerning the accuracy, completeness, suitability, or utility of any information, apparatus, product, or process discussed in this publication, and assumes no liability therefore. This information should not be used without first securing competent advice with respect to its suitability for any general or specific application. Anyone utilizing this information assumes all liability arising from such use, including but not limited to infringement of any patent or patents.
Photocopies. Authorization to photocopy material for internal or personal use under circumstances not falling within the fair use provisions of the Copyright Act is granted by ASCE to libraries and other users registered with the Copyright Clearance Center (CCC) Transactional Reporting Service, provided that the base fee of $8.00 per chapter plus $.50 per page is paid directly to CCC, 222 Rosewood Drive, Danvers, MA 01923. The identification for ASCE Books is 0-7844-0388-0/98/ $8.00 + $.50 per page. Requests for special permission or bulk copying should be addressed to Permissions & Copyright Dept., ASCE.

Copyright © 1998 by the American Society of Civil Engineers,
All Rights Reserved.
Library of Congress Catalog Card No: 98-39144   ISBN 0-7844-0388-0
Manufactured in the United States of America.

Geotechnical Special Publications

1. Terzaghi Lectures
2. Geotechnical Aspects of Stiff and Hard Clays
3. Landslide Dams: Processes, Risk, and Mitigation
4. Tiebacks for Bulkheads
5. Settlement of Shallow Foundation on Cohesionless Soils: Design and Performance
6. Use of In Situ Tests in Geotechnical Engineering
7. Timber Bulkheads
8. Foundations for Transmission Line Towers
9. Foundations & Excavations in Decomposed Rock of the Piedmont Province
10. Engineering Aspects of Soil Erosion, Dispersive Clays and Loess
11. Dynamic Response of Pile Foundations–Experiment, Analysis and Observation
12. Soil Improvement: A Ten Year Update
13. Geotechnical Practice for Solid Waste Disposal '87
14. Geotechnical Aspects of Karst Terrains
15. Measured Performance Shallow Foundations
16. Special Topics in Foundations
17. Soil Properties Evaluation from Centrifugal Models
18. Geosynthetics for Soil Improvement
19. Mine Induced Subsidence: Effects on Engineered Structures
20. Earthquake Engineering & Soil Dynamics II
21. Hydraulic Fill Structures
22. Foundation Engineering
23. Predicted and Observed Axial Behavior of Piles
24. Resilient Moduli of Soils: Laboratory Conditions
25. Design and Performance of Earth Retaining Structures
26. Waste Containment Systems: Construction, Regulation, and Performance
27. Geotechnical Engineering Congress
28. Detection of and Construction at the Soil/Rock Interface
29. Recent Advances in Instrumentation, Data Acquisition and Testing in Soil Dynamics
30. Grouting, Soil Improvement and Geosynthetics
31. Stability and Performance of Slopes and Embankments II
32. Embankment of Dams–James L. Sherard Contributions
33. Excavation and Support for the Urban Infrastructure
34. Piles Under Dynamic Loads
35. Geotechnical Practice in Dam Rehabilitation
36. Fly Ash for Soil Improvement
37. Advances in Site Characterization: Data Acquisition, Data Management and Data Interpretation
38. Design and Performance of Deep Foundations: Piles and Piers in Soil and Soft Rock
39. Unsaturated Soils
40. Vertical and Horizontal Deformations of Foundations and Embankments
41. Predicted and Measured Behavior of Five Spread Footings on Sand
42. Serviceability of Earth Retaining Structures
43. Fracture Mechanics Applied to Geotechnical Engineering

44	*Ground Failures Under Seismic Conditions*
45	*In Situ Deep Soil Improvement*
46	*Geoenvironment 2000*
47	*Geo-Environmental Issues Facing the Americas*
48	*Soil Suction Applications in Geotechnical Engineering*
49	*Soil Improvement for Earthquake Hazard Mitigation*
50	*Foundation Upgrading and Repair for Infrastructure Improvement*
51	*Performance of Deep Foundations Under Seismic Loading*
52	*Landslides Under Static and Dynamic Conditions–Analysis, Monitoring, and Mitigation*
53	*Landfill Closures–Environmental Protection and Land Recovery*
54	*Earthquake Design and Performance of Solid Waste Landfills*
55	*Earthquake-Induced Movements and Seismic Remediation of Existing Foundations and Abutments*
56	*Static and Dynamic Properties of Gravelly Soils*
57	*Verification of Geotechnical Grouting*
58	*Uncertainty in the Geologic Environment*
59	*Engineered Contaminated Soils and Interaction of Soil Geomembranes*
60	*Analysis and Design of Retaining Structures Against Earthquakes*
61	*Measuring and Modeling Time Dependent Soil Behavior*
62	*Case Histories of Geophysics Applied to Civil Engineering and Public Policy*
63	*Design with Residual Materials: Geotechnical and Construction Considerations*
64	*Observation and Modeling in Numerical Analysis and Model Tests in Dynamic Soil-Structure Interaction Problems*
65	*Dredging and Management of Dredged Material*
66	*Grouting: Compaction, Remediation and Testing*
67	*Spatial Analysis in Soil Dynamics and Earthquake Engineering*
68	*Unsaturated Soil Engineering Practice*
69	*Ground Improvement, Ground Reinforcement, Ground Treatment: Developments 1987-1997*
70	*Seismic Analysis and Design for Soil-Pile-Structure Interactions*
71	*In Situ Remediation of the Geoenvironment*
72	*Degradation of Natural Building Stone*
73	*Innovative Design and Construction for Foundations and Substructures Subject to Freezing and Frost*
74	*Guidelines of Engineering Practice for Braced and Tied-Back Excavations*
75	*Geotechnical Earthquake Engineering and Soil Dynamics III*
76	*Geosynthetics in Foundation Reinforcement and Erosion Control Systems*
77	*Stability of Natural Slopes in the Coastal Plain*
78	*Filtration and Drainage in Geotechnical/Geoenvironmental Engineering*
79	*Recycled Materials in Geotechnical Applications*
80	*Grouts and Grouting: A Potpourri of Projects*
81	*Soil Improvement for Big Digs*

PREFACE

It is more challenging to modernize a historical city than to build a new one at a new location. However, the latter one is seldom an acceptable alternative due to the cultural, political, and economical heritage inherent in an old city. Three major challenges encountered in modernizing a historical city are: 1) Vicinity to existing structures, 2) Soil and site conditions, and 3) Maintaining the city in function during construction. To overcome these challenges, numerous innovative geosystems technologies have been developed to maintain the integrity of the existing buildings and structures or to modify the soft ground to bear the loads, which were never expected in the past. One example from overseas is the dynamic innovation and development of the deep mixing technologies. They make the deep excavation near existing facilities safe and economical and make the construction of mega structures on soft ground or under sea water possible. Another example in the United States is the "big dig"- Boston, MA Central Artery/Tunnel project. The difficult construction conditions turn Boston into the showroom of innovation subsurface construction technologies.

This special technical publication constitutes the proceedings of two sessions at the ASCE National Convention held in Boston, Massachusetts in October 1998 and also includes additional papers related to the soil improvement and deep excavation. The two technical sessions entitled "Soil Improvement for Big Digs I: Deep Soil Mixing" and "Soil Improvement for Big Digs II: Ground Freezing and Minipiles" were sponsored by the Soil Improvement and Geosyntetics Committee of ASCE. The objective of these sessions is to share information gained from recent experiences involving soil improvement and deep excavation with practicing engineers, researchers, and specialty contractors.

Each paper included in this volume has received at least two positive peer reviews and authors were given the opportunity to modify their papers based on reviewers suggestions prior to final submittal of the papers. Reviews of papers for publication were conducted by members of the Soil Improvement and Geosyntetics committee and by other ASCE members with expertise in the subject areas. All papers published are eligible for discussion in the *Journal of Geotechnical Engineering* and are also eligible for ASCE awards.

Ali Maher and David S. Yang
(Editors)

ACKNOWLEDGEMENTS

The editors of this publication express appreciation to all authors and reviewers who made this publication possible. The editors would like to thank the following individuals for providing the technical reviews of the papers presented in this report: Lee Abramson, Eric Bahner, Don Bruce, John F. Donohoe, Tuncer Edil, Dave Elton, Nenad Gucunski, Robert Holtz, James Hussin, Zia Islam, Ali Maher, George Munfakh, Kyle Rollins, Vernon Schaefer, Kevin Sharp, Nassef Soliman, Chris Swan, Joe Wang, and David S. Yang. Furthermore, the editors would like to thank Laura Picone and Patrick J. Szary who provided assistance in organizing the papers and preparing the layout of this report in a timely and efficient manner.

Cover photo courtesy of SWING Corporation.

Contents

A. Deep Soil Mixing

Deep Mixing Method: A Global Perspective 1
Donald A. Bruce, Mary Ellen C. Bruce, and Albert F. DiMillio

Design of a Soil Mixed Composite Gravity Wall 27
Peter J. Nicholson, James K. Mitchell, Eric W. Bahner, and Yoshihara Moriwaki

Design and Construction of a Deep Soil Mix Retaining Wall for the Lake Parkway Freeway Extension 41
Eric W. Bahner and Aiman M. Naguib

Soil–Cement Pile/Column: A System of Deep Mixing 59
Osamu Taki and Roy A. Bell

Experiences with GeoJet Piles in Sandy Clays 72
William M. Isenhower, Jose A. Arrellaga, Shin-Tower Wang, and J.O. Johnson

The Application of Deep Mixing Pile Walls for Retaining Structures in Excavations 84
Yong Shao, Chunming Zhang, and Emir Jose Macari

Dry Jet Mixing for Stabilization of Very Soft Soils and Organic Soils 96
David S. Yang, Jack N. Yagihashi, and Steve S. Yoshizawa

SWING Method for Deep Mixing 111
David S. Yang, Jack N. Yagihashi, and Steve S. Yoshizawa

Pre-Construction Aspects of Deep Soil–Cement Mixing for CA/T Project 122
Prabir K. Das, Justice J.G. Maswoswe, and Edward Y.P. Yin

Strength Gain of Organic Ground with Cement-Type Binders 135
Melanie B. Hampton and Tuncer B. Edil

B. Ground Freezing

The Freezing of Soil Masses as an Aid to Engineering Construction 149
John F. Donohoe, Derek Maishman, and Paul C. Schmall

Freeze–Thaw Effects on Boston Blue Clay 161
Christopher Swan and Christopher Greene

C. Minipiles and Soil Anchors

A Case Study of Timber Pile In Situ Soil Reinforcement 177
Ching L. Kuo, Wing Heung, Francisco J. Tejidor, and John Roberts

Augered Minipiles: A Cost-Effective Foundation for Light Structures 190
Bashar S. Qubain and Jianchao Li

An Underpinning Solution to a Persistent Foundation Settlement Problem 202
Eric J. Seksinsky and Bashar S. Qubain

Soil Nail Walls in Residual Soils..214
Mike Khalil, Mark Rhodes, Jim Daly, and Jeanine Ferris

Monitoring the Performance of a Soil Nailed Wall 226
Nassef N. Soliman and Kwang Ro

Seismic Behavior of Micropile Systems..239
Aomar Benslimane, Ilan Juran, Sherif Hanna, and Serguey Drabkin

Recent Developments in Soil Nailing: Design and Practice..259
Sherif Hanna, Ilan Juran, Ofer Levy, and Aomar Benslimane

D. Other Methods

Chemical Stabilization of Kaolinite by Electrochemical Injection 285
Senda Ozkan, Robert J. Gale, and Roger K. Seals

Surcharge of Phosphatic Waste Clay with Strip Drains 298
Wing Heung, Ching L. Kuo, and John Roberts

Seismic Retrofit of the Fourth Street & Riverside Viaducts with Micropiles 313
Gary E. Taylor, Francis B. Gularte, and Greg G. Gualarte

Design & Construction of the Runway 13–31 Overrun Area at LaGuardia Airport.....326
Raymond E. Sandiford, Arnold Aronowitz, and Stephen Law

Subject Index 339
Author Index 341

Deep Mixing Method: A Global Perspective

Donald A. Bruce, M.ASCE[1]
Mary Ellen C. Bruce, M.ASCE[2]
Albert F. DiMillio, M.ASCE[3]

Abstract

Various types of contemporary Deep Mixing Method (DMM) techniques have been used in the United States since 1986. Such techniques owe their origins to Japanese and Scandinavian developments, which began almost three decades ago. The growing demands of urban infrastructure development and rehabilitation have created a very active and rapidly expanding market demand in the United States especially since the early 1990s, and there is a clear need for a new and fundamental review of the surprisingly large number of DMM techniques that are being used domestically, or are available in other parts of the world. Following a summary tracing the historical development of DMM, and a generic classification of applications, the paper provides a review of each of the many different proprietary methods, which the authors have identified during preparation of an international survey funded by the Federal Highway Administration (FHWA). Data are also provided on commercial aspects of the various DMM techniques worldwide.

Introduction

The Deep Mixing Method (DMM) encompasses a group of technologies that provide in situ soil treatment. Materials of various types, but usually of cementitious nature, are introduced and blended into the soil through hollow, rotated shafts equipped with

---

[1]Principal, ECO Geosystems, Inc., P.O. Box 237, Venetia (Pittsburgh), PA, 15367
Phone: (724) 942-0570, email: eco111@aol.com
[2]President, geotechnica, s.a., Inc., Venetia (Pittsburgh), PA.
[3]Geotechnical Research Manager, U.S. Department of Transportation, Federal Highway Administration, McLean, VA.

cutting tools, and mixing paddles or augers that extend for various distances above the tip. The materials may be injected in either slurry (wet) or dry form. The treated soil or fill mass that results generally has a higher strength, lower compressibility and (usually) lower permeability than the virgin soil, although the exact properties obtained will reflect both the characteristics of the native soil, and the construction techniques and variables that are selected.

Although the original concept appears to have been developed in the United States in 1954, current practice reflects the intense efforts of researchers, backed by strong federal resources and demand, in both Japan and Scandinavia since 1967. During the last decade, however, domestic challenges to the specialty ground engineering community in the arenas of urban infrastructure development, seismic mitigation and environmental remediation, have led to a rapid growth in the use of such techniques in the United States also.

Recent international conferences, such as in Tokyo in 1996 and Logan, UT in 1997 have highlighted that there exist a surprisingly large number of different DMM techniques, each one typically proprietary to one, or a group of, specialty contractors. It is also clear that each technique has its own particular advantages and limitations, technically, logistically, and environmentally.

Given the rapid growth in the usage of DMM in the United States, it is therefore timely to present a global overview of these various different techniques so that they can be better understood by the engineering community and so more appropriately used. This paper first provides a summary of the historical evolution of DMM so that the reader can appreciate the lineage of the respective methods. It continues by briefly summarizing the main groups of applications, and comparing DMM with competitive methods in each category. The main focus of the paper, however, is the tabulation, within a new generic classification framework, of the different techniques, their characteristics, equipment, performance and so on. The paper concludes with a presentation of some commercial data relating to both domestic and international usage. Space restrictions prevent a more detailed review of treated soil properties, or a discussion of the advantages and disadvantages of DMM relative to alternative or competitive technologies. The interested reader is referred to the FHWA study (1998).

Historical Evolution

Table 1 provides a chronology of the major events in the ongoing development of the DMM techniques. It refers to a large number of these techniques by name, bearing in mind that the details of these techniques themselves are provided in Appendices 1 and 2. Table 1 highlights the commitment and energy of engineers in Scandinavia and Japan for over 30 years, initially pursuing similar paths, but soon following different directions in response to particular national demands. Also apparent is the accelerating rate of progress in other regions, principally the United States and Western Europe, over the last 10 years.

Year	Event
1954	Intrusion Prepakt Co. (U.S.) develop the Mixed in Place (MIP) Piling Technique (single auger), and see sporadic use in the U.S., although widespread use continues in Japan till early 1970s.
1967	Port and Harbor Research Institute (PHRI, Ministry of Transportation, Japan) begin laboratory tests, using granular lime for treating soft marine soils (DLM). Research continued by Okumura, Terashi et al. through early 1970s to a) investigate lime-marine clay reaction and b) develop appropriate mixing equipment (U.C.S. of 0.1 to 1 MPa achieved) Early equipment used on first marine trial near Hareda Airport (10m below water surface).
1967	Laboratory and field researches begin on Swedish Lime Column method for treating soft clays under embankments using unslaked lime (Kjeld Paus, Linden - Alimak AB, in cooperation with Swedish Geotechnical Institute (SGI), EurocAB and ByggproduktionAB).
1974	PHRI report that the Deep Lime Mixing method (DLM) has commenced full scale application in Japan. First applications in reclaimed soft clay at Chiba. (DLM continues to be popular until late 1970s when CDM and DJM supersede it.)
1975	Following researches from 1973 to 1974, PHRI develop the forerunner of the Cement Deep Mixing method (CDM) using cement grout and employ it for the first time in large scale projects in soft marine soils offshore. (Original variants include DCM, CMC (still in use from 1974), then DCCM, DECOM, Demic, etc.)
1976	Public Works Research Institute (PWRI) (Ministry of Construction, Japan) begins researches on the Dry Jet Mixing method (DJM) using dry powdered cement (or less commonly, quick-lime); "first practical stage" completed in late 1980. Representatives of PHRI also participate.
1976	SMW method used commercially for first time in Japan.
1977	First "practical use" of CDM in Japan (marine and land uses).
1980	First commercial use in Japan of DJM (land use only).
1981	Prof. Mitchell presents General Report at ICSMFE in Stockholm on lime-cement columns.
1985	SGI (Sweden) publishes 10 year progress review. (Åhnberg and Holm).
1986	SMW Seiko Inc. commence operations in U.S. under license from Seiko Kogyo.
1987 - 1988	SMW method used in massive ground treatment and improvement program for seismic retrofit at Jackson Lake Dam, WY.

Table 1. Highlights of Historical Development of DMM (continues).

Year	Event
1988 - 1989	Development by Geocon, Inc. in U.S. of DSM (Deep Soil Mixing) and SSM (Shallow Soil Mixing) techniques.
1989	Start of exponential growth in use of Lime Cement Columns in Sweden and Finland.
1992 - 1994	SMW method used for massive earth retention and ground treatment project at Logan Airport, Boston.
1992 - 1993	First SCC installation in U.S. (Richmond, CA).
1993	CDM and DJM Research Institutes publish Design and Construction Manuals (in Japanese).
1994	First commercial application of original Geojet system in the U.S. (Texas) following several years of development by Brown & Root.
1995	Swedish government sets up new "Swedish Deep Stabilization Research Center" at SGI (1995-2000: $8 -10 M): "Svensk Djupstabilisering". Consortium includes owners, Government, contractors, universities, consultants, research organizations, co-ordinated by Holm of SGI.
1995	Finnish government sets up new research consortium for the ongoing Road Structures Research Programme ("TPPT") - till 2001.
1995	Swedish Geotechnical Society publishes new design guide for lime and lime-cement columns (P. Carlsten).
1996	SGI (Sweden) publish 21-year experience review.
1996	First commercial use of lime columns in the U.S. (by Stabilator in Queens, NY).
1996 - 1997	Hayward Baker install 1.2 to 1.8m diameter columns for foundations, earth retention and ground improvement.
1997	SMW method used for huge soil treatment project at Fort Point Channel, Boston, MA (Largest DMM project to date in North America), and other adjacent projects. Input at design stage to U.S. consultants by Dr. Terashi (Japan).
1997	First commercial use in U.S. of modified Geojet system (by Condon Johnson Associates at San Francisco Airport, CA).
1997	Major lime-cement column application (I-15, Salt Lake City) proposed by Swedish contractor, Stabilator.
1997	Raito Kogyo (Japan) establish U.S. subsidiary in California.
1997 - 1998	Master Builders Technologies develop families of dispersants for soil (and grout) to aid DMM penetration and mixing efficiency.

Table 1. Highlights of Historical Development of DMM (concluded).

Applications and Applicability of DMM

Six basic groups of applications can be identified for contemporary deep mixing methods:

1. Hydraulic cutoffs: DMM walls to prevent water movement through or under water retaining structures, such as dams or levees (e.g., Sacramento, CA) and into deep basements excavated below the water table.
2. Structural walls: DMM walls containing steel elements to resist lateral earth pressures in the construction of deep excavations, such as for cut and cover tunnels (Ted Williams Tunnel, Boston, MA) and deep basements.
3. Ground treatment: Block treatment to strengthen in a uniform manner large volumes of foundation soil in conjunction with deep excavations (Fort Point Channel, Boston, MA), and structural foundations.
4. Ground improvement: Discrete DMM elements (columns or panels) used as reinforcing elements to improve the overall performance of large, compressible soil masses under relatively lightly loaded structures, such as road (e.g., I-15, UT) or railway embankments.
5. Liquefaction mitigation: Interlocking DMM box or cellular structures to reduce the tendency for mass liquefaction and lateral spreading during seismic events under large embankments (e.g., Jackson Lake Dam, WY) or buildings.
6. Hazardous materials: DMM walls to contain, or DMM block treatment to fix, environmentally unacceptable materials.

Classification of Methods

A generic classification of the numerous methods used internationally can be made on the following simple basis:

- Is the cementitious material injected in a slurry or wet (W) form, or in a dry (D) state?
- Is this binder mixed with the soil via rotary energy only (R) or is the mixing enhanced/ facilitated by high pressure jet (J) grout type methods?
- Is the mixing action only occurring near to the drilling tool (E), or is it continued along the shaft (S) for a significant distance above it, via augers and/or paddles?

The classification shown on Figure 1 has therefore been developed by the authors, and four categories of methods - WRS, WRE, WJE and DRE - have been identified. No methods have been found in the DRS, DJE, or DJS categories since dry injection methods only feature end mixing with relatively low pressure binder injection pressures via compressed air, and jetted methods only feature end mixing (hence no WJS).

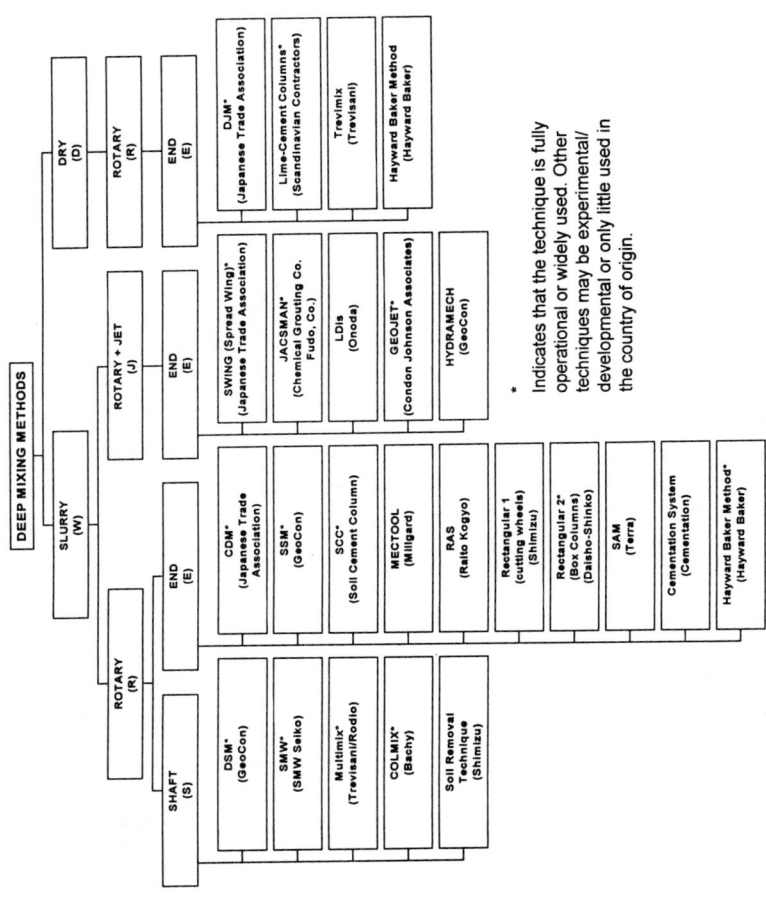

Figure 1. Classification of DMM techniques.

Features of Methods

The FHWA Report (1998) provides extensive data on each of the methods identified in Figure 1. Space restrictions prevent more than a brief summary of the more significant methods being provided in this publication (Appendices 1 and 2). The following general points must be borne in mind while considering these data:

- New methods, refinements of existing methods, and developments in materials (e.g., use of flyash, gypsum and slag in slurries; clay dispersants to aid penetration and improve mixing efficiency) are continually underway.
- As noted by Taki and Bell (1997), the technical goal of any DMM technique is to provide a uniformly treated mixed body, with no discrete lumps of binder or soil, a uniform moisture content, and a uniform distribution of binder throughout the mass. The most important requirements for installation are therefore: thorough and uniform mixing of the soil and binder; appropriate water/cement ratio; and appropriate grout injection ratio (e.g., volume of grout: volume of treated soil).
- The table includes methods not conventionally or nationally regarded as DMM, for example, the SMW Method (used only for walls in Japan), while the Scandinavian practitioners do not conventionally address their Lime-Cement Column Method as DMM.
- Despite their generic similarity, there are major and significant regional and procedural variations. For example, unconfined compressive strengths (U.C.S.) of treated soil using WRE, WRS, and WJE are typically higher than 1 MPa, except (e.g., FGC-CDM) where lower strengths are deliberately engineered. For DRE methods in Japan (e.g., DJM) a minimum U.C.S. of 0.5 MPa is obtained, whereas for the comparable DRE Scandinavian method (Lime-Cement Columns), rarely are strengths in excess of 0.15 MPa designed and/or achieved. Furthermore, treated soils in Scandinavia may be considered as providing vertical drainage, whilst similar soils in other countries, by other methods, may be regarded as relatively impermeable.
- Table 2 (Terashi, 1997) summarizes the factors influencing the strength of treated soil. In laboratory testing, there is no way to simulate factors III and IV except for the amount of binder and the curing time. Thus laboratory testing features standardization of these factors, and so it must be realized that the strength data provided by such tests is "not a precise predication" (Terashi, 1997) but only an "index" of the actual strength. Field testing is essential, and invariably appears to provide, for a number of reasons, inferior and more variable strength data.
- Deep mixing is, of course, not a panacea for all soft ground treatment, improvement, retention and containment problems, and in different applications it can be more or less practical, economic or preferable than competitive technologies. In the most general terms, DMM may be most attractive in projects where the ground is neither very stiff nor very dense, nor contains boulders or other obstructions; to depths of less than about 30m;

where there is relatively unrestricted overhead clearance; where a constant and good supply of binder can be assured; where a significant amount of spoil can be tolerated; where a relatively vibration-free technology is required; where treated or improved ground volumes are large; where "performance specifications" are applicable; or where treated ground strengths have to be closely engineered (typically 0.1 to 5 MPa). Otherwise, and depending always on local conditions, it may prove more appropriate to use jet grouting, diaphragm walling, sheet piling, caissons, beams and lagging, driven piles, wick drains, micropiles, soil nails, vibrodensification, compaction grouting, deep dynamic consolidation, bioremediation, or vapor extraction.

- The materials injected are tailored to the method used, their local availability, the ground to be treated and the desired or intended result. Generally, for the methods using a fluid grout, the constituents include cements, water, bentonite, clay, gypsum, flyash, and various additives. Water cement ratios typically range from less than 1 to over 2, although the actual in place w/c ratio will depend on any "predrilling" activities with water, or other fluids. Most recently, dispersants (Gause, 1997) can be used, both to breakdown cohesive soils, and also to render more efficient the grout injected. For dry injection methods, cement and/or unslaked lime are the prime materials used.
- For wet methods (mechanically simpler and so preferable in "difficult" geographic locations), the cement injected is typically in the range of 100 to 500 kg per cubic meter of soil to be treated. The ratio of volume of fluid grout injected to soil mass treated is typically about 20 to 40%. (A lower injection ratio is preferable, to minimize cement usage and spoil).
- For dry methods, (in soils of 60 to over 200% moisture content), typically 100 to 300 kg of dry materials per cubic meter of treated soil are used, providing strengths of 0.2 to 20 MPa, depending very much on soil type (low strengths and solids contents in Scandinavia), with minimal spoil or heave potential.
- Treated soil properties (recalling that cohesive soils require more cement to give equivalent strengths than cohesionless soils) are usually in the ranges shown in Table 3.
- It must be remembered that different techniques are intended specifically to provide higher strengths, or lower permeabilities and so the figures cited above are gross ranges only, and that the data provided by the individual corporations supersede those presented above for specific applications.

I	Characteristics of hardening agent
	1. Type of hardening agent
	2. Quality
	3. Mixing water and additives
II	Characteristics and conditions of soil (especially important for clays)
	1. Physical chemical and mineralogical properties of soil
	2. Organic content
	3. pH of pore water
	4. Water content
III	Mixing conditions
	1. Degree of mixing
	2. Timing of mixing/re-mixing
	3. Quality of hardening agent
IV	Curing conditions
	1. Temperature
	2. Curing time
	3. Humidity
	4. Wetting and drying/freezing and thawing, etc.

Table 2. Factors affecting the strength increase (Terashi, 1997).

U.C.S.	0.2 - 5.0 MPa (0.5 - 5 MPa in granular soils) (0.2 - 2 MPa in cohesives)
k	10^{-6} - 10^{-9} m/s (lower if bentonite is used)
E	350 to 1000 times U.C.S. for lab samples and 150 to 500 times U.C.S. for field samples
Shear strength (direct shear, no normal stress)	40 to 50% of U.C.S. at U.C.S. values < 1 MPa, but this ratio decreases gradually as U.C.S. increases.
Tensile strength	Typically 8 - 14% U.C.S.
28-day U.C.S.	1.4 to 1.5 times the 7-day strength for silts and clays 2 times the 7-day strength for sands
60-day U.C.S.	1.5 times the 28-day U.C.S., while the ratio of 15 years to 60 days U.C.S. may be as high as 3 to 1. In general, grouts with high w/c ratios have much less long term strength gain beyond 28 days, however.

Table 3. Typical data on soil treated by deep mixing.

Commercial Aspects

In the United States, there are at least nine companies who offer, or claim to offer, deep mixing services. Four (GeoCon, Condon Johnson, Terra, and Millgard) appear to have no links with foreign ownership or licensees, having developed their own systems. The others (Hayward Baker, Raito, Seiko, Stabilator, and SCC) are either U.S. operations with foreign ownership or use methods under foreign license. Based on the authors' investigations, it would seem that from 1986 to 1992, the annual value of deep mixing work conducted was in the range of $10 to 20 million, increasing by over 50 percent to 1996. Since then, as a result of massive works in Boston, Salt Lake City and the West Coast, this annual volume is probably now in the range of $50 to 80 million. For DMM used in environmental applications, the annual market may be around $20 to 30 million, increasing at about 5 to 10 percent annually.

Large scale systems may cost $80,000 to $200,000 to mobilize (much lower for methods such as Lime-Cement Columns). Typical prices for treatment are $100 to 250/m^2, or $50 to 100/m^3.

In Japan, the CDM Association claims to have treated over 26 million m^3 of soil from 1977 to 1995 (30 percent in the period 1992 to 1995) with about 60 percent being offshore (Figure 2). The DJM Association records 16 million m^3 of soil treatment from 1980 to 1996, involving 2345 separate projects, and an annual volume now approaching 2 million m^3 (Figure 3). By 1994, SMW Seiko, referring to their deep mixing wall system, had recorded 4,000 projects worldwide for a total treatment of 12.5 million m^2 (7 million m^3).

In Scandinavia, Åhnberg's data (Figure 4) illustrate the rapid growth in Swedish applications, while similar data illustrate a strong but smaller and steadier market in Finland (about 250,000 m^3 per year, 80 percent of which is lime-cement columns.) Markets in Norway and the Baltic countries are much smaller but have considerable growth potential. Selling prices in Scandinavia are typically in the range of $7 to 12/lin. m.

Similar data have not been found for other European countries, but there is no evidence that levels of activity in countries like U.K., France, Germany and Italy currently approach those in the U.S.

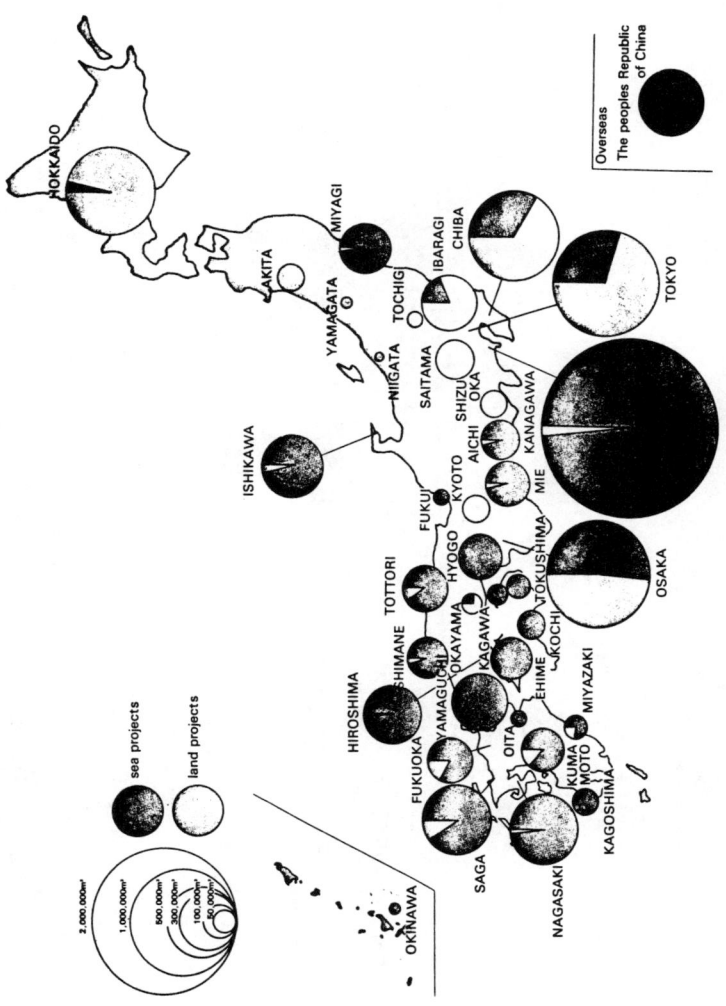

Figure 2. Data on use of CDM in Japan and China till 1993 (CDM Association, 1994).

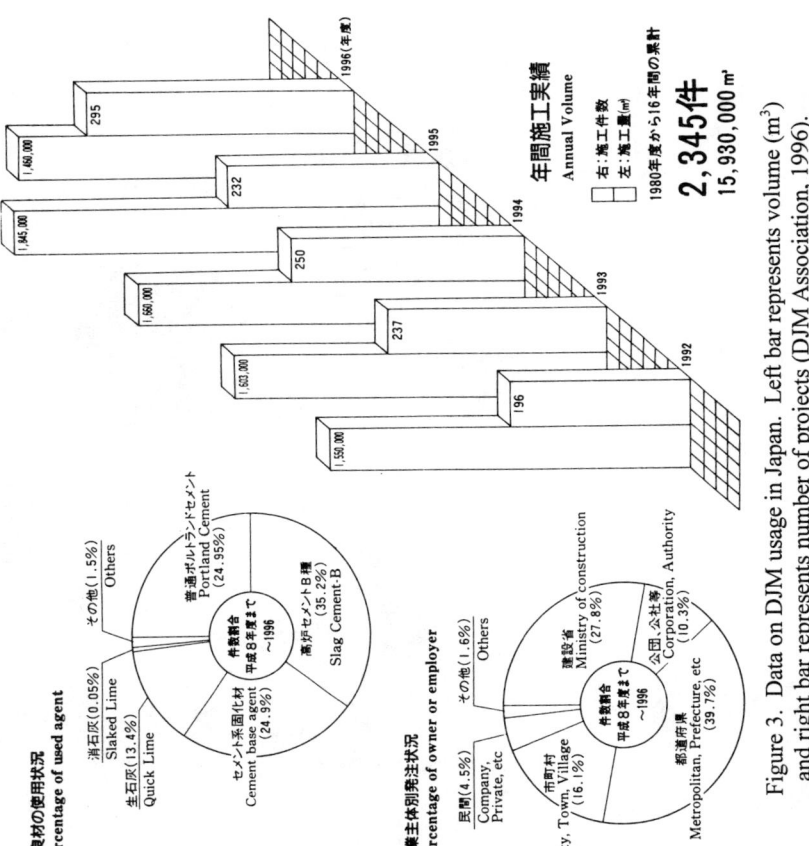

Figure 3. Data on DJM usage in Japan. Left bar represents volume (m³) and right bar represents number of projects (DJM Association, 1996).

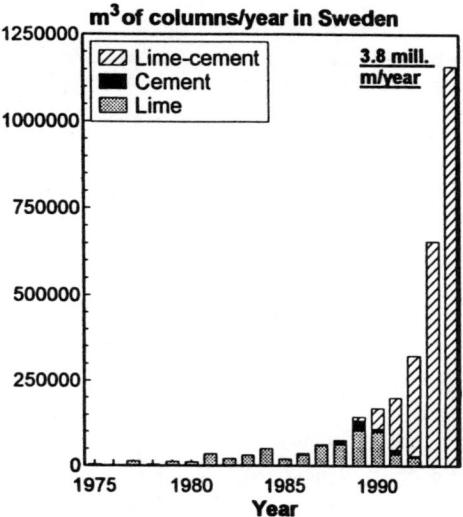

Figure 4. Use of different stabilizing agents for deep stabilization of soils in Sweden 1975-1994 (Åhnberg, 1996).

Final Remarks

Deep mixing is one of the world's most attractive and fastest growing specialty geotechnical construction processes. It offers solutions in a wide range of applications, in softer soils to moderate depths. Levels of knowledge and expertise are exceptionally high in Scandinavia and Japan, and rapidly improving in North America, where its technical and commercial potential is now becoming fully recognized. It is hoped that this paper will provide a structured introduction to the many techniques and systems that currently exist and a basis for understanding the innovations that will doubtless follow in quick order.

Acknowledgments

The authors have been singularly fortunate to have received input from specialist colleagues throughout the world from a variety of institutions and corporations. These individuals include Peter Nicholson, Dan Himick, Osamu Taki, Dave Druss, Tom O'Rourke, George Burke, Maasaki Terashi, James Johnson, Goran Hölm, Johan Hagblom, David Yang, and Seth Pearlman. To all, our thanks.

References

Åhnberg, H. (1996). "Stress dependent parameters of cement and lime stabilized soils."*Grouting and Deep Mixing*, Proceedings of IS-Tokyo'96, The Second International Conference on Ground Improvement Geosystems, Tokyo, May 14-17. pp. 387-392.

American Society of Civil Engineers. (1997). *In Situ Ground Improvement, Reinforcement and Treatment: A Twenty Year Update and a Vision for the 21st Century*, Ground Reinforcement Committee, Geo-Institute Conference, Logan, UT, July 16-17.

Burke, K., A. Furth, and D. Rhodes. (1996). "Site remediation of hexavalent chromium in a plater's sump: heavy metal chemical fixation - A case history." Proc. 17th Annual AESF/EPA Conference on Pollution Prevention and Control. February 1. Orlando, FL. 10 p.

CDM Association. (1996). Brochure.

Day, S.R. and C.R. Ryan. (1995). "Containment, stabilization and treatment of contaminated soils using in-situ soil mixing." Proc. ASCE Specialty Conference, New Orleans, LA. February 22-24. 17 p.

DJM Association. (1994). Brochure.

Federal Highway Administration (1998). "An Introduction to the Deep Mixing Methods used in Geotechnical Applications." Prepared by ECO Geosystems, Inc.

Fujita, T. (1996). "Application of DJM Method under special conditions." *Grouting and Deep Mixing*, Proceedings of IS-Tokyo'96, The Second International Conference on Ground Improvement Geosystems, Tokyo, May 14-17. pp. 591-594.

Gause, C. (1997) Personal Communication.

Hölm, G. (1994). "Deep stabilization by admixtures," 13th International Conference on SMFE, New Delhi, *Proceedings*, Vol. 3, pp. 1123-1126.

Miyoshi, A. and K. Hirayama. (1996) "Test of solidified columns using a combined system of mechanical churning and jetting." *Grouting and Deep Mixing*, Proceedings of IS-Tokyo'96, The Second International Conference on Ground Improvement Geosystems, Tokyo, May 14-17. pp. 743-748.

Okumura, T. (1996) "Deep mixing method of Japan." *Grouting and Deep Mixing*, Proceedings of IS-Tokyo'96, The Second International Conference on Ground Improvement Geosystems, Tokyo, May 14-17. Vol. 2. pp.879-887.

Pagliacci, F. and G. Pagotto. (1994). "Soil improvement through mechanical deep mixing treatment in Thailand." Proc. 5th DFI International Conference, Bruges, Belgium. June 13-15, pp. 5.11-5.17.

Rathmeyer, H. (1996). "Deep mixing methods for soft subsoil improvement in the Nordic Countries." *Grouting and Deep Mixing*, Proceedings of IS-Tokyo'96, The Second International Conference on Ground Improvement Geosystems, Tokyo, May 14-17. Vol. 2, pp. 869-877.

Reavis, G. and F.C. Freyaldenhoven. (1994). "GEOJET Foundation System." *Geotechnical News*. Vol. 12, No. 4, December. pp. 56-60.

Ryan, C.R. and B.H. Jasperse. (1989). "Deep soil mixing at Jackson Lake Dam." Proc. ASCE 1989 Foundation Engineering Congress, *Foundation Engineering: Current Principles and Practices*. Vol. 1 and 2, Evanston, IL. June 25-29, pp. 354-367.

Ryan C.R and Jasperse, B.H. (1992). "Stabilization and fixation using soil mixing," ASCE Geotechnical Special Publication No. 30. *Grouting Soil Improvement and Geosynthetics*, New Orleans. pp. 1273-1284.

Taki, O. and R. Bell. (1997). Booklet on soil-cement pile column. 23 p.

Taki, O. and D.S. Yang. (1989). "Excavation support and groundwater control using soil-cement mixing wall for subway projects." Proceedings Rapid Excavation and Tunneling Conference. Los Angeles, CA. June 11-14. pp. 156-175.

Taki, O. and D.S. Yang. (1991). "Soil-cement mixed wall technique." American Society of Civil Engineers, Proceedings, Geotechnical Engineering Congress, Denver, CO. pp. 298-309.

Terashi, M. (1997). "Deep mixing method - Brief state-of-the-art." 14th International Conference on Soil Mechanics and Foundation Engineering, 4 p.

Tokyo Conference. (1997). Grouting and deep mixing. Proceedings of IS-Tokyo'96, The Second International Conference on Ground Improvement Geosystems, A.A. Balkema, Rotterdam, 2 vols., Tokyo, May 14-17.

Yang, D.S. (1997). "Deep mixing." *In Situ Ground Improvement, Reinforcement and Treatment: A Twenty Year Update and a Vision for the 21st Century*, Ground Reinforcement Committee, American Society of Civil Engineers, Geo-Institute Conference, Logan, UT, July 16-17. pp. 130-150.

Classification Name Company Geography	WRS DSM GeoCon N. America	WRS SMW SMW Seiko S.E. Asia, U.S.
General Description of Method	Adjacent discontinuous augers rotate in alternate directions. Most of grout injected mainly on downstroke	Adjacent discontinuous augers rotate in alternate directions. Water air or grout used on downstroke and grout on upstroke
Special Features/Patented Aspects	Lower 3m usually double stroked. Strong QA/QC by electronic methods	Special electric head patented Double stroking "oscillation" common especially in cohesive soils. Discontinuous auger flights are positioned at intervals to reduce torque requirements.
Details of Installation Shafts Diameter Depth RPM Productivity/output	1-4, usually 4 0.6 to 1.0m, usually 0.8m 35m 20 - 50 0.6 m/min penetration 2 m/min withdrawal/mixing	1-5, usually 3 0.55 to 1.5m, usually 1m 60m claimed, 35m practical 15-20 0.5 m/min penetration 2 m/min withdrawal/mixing 100-200 m^3 per shift i.e., 100-250 m^2 per shift
Mix Design *(depends on soil type and strength requirements)* Materials w/c ratio Cement ratio (k_{cement}/m^3_{soil}) Volume ratio(Vol_{grout}:Vol_{soil})	Cement grout ± bentonite and other additives w/c = 1.2-2.5 300-500kg/m^3 soil 30-40%	Cement grout ± bentonite and other additives w/c = 1.3-2.5 250-750 kg/m^3 soil 50-100%
Reported Treated Soil Properties U.C.S. k E	0.3-7 MPa (clay strengths approx. 40% those in sands) 1×10^{-8} - 1×10^{-9} m/s 500 to 1350 x U.C.S.	0.3-1.3 MPa (clay) 1.4-4.2 MPa (sands) 1×10^{-7} - 1×10^{-10} m/s 500 to 1350 x U.C.S.
Specific Relative Advantages and Disadvantages	Economic, proven systems Mixing efficiency can be poor in stiff cohesives Can generate large spoil volumes	
Notes	Developed in late 1980s	Developed in 1972: first used 1976 in Japan, 1986 in U.S.
Representative References	Ryan and Jasperse (1989, 1992) Day and Ryan (1995)	Taki and Yang (1989, 1991) Yang (1997)

Appendix 1. Details of major fully operational deep mixing techniques (continues).

Classification	WRS	WRE
Name	Trevimix (Multimix)	CDM
Company	Trevisani, Rodio	48 members of CDM Association
Geography	Italy	Japan, China
General Description of Method	Adjacent augers rotate in opposite directions. Grout injected during penetration. Prestroked with water in clays. Auger rotation reversed during withdrawal. Mixing occurs over 8-10m length of shaft.	Shafts have 4-6 mixing blades above cutting tool. Grout injected mainly during penetration. Also a 2 min mixing period at full depth.
Special Features/Patented Aspects	Pre-drilling with water ± additives in very resistant soils. Process is patented by Trevisani	Comprises numerous subtly different methods all protected under CDM Association
Details of Installation		
Shafts	1-3, typically 3	2-8m (marine): 1-2m (land)
Diameter	550-800mm at 450-600mm spacings	1-2m (marine): 0.7-1.5 (land)
Depth	25m	70m (marine): 40m (land)
RPM	12-30	20-30 (penetration):60 (withdrawal)
Productivity	0.35-1.1 m/min penetration 0.48-2 m/min withdrawal	0.5-2 m/min (avg. 1/m/min) 1-2 m/min (withdrawal) (1000 m^3/shift for marine; 100-200 m^3/shift on land)
Mix Design *(depends on soil type and strength requirements)*		
Materials	Cement, water mainly plus bentonite used in clays, additives common	Wide range of materials including cement, bentonite, gypsum, flyash
w/c ratio	Typically low, i.e., 0.6-1.0	0.6-1.3 (usually low)
Cement ratio (k_{cement}/m^3_{soil})	200-250 kg/m^3	100-200 kg/m^3
Volume ratio ($Vol_{grout}:Vol_{soil}$)		15-30%
Reported Treated Soil Properties		
U.C.S.	0.5-5 MPa (sands) 0.2-1 MPa (silts, clays)	Strengths can be closely controlled by varying grout composition from < 0.5-4 MPa
k	< 1 x 10^{-7} m/s	10^{-8} - 10^{-9} m/s
E		350-1000 x U.C.S.
Specific Relative Advantages and Disadvantages	Goal is to minimize soil removal (10-20%) and enhance mixing efficiency	Vast amount of R&D information available. Specifically developed for softer marine deposits and fills
Notes	Developed jointly in late 1980s	Association founded in 1977 Research initiated under Japanese Government (1967). Offered in the U.S. by Raito.
Representative References	Pagliacci and Pagotto (1994)	CDM (1996) Okumura (1996)

Appendix 1. Details of major fully operational deep mixing techniques (continues).

Classification Name Company Geography	WRE SCC SCC Technology SCC (U.S.A.): Tenox (Japan)	WRE HBM Hayward Baker Inc: A Keller Co. (U.S.A. but with opportunities for sister companies worldwide)
General Description of Method	Grout is injected during penetration. A non-rotated "share blade" is located above tip. At target depth, 1 minute of additional injection plus oscillation for 1.5-3m Withdrawal with counter rotation	Grout injected during penetration followed by 5 minutes mixing and oscillation at full depth, and rapid extraction with injection of "backfill grout" only (1-5% total)
Special Features/Patented Aspects	Very thorough mixing via "share blade" action which is patented.	Method proprietary to Keller.

Details of Installation   Shafts
    Diameter
    Depth
    RPM
    Productivity
	Single with 3 rotated mixing blades plus "share blade"
	0.6m
	18m max
	30-50
	1 m/min in and out
		Single with 2 or 3 pairs of paddles
		1.2-2.6m, typically 2.1m
		15m max
		20-25 (penetration): higher upon withdrawal
		0.3-0.5 m/min (penetration): faster upon withdrawal. In excess of 500 m^3/shift

Mix Design
(depends on soil type and strength requirements)   Materials
  w/c ratio
  Cement ratio (k_{cement}/m^3_{soil})
  Volume ratio ($Vol_{grout}:Vol_{soil}$)
	Typically cement, but others e.g., ash, bentonite possible.
	0.6 (clays) to 1.0 (sands)
	100-450 kg/m^3 cement
	30-35%
		Varied in response to soil type and needs
		1-2
		150 kg/m^3 cement
		15-30%

Reported Treated Soil Properties U.C.S. k	3.5-7 MPa (sands) 1.3-7 MPa (cohesives) 10^{-8} m/s	3-5-10 MPa (sands) 0.2-1.4 MPa (clays) 10^{-10} m/s
Specific Relative Advantages and Disadvantages	Low spoil with minimal cement loss claimed, due to low w/c and minimized injected volume	Good mixing: moderate penetration: low spoils volume
Notes	Used since 1979 in Japan and 1993 in U.S.A.	Developed since 1990
Representative References	Taki and Bell (1997)	Burke et al. (1996)

Appendix 1. Details of major fully operational deep mixing techniques (continues).

SOIL IMPROVEMENT FOR BIG DIGS

Classification Name Company Geography	WJE JACSMAN Chemical Grout Co., Fudo Co. Japan	WJE GeoJet Condon Johnson Associates Western U.S.A.
General Description of Method	Grout injected at low pressure via blades during penetration. During withdrawal, obliquely inclined jets are used at high pressures to increase diameter and enhance mixing efficiency	Grout is jetted via ports on a pair of wings during penetration
Special Features/Patented Aspects	The combination of DM and jet grouting ensures good joints between adjacent columns, and columns of controlled diameter and quality. Column formed is 1.9m x 2.7m in plan. Patented process.	Combination of mechanical and hydraulic cutting/mixing gives high quality mixing and fast penetration. Licenced by CJA.
Details of Installation Shafts Diameter Depth RPM Productivity	 2 shafts each with 3 blades. 1m (blades at 0.8m spacing) 20m 20 1 m/min penetration 0.5-1 m/min withdrawal	 1 shaft with pair of wings 400-1200mm 40m max (20m typical) 150-200 (recent developments focusing on slower rpm) 2-12 m/in (penetration) 10 m/min (withdrawal) 150m of piles/hr possible
Mix Design *(depends on soil type and strength requirements)* Materials w/c ratio Cement ratio (k_{cement}/m^3_{soil}) Volume ratio ($Vol_{grout}:Vol_{soil}$)	 Cement 1.0 200 kg/m³ (jetted) 320 kg/m³ (DM) Air also used to enhance jetting 200 l/min per shaft during DM penetration 300 l/min during withdrawal (jetting)	 Cement, additives 1.0 570 kg cement/m of column (typical) 30-40%
Reported Treated Soil Properties U.C.S.	1-6 MPa (silty sand and clay)	4.8-10.3 MPa (clay)
Specific Relative Advantages and Disadvantages	New system combining DM and jet grouting principles to enhance volume and quality of treatment	Computer control of penetration parameters excellent. High strength. Low spoil volumes. Excellent mixing.
Notes	Name is an acronym for Jet and Churning System Management. Not yet fully operational	Developed since early 1990s. Fully operational in Bay Area.
Representative References	Miyoshi and Hirayama (1996)	Reavis and Freyaldenhoven (1994)

Appendix 1. Details of major fully operational deep mixing techniques (continues).

Classification	DRE	DRE
Name	Dry Jet Mixing	Lime-Cement Columns
Company	DJM Association (64 companies)	Various (Scandinavia and Far East); Stabilator (U.S.A.)
Geography	Japan	Scandinavia, Far East, U.S.A.
General Description of Method	Soil is penetrated while injecting compressed air from the lower blades. Dry materials are injected during withdrawal via compressed air, and reverse rotation.	Soil is penetrated while injecting compressed air below mixing tool. Dry materials are injected during withdrawal via compressed air, and reverse rotation.
Special Features/Patented Aspects	System is patented and protected by DJM Association. Two basic patents (blade design and control system)	Very low spoil. High productivity. Efficient mixing. Patents are held by the contractors. Strong reliance on computer control.
Details of Installation		
Shafts	1-2 shafts spaced at 0.8 to ~1.5m, each with 2 pairs of blades	Single shaft, various types of cutting/mixing blades.
Diameter	1m	500-1200mm, typically 600, 800mm
Depth	33m max	25m max
RPM	5-50 typically	100-200, usually 130-170
Productivity	50-55 mins for 9m column 0.5-3 m/min withdrawal	0.6-0.9 m/min (withdrawal) 400-1000 lin m/shift
Mix Design *(depends on soil type and strength requirements)* Materials	Usually cement, but quicklime in clays of very high moisture contents	Cement and lime in various percentages (typically 50:50 or 75:25)
Cement ratio (k_{cement}/m^3_{soil})	100-200 kg/m³ Range is 100-500 kg/m³ but typical is 100-300 kg/m³	23-28 kg/m (600mm diameter); 40 kg/m (800mm diameter) i.e., 80-150 kg/m³
Reported Treated Soil Properties U.C.S.	Greatly varies depending on soil and binder, 1-10 MPa	Varies but typically 0.2-0.3 MPa (0.2-2 MPa, possible)
Specific Relative Advantages and Disadvantages	Very little spoils; efficient mixing. Great deal of R&D experience.	Same as for DJM. Excellent Swedish/Finnish research continues.
Notes	Developed by Japanese Government and fully operational in 1980. Offered in the U.S. by Raito.	Developed by Swedish industry and Government, with first commercial applications in mid 1970s.
Representative References	DJM Brochure (1994); Fujita (1996)	Holm (1994); Rathmeyer (1996)

Appendix 1. Details of major fully operational deep mixing techniques (concluded).

DSM system (Courtesy: GeoCon, Inc.)

Appendix 2. Illustrations of various deep mixing methods described in Appendix 1 (continues).

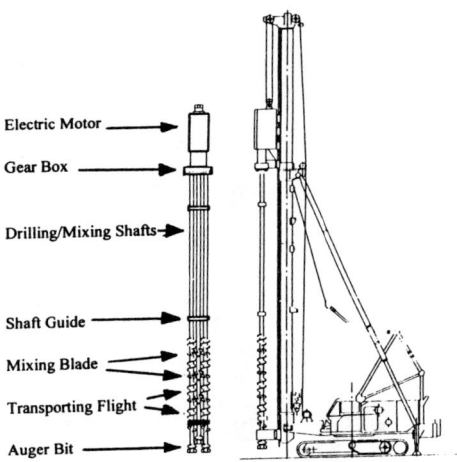

Schematic of SMW system (Taki and Bell, 1997).

Title	CDM Cement Deep Mixing	SMW Soil Mix Wall
Sketches of Representative Mixing Mechanisms	⌀ = 39" to 63" available 1 ft = 0.305 m	⌀ = 22" to 40" available 1 ft = 0.305 m
Descriptions	Rotation of multiple axis shafts create relative movement and shear in soil for soil-reagent mixing.	Uses multiple auger, paddle shafts rotating in alternating directions to mix in situ soil with cement grout or other reagents to form continuous soil-cement walls.

Comparisons of SMW and CDM shaft arrangements (Yang, 1997).

Appendix 2. Illustrations of various deep mixing methods described in Appendix 1 (continues).

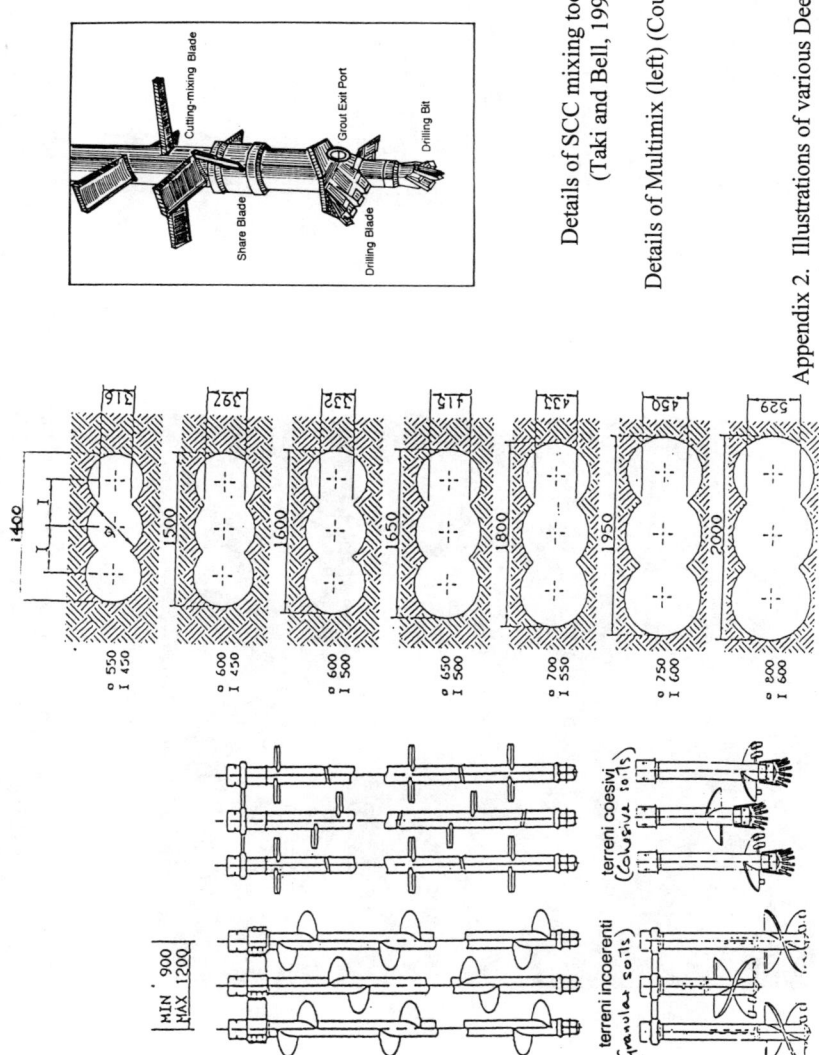

Appendix 2. Illustrations of various Deep Mixing Methods described in Appendix 1 (continues).

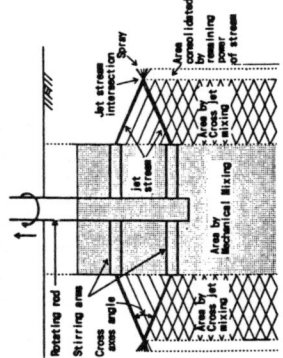

Details of JACSMAN system
(Miyoshi and Hirayama, 1996)

HBM tooling
(Courtesy: Hayward Baker, Inc.)

Appendix 2. Illustrations of various Deep Mixing Methods described in Appendix 1 (continues).

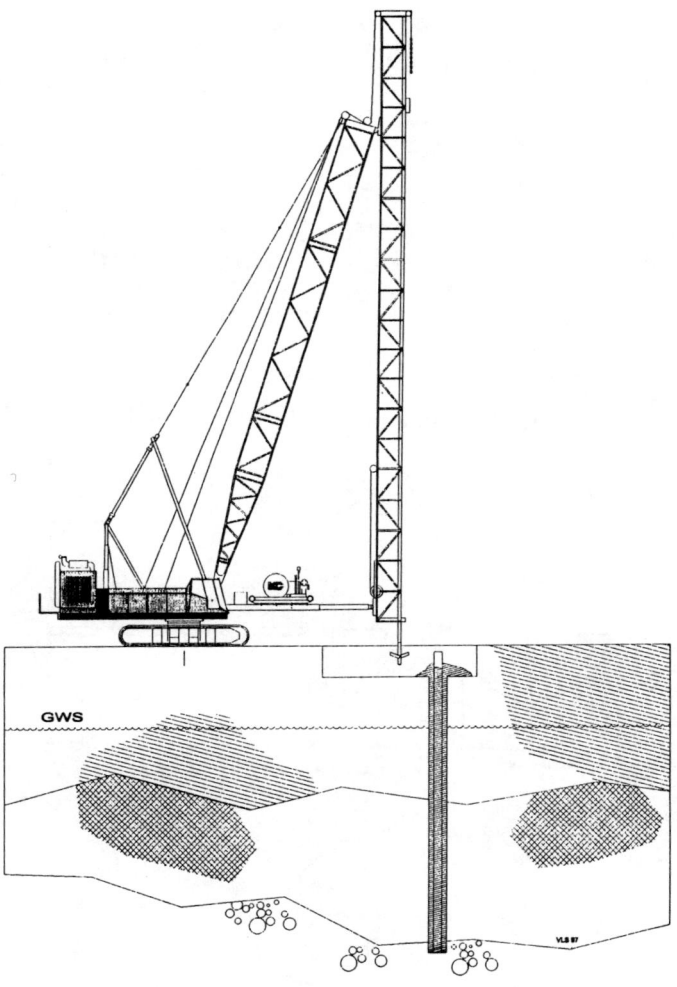

GeoJet system (Courtesy: Condon Johnson Associates).

Appendix 2. Illustrations of various deep mixing methods described in Appendix 1 (continues).

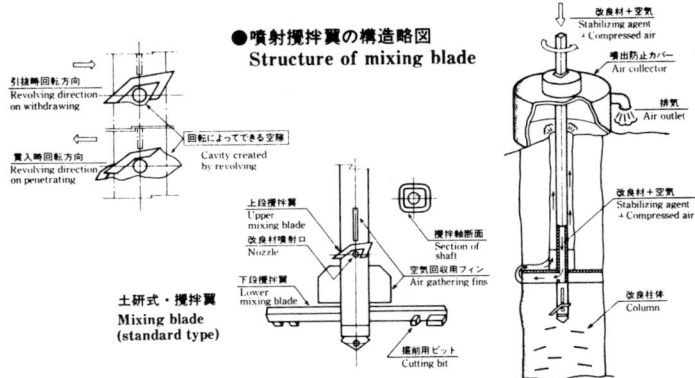

Details of DJM system (DJM Association, 1994).

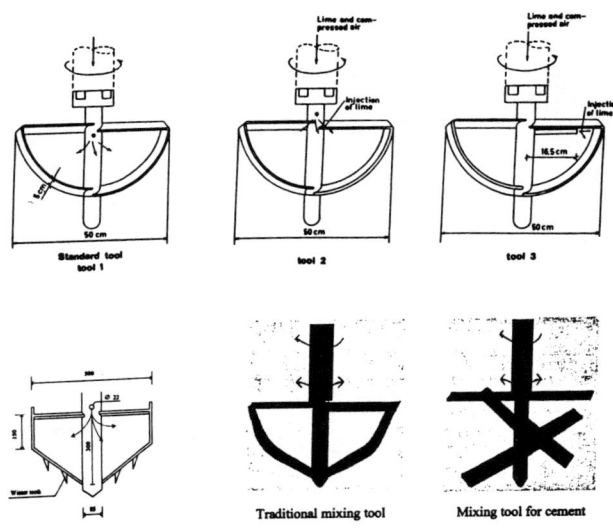

Details of mixing tools for Lime-Cement Columns (Courtesy: Stabilator).

Appendix 2. Illustrations of various deep mixing methods described in Appendix 1 (concluded).

DESIGN OF A SOIL MIXED COMPOSITE GRAVITY WALL

P.J. Nicholson[1], Member, ASCE; J.K. Mitchell[2], Hon. Member ASCE; E.W. Bahner[3], Member ASCE; and Y. Moriwaki[4], Member, ASCE

ABSTRACT

One application, only in-situ soil mixing (or deep mixing method, DMM) is becoming increasingly common as an economical construction method for temporary support of deep foundation excavations and is also used in the construction of permanent retaining walls. Steel beams placed into the fresh soil/cement are generally used as the primary structural support, later supplemented with bracing or tie-back anchors as the excavation depth requires. These additional elements of bracing increase the cost of the support or retaining wall system, add complexity and delay the construction schedule. This paper presents the design for a new, patent pending, DMM wall system using the DSM technique that utilizes an array of soil/cement mixed columns arranged in the native soil in a manner that creates a gravity block. This gravity block resists the sliding and overturning loads and thus eliminates the need for bracing or tie-back anchors for stability. All the support elements are placed prior to any excavation, thus speeding the construction schedule. Because it is a gravity block, no steel reinforcement is needed.

The analysis of this novel system was performed by two independent teams of researchers to determine the potential failure mechanism of the composite wall system. Finite element and pile group analysis were used to estimate column spacing, required soil-cement strength and variations in pattern geometry. The conclusion of the design teams is that the composite gravity wall system is a viable and potentially cost-effective method for both temporary support of excavation and permanent wall construction.

---

[1] President, Geo-Con, Monroeville, PA 15146

[2] University Distinguished Professor, Department of Civil and Environmental Engineering, Virginia Tech University, Blacksburg, VA 24061-0105

[3] Geotechnical Engineering Manager, Woodward-Clyde International Americas, Waukesha, WI 53188

[4] Senior Principal, Woodward-Clyde International Americas, Santa Ana, CA 92705

INTRODUCTION

Soil/cement mixing (DMM) for support of excavation has been known for over twenty years in Scandinavia and Japan and has become more known in the United States over the last 10 years. Starting with a major project at Jackson Lake Dam, WY in 1987, the use of the DMM has expanded with recent projects in Milwaukee, San Francisco, Boston and other places. Typically, soil cement columns, reinforced with steel beams placed down the center of the columns, have been used for temporary support of excavation (Fig. 1).

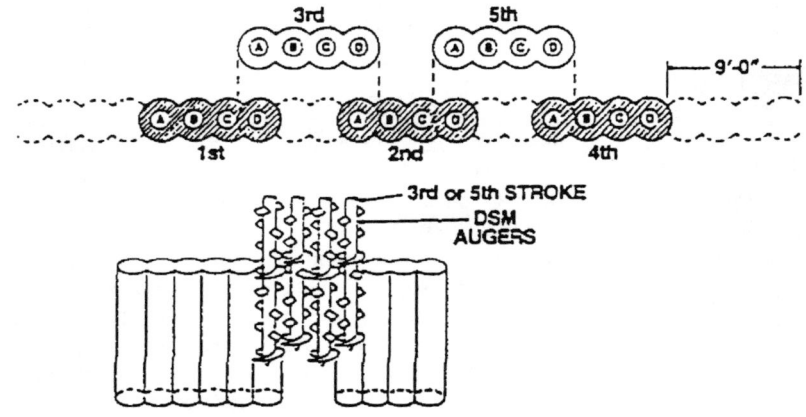

Figure 1. DSM With Steel Beams

Permanent walls have been constructed with this system by adding a concrete facing to the DSM wall. As a next step, Geo-Con and Woodward-Clyde in conjunction with the Geotechnical Engineering Department of Virginia Tech University, have developed a concept of reinforcing the earth to provide a gravity wall system that is self supporting, without bracing or tie-back anchors. This paper presents the results of the initial studies completed by the team in the development of this new technology.

HISTORY OF GRAVITY EARTH RETENTION STRUCTURES

Gravity walls are commonly used civil engineering structures and the basic concepts of their design are widely understood. The Romans used similar structures made of stone and in some cased enhanced with rudimentary forms of cement. In a parallel manner, there occurred development of methods to reinforce ground to improve its strength. Straw or reeds were laid with soil to provide steeper slopes than natural soil alone could attain. Nature has even devised some reinforcing techniques such as the root structures of trees that vegetate and reinforce steeply sloping hills. Manmade ground reinforcement dates back to the time when man first decided to build enclosed areas in which to live – witness the use of adobe and clay, reinforced with straw. History also records the use of structures called "ziggurats" over three thousand years ago. These were constructed of clay bricks reinforced with woven mats of reeds laid horizontally on a layer of sand and gravel at vertical spacing varying between 0.6 to 2 meters. The Romans were known to have used earth reinforcing techniques as evidenced by the reed reinforced earth levees constructed along the Tiber River and elsewhere. (Jones, 1985)

In modern times, man had used massive concrete, reinforced or un-reinforced; bin type structures, usually earth or rubble filled; and more recently, gravity structures constructed by means of "Reinforced Earth" ™ or "Mechanically Stabilized Earth" (MSE) that uses strips of steel or fabric placed horizontal at a predetermined spacing with granular backfill to form a gravity retaining structure. Later came structures created by "Soil Nailing", where sub-horizontal bars were grouted into the ground at a spacing of 1-2 meters to "stitch" or "nail" the ground into a gravity wall mass. These two types of structures, both MSE and Soil Nailing, work on the principle that closely spaced reinforcing elements will behave in a composite manner with the surrounding soil, creating a single mass or gravity block. This gravity block is then designed to resist sliding, overturning and slope stability failure from the external soil, water and other loads imposed on it.

Gravity walls have been constructed from soil mixing elements since 1990 in the United States. In these three previous cases, the walls were composed of soil that was 100% treated with cement throughout the designed gravity mass. At least three projects have been completed in the United States using this concept, two for temporary support and the other for both temporary and permanent support. This latter project was in Columbus, GA in 1991, where a gravity mass of cemented soil was used to provide temporary and permanent support of a river front wall. After the soil cement had set, the river bank was excavated (Fig. 2) and the cemented soil mass was provided with a concrete facing. It currently serves as a river-front park (Fig. 3).

Figure 2. Columbus, GA River Front Wall Under Construction

Figure 3. Columbus, GA River Front Park, Retaining Wall Completed

SOIL CEMENT COLUMNS

As stated above, soil/cement mixing has become common in the United States over the past 10 years. A typical installation may consist of multiple columns, usually reinforced with beams to provide a shoring wall with characteristics similar to sheet piling (Fig. 1) (Jasperse, 1990).

At least 30 projects of this type have been successfully completed in the United States – thousands in Japan. Additionally, deep mixing has been used as underground reinforcement (Nicholson, 1994) and, as stated above, as cemented gravity structures. Currently a large soil/cement mixing project is underway in Boston, MA as part of the Central Artery/Tunnel project. There, approximately 500,000 cubic meters of soil will be mass treated to improve their structural characteristics and allow excavation for the submerged Fort Point Channel tubes to proceed (Druss, 1997).

Soil-cement is formed by mixing a cement-based slurry with the existing soils, which can vary from fine grained clays to coarse grained sand and gravels*. In this process, single or multiple stem, or shafts with bottom augers and paddles along the shaft, are rotated into the ground. As the tools penetrate the ground, a mixture of cement and water slurry is pumped down the hollow center of the stems. The rotation of the auger and mixing paddles blends the cement slurry with the existing soil. After the desired depth is reached, the rotation is reversed and the tools are pulled while additional cement slurry is added. The resultant column of mixed soil/cement has the properties of a very lean 345-3450 kPa (50-500 psi) concrete. An array of such columns, 0.5 to 3 m in diameter, installed into the ground to the appropriate depth is used to construct the "composite gravity wall" known as "VERT" (Vertically Earth Reinforced Technology) (developmental/experimental stage) (Fig. 4).

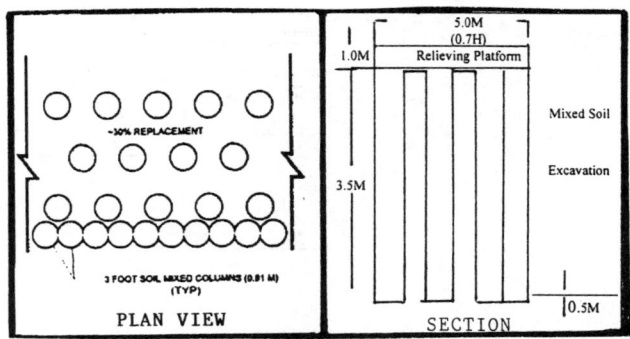

Figure 4. VERT Prototype Wall at Texas A&M University

*   Other processes are also used that involve injection of dry cement.

In the current design concept, a combination of the two popular forms of soil/cement mixing are to be utilized. First a row of tangent or secant columns is formed along the intended line of the proposed excavation. Then additional vertical reinforcing soil/cement columns are added in a pattern that will ensure composite action of the mass of soil encompassed by the soil cement columns. In a typical installation, there are either 3 or 4 rows of staggered, circular soil/cement elements forming a gravity structure that typically has a width of 0.6 to 0.8 times the exposed wall height. In the current design, the DMM elements are capped by a 1 m thick layer of mixed cement and soil spoil from the construction of the soil/cement elements. This cap is used to tie the tops of individual elements together and also act as a relieving platform for surcharge loads due to construction traffic, material and building surcharges.

SUMMARY OF ANALYTICAL STUDIES DONE AT VIRGINIA TECH

COMPOSITE ACTION AND STABILITY

A major concern in the use of DMM panels, walls, and columns for support of embankments is whether the soil-cement elements and untreated soil will act as a composite material, or the untreated soil will extrude between the elements under the influence of lateral loading. Finite element analyses of the DMM supported embankment and MSE wall shown in transverse section in Fig. 5(a) were made using finite element codes SOILSTRUCT and SAGE.

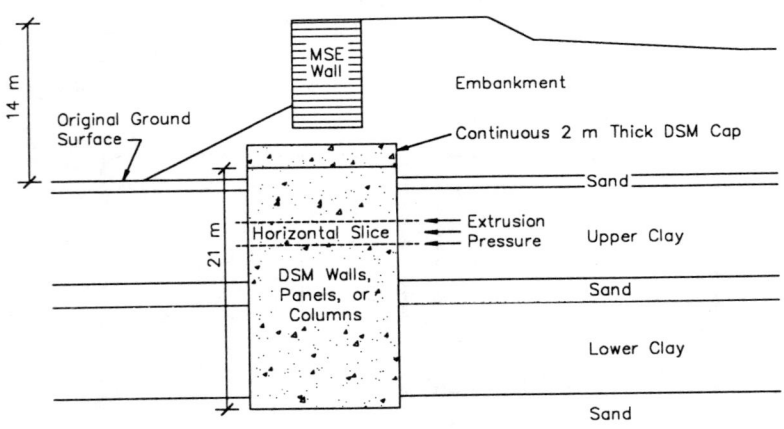

a) Cross-Section

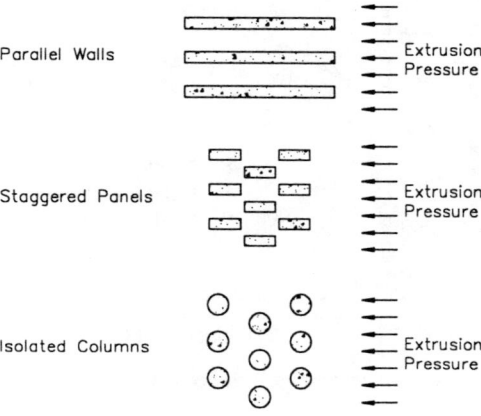

b) Horizontal Slices

Figure 5. DMM Configurations Analyzed for Embankment Support and Stabilization

A horizontal slice through three DMM wall configurations -- parallel walls, staggered panels, and circular columns -- as shown in plan view in Fig. 5(b) were analyzed. The configurations were compared on the basis of the same area replacement ratio and equal amounts of DMM per unit length of embankment. The undrained shear strength of the clay was assumed to be 58 kPa, and a hyperbolic representation was made of the stress-strain behavior. For the horizontal slice analyses the DMM elements were assumed fixed in place and of sufficient strength to resist any lateral pressure applied by the clay.

Monotonically increasing horizontal pressure was applied to each horizontal slice as shown in Fig. 5(b) and the limiting value determined at which the computed deformations began to increase rapidly. The onset of large deformations would be indicative of clay squeezing between the DMM elements. The results are shown in Fig. 6.

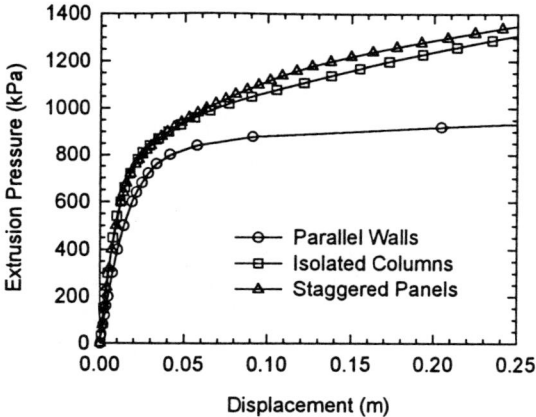

Figure 6. Extrusion Pressure in Horizontal Slice Analyses of Three DMM Configurations

Similar values for the limiting horizontal pressure were obtained using an approximate bearing capacity analysis. Essentially the same pressure-displacement relationships were found for the staggered panel and circular column DMM patterns. Both of these configurations provided greater resistance to clay squeezing than did the parallel wall arrangement. By this method, the limiting value of horizontal pressure can be determined for any clay strength and DMM configuration.

Whether the DMM elements themselves are capable of resisting the horizontal pressure depend upon the shear strength of the DMM. From knowledge of this strength, DMM-soil area ratio, and the pressure to extrude the soil, a profile of overall shear resistance vs. depth may be developed. The extrusion failure mode may control over one certain part of the depth, and a composite strength may control over another. The composite shear strength is defined by the sum of the soil shear strength and the allowable shear strength of the DMM material, each taken in proportion to its area ratio in the cross section. However, in some cases, the allowable shear on the DMM may be limited by the tensile capacity of the DMM elements and this should be evaluated upon the basis of the load distribution along the column.

Once a limiting strength profile has been developed, the overall stability of a proposed configuration, such as that in Fig. 7a), may be evaluated using a limiting equilibrium method. For the actual case used in the analyses outlined above, a design section of the I-15 reconstruction in Salt Lake City, Utah, factors of safety of 1.4 to 2.4 were obtained for DMM shear strengths ranging from 300 to 1000 kPa.

DMM GRAVITY WALL ANALYSIS

The situation is somewhat different when DMM is proposed for use as a faced gravity wall in that the squeezing mechanism is no longer a controlling criterion. A study was made of the gravity wall layout shown in transverse section in Fig. 7(a), proposed for another section along the I-15 reconstruction in Salt Lake City. In this case, the DMM column layout shown in plan in Fig. 7(b) includes a row of DMM tangent piles along the excavation face. Finite element analyses of the section shown in Fig. 7(a) were made using the unit weight and strength properties shown on the drawing, and hyperbolic representation of the stress-strain behavior.

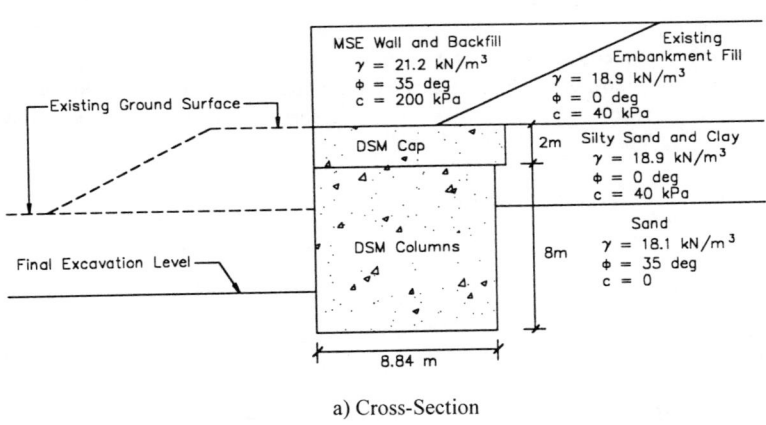

a) Cross-Section

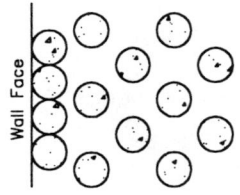

b) Horizontal Slice through DMM Columns

Figure 7. DMM System for Support of Excavation Face and Overlying

The maximum tensile stresses in the DMM columns and in the DMM cap were about 80 kPa as compared to the design compressive strength of 1200 kPa, which might provide a tensile strength of about 120 kPa in un-reinforced DMM. The maximum computed deformations at the wall face towards the excavation were about 55 mm if the MSE wall was placed prior to excavation and about 65 mm if the excavation was carried out prior to placement of the MSE wall above the DMM. It was found that the overall factor of safety against stability failure was 2.1 using Spencer's method and UTEXAS3.

Overall, the analyses showed that DMM VERT walls should provide stable support for embankments and excavations, with deformations that are well within acceptable limits.

SUMMARY OF ANALYTICAL STUDIES -- WOODWARD-CLYDE
A model similar to the one used by Virginia Tech University was used but one significant difference should be noted: all the DMM elements used in the Woodward-Clyde study were 0.9 m by 3 m panels of DMM that were arranged in staggered rows four meters apart. The overall interaction behavior of these DMM panels and the surrounding soil was evaluated using a transverse cross section as shown on Fig. 8.

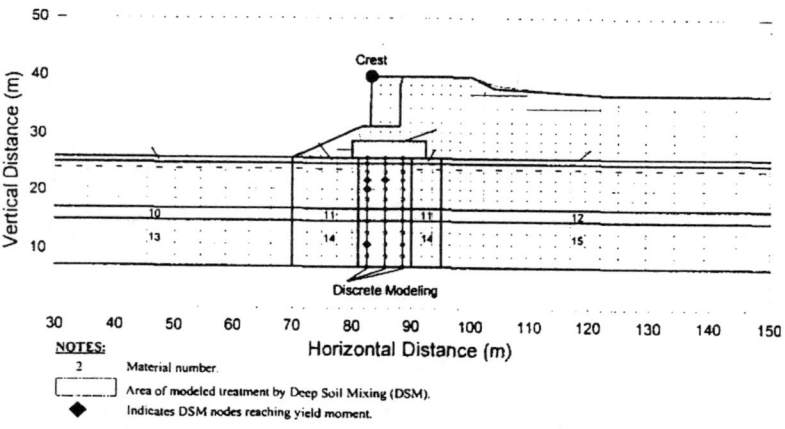

Figure 8. Analysis Cross Section - Proposed I-15 Wall

The analyses were performed using the computer program FLAC (Fast Lagrangian Analysis of Continua, Itasca Group, Inc.). FLAC is a two-dimensional explicit finite difference program for evaluating the response of earth and earth-structure systems under various loading conditions including a very large deformation behavior undergoing plastic flow. Since FLAC can also incorporate simple structural members, it was decided to model the DMM panels or column as "piles".

In FLAC, pile elements are modeled as beam elements connected to soil elements by bilinear p-y and t-z curves to represent pile-soil interaction effects. Through these connections, it is possible for piles to cut through surrounding soils and vice versa. The three-dimensional nature of piles embedded into surrounding soil is modeled by appropriately scaling down the pile stiffness and strength of the associated p-y and t-z curves. The maximum or yield moment values can be assigned to pile nodes allowing a formation of a hinge once the yield moment is reached at a pile node.

MODELING AND ANALYSIS RESULTS

The DMM panels were modeled in two ways. In one way, as indicated on Figure 8, the three rows of DMM panels were modeled using three rows of pile elements in FLAC. Because the loading conditions imposed by the transverse section primarily induced lateral deformation, the pile elements were connected to the surrounding soils only by horizontal non-linear springs or p-y curves. These p-y curves were specified in such a way that the resulting lateral resistance would be consistent with the results of the local interaction analysis. In another way, the DMM panels in the transverse section were modeled as a "composite" material where the material behavior corresponds to a weighted average of the shear strength between the soil and the panel strengths. For comparison purposes, the first modeling is referred to as the "discrete" model, and the second as the "composite" model.

Table 1
Summary of Material Properties Used in FLAC Deformation Analysis

Material Numbers	Description	Υ (kN/m^3)	c (kPa)	S_u (kPa)	$\varnothing$ (degrees)	G_{max} (kPa)	v (Poisson's Ratio)	V_a (m/s)
1	MSE Wall	21.2	200	-	35	$K_{2max}^{(2)}=75$	0.30	-
2	Fill	21.2	0	-	35	$K_{2max}^{(2)}=75$	0.30	-
3	DMM$^{(3)}$ Pad & Panels	16.5	-	600	-	1.1×10^6	0.30	-
4	Sand	18.5	-	-	35	75,000	0.30	200
5	Sand	18.5	-	-	35	140,000	0.30	275
6	Sand	18.5	-	-	35	170,000	0.30	300
7	Clay	18.5	-	46	-	83,000	0.48	210
8	Clay	18.5	-	58	-	140,000	0.48	285

Material Numbers	Description	Υ (kN/m³)	c (kPa)	S_u (kPa)	$\varnothing$ (degrees)	G_{max} (kPa)	v (Poisson's Ratio)	V_a (m/s)
9	Clay	18.5	-	70	-	187,000	0.48	315
10	Sand	18.5	-	-	35	99,800	0.48	230
11	Sand	18.5	-	-	35	153,000	0.48	285
12	Sand	18.5	-	-	35	193,000	0.48	320
13	Clay	18.5	-	63	-	118,000	0.48	250
14	Clay	18.5	-	72	-	170,000	0.48	300
15	Clay	18.5	-	84	-	205,000	0.48	330

NOTES:
(1) Groundwater is based on an artesian condition, represented by a unit weight of water of 12.3 kN/m³.
(2) Values of shear modulus, Gmax, for the MSE wall and fill materials are calculated using:
$G_{max}(psf) = 1,000 \, K_{2max} \, (o_v[psf])^{0.5}$
(3) DMM: Deep Mixing Method
Υ Unit weight
c   Cohesion
S_u Shear strength
ϕ Friction angle
G_{max} Shear modulus
V_s Shear wave velocity

The results of FLAC analyses using the material properties shown in Table 1 indicate that either model is stable, resulting in very small calculated deformation under the gravity loading conditions, indicating higher-than-one factors of safety against instability of the DMM system. The final results will not provide the system's margin of safety nor provide the performance mechanisms of the DMM system under limiting conditions.

An attempt was made to approximate margins of safety by assigning a "shear strength ratio" to each material. This ratio varied from 1 (no strength reduction) to 0.6 (40 percent strength reduction). The analysis case shown on Fig. 9 corresponds to a longitudinal panel spacing of 4 m and panel shear strength of 66 kPa. In this analysis, the DMM panels start to reach yield moments when the shear strength ratio is about 0.7. These yielding nodes, in spite of very small horizontal displacement values shown on Fig. 2, indicate the one mode of system failure may be the panels yielding in bending. The results of the FLAC analysis shown on Fig. 9 indicate that the computed relative movement between DMM panels and the surrounding soils for the discrete model case are very small. This indicates that the DMM panel is not cutting through the surrounding soils.

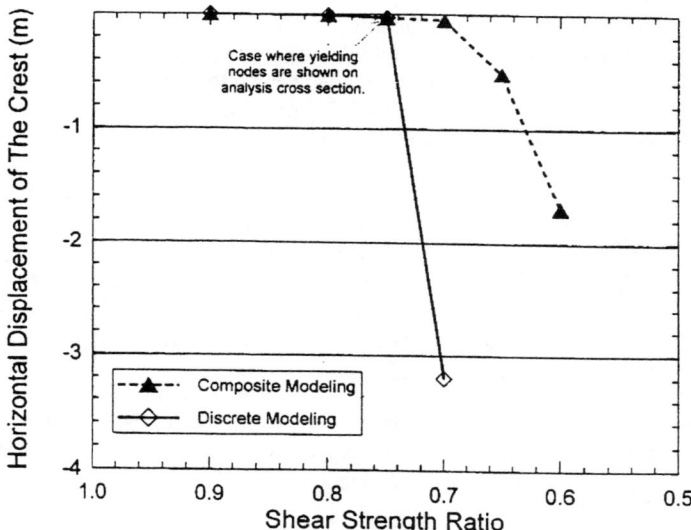

Figure 9. Relationship Between Horizontal Displacement and Shear Strength Ratio for I-15 Wall

FIELD TESTING

In order to verify the results obtained in the above desk studies, a full scale field trial of the VERT Composite Gravity Wall is currently underway at Texas A&M University. This test wall will be constructed 8.5 m high with a full 16 meter wide level bottom section of wall plus two 12 meter wing walls (Fig. 10).

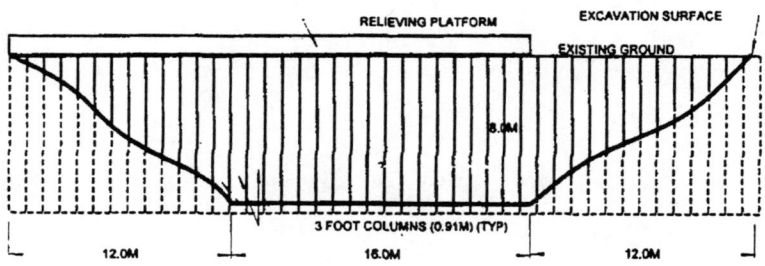

Figure 10. Elevation View of Gravity Wall - Texas A&M University

Inclinometers, extensometers and optical surveying will be used to monitor movement in the wall as it is excavated in three stages to the full 8.5 meter depth. After several weeks of monitoring, it is planned that if movements are not excessive (greater than 50 mm), additional loading will be placed on the wall by means of an earthen embankment placed just behind the gravity wall. A decision of whether or not to fail the wall will be made at that time based upon the recommendation of the Peer Review Committee.

The Texas A&M site is also the location of another type of wall installed in 1993 by Schnable Foundation Co. with the help of the Federal Highway Administration. This is a soldier pile and lagging wall, tied-back with anchors (5 Briaud). This wall has been the subject of numerous papers and is still being monitored for movement. Since the walls are located in the same soils, within 40 meters of one another, it will be possible to compare the reaction of the two walls to the loadings during excavation.

It is hoped that this full scale test will establish the composite gravity wall as a viable method of installing temporary excavation support as well as providing a permanent part of the final structural support of walls and basements.

CONCLUSIONS

Gravity retaining walls constructed by the use of DMM have been used since 1990. Three walls, all constructed using 100% cement treatment of the gravity mass, have been constructed in the United States. In order to make this type retaining structure more economical, less than 100% treatment of the gravity mass has been proposed. Treatment of the gravity mass by use of a discrete array of soil cement columns has been modeled by two separate researchers. A prototype wall has been constructed at Texas A&M University utilizing a replacement ratio of approximately 35% of the gravity mass. This wall is currently undergoing monitoring of its behavior after excavation of 8.5 meters at the face of the wall was completed and a surcharge of 3 meters was placed at the rear of the wall. Recent measurements show that the wall is performing in line with predictions of the Virginia Tech University model. Maximum recorded movements to date approximate 18 mm.

REFERENCES

Druss, David L., "Central Artery/Tunnel Project", University of Wisconsin, 1997.

Jasperse, Brian H., *Installation of Vertical Barriers Using Deep Soil Mixing*, HazMat Central 1990, Rosemont, IL, 1990.

Jones, C.I.F.P., Earth Reinforcement & Soil Structures, 1985.

Nicholson, Peter J., and Eddy Chu, "Excavation Support Remediation By Ground Treatment", ASCE Hershey PA Proceedings, 1994.

DESIGN AND CONSTRUCTION OF A DEEP SOIL MIX RETAINING WALL FOR THE LAKE PARKWAY FREEWAY EXTENSION

By E.W. BAHNER[1] Member, ASCE, A.M. NAGUIB[2]

ABSTRACT: The extension of a new freeway in an urban area may require advanced construction methods and design to minimize land acquision, disruption to existing neighborhoods, and costs. The extension of the Lake Parkway Freeway in suburban Milwaukee, Wisconsin faced all of these and many more challenges. The State of Wisconsin solicited bids for the construction on a design/build (D/B) basis to take maximum advantage of contractor innovations and to minimize schedule and costs. The construction contract was awarded to the winning team which proposed the Deep Mixing Method (DMM) of wall construction. The freeway extension is about 1000 meters long and runs through a residential area of St. Francis, Wisconsin 1000 meters west of Lake Michigan. Most of this section of the freeway will be depressed below grade in a trench about 9 meters deep. The freeway alignment also involves rerouting a railroad line, surface streets, and overhead electrical transmission lines. There are also buried high pressure sludge lines, water lines, and other underground utilities which needed to be protected and/or avoided during construction of the freeway. In order to minimize the impact to the neighborhood, the State Department of Transportation specified that the completed wall be waterproof, minimize construction noise, and minimize schedule impacts to the railroad and electrical transmission lines. The ability to provide a flexible schedule was critical to coordinating with the utility companies and the neighborhood. The retaining wall was composed of a 12 to 18 m deep soil-cement DMM wall which was seated in clay till and low permeability glacial sediments. The DMM wall was supported with walers and permanently tied-back with earth anchors. A 610 mm thick facade of cast-in-place concrete was used as an architectural covering. At the ends of the trench the freeway gradually returns to grade, so in these areas the wall is purely for groundwater cutoff and is not exposed. The middle one third of the wall is covered with a bridge which supports the railroad and surface street crossing. A reinforced DMM wall was used at the bridge abutments to take the full lateral loads of the excavation, and eliminate the need to design the piles of the new bridge foundations to take these loads. The construction of this structure using DMM and the D/B contracting method proved successful for the owner and contractors.

INTRODUCTION

The Lake Parkway Freeway is an extension of Interstate 794 designed to connect the Hoan bridge to Layton Avenue north of General Mitchell International Airport. Where this freeway passes through the city of St. Francis, the Wisconsin Department of Transportation (WDOT) initially proposed to construct an elevated freeway. However, public outcry in the city of St. Francis, especially over the potential division of the city by the bridge forced WDOT and their consultant to consider other options including a tunnel and depressed roadway section. A depressed roadway was ultimately selected as the more cost-effective of the two alternatives. To construct

---

By E.W. BAHNER[1] Geotechnical Engineering Manager, Woodward-Clyde International, Milwaukee, WI, USA, Email: ewbahneØ@wcc.com, Ph..414.513.0577. AND A.M. NAGUIB[2], Office Manager, Geo-Con Inc., Denton, TX, USA, Email: amnaguiØwcc.com Ph. 940.383.1400.

the depressed roadway, WDOT decided on the design/build (D/B) approach to take advantage of contractor innovation, and minimize both schedule and cost. The D/B team of Schnabel Foundation Company, Geo-Con, Inc. and Woodward-Clyde

International proposed a permanently tiedback deep mixing method (DMM) wall, the approach that was ultimately deemed the quickest and most cost-effective method. An example section of the DMM wall is shown in Figure 1. The anchors and wales are not shown for clarity. This project represented the first time that this innovative earth retention method was used in the upper midwest. The construction of the DMM wall was completed 1 month ahead of schedule and below budget.

PROJECT DESCRIPTION

The project consisted of approximately 912 meters of depressed roadway beginning at East Elizabeth Avenue on the south end of the alignment, and ending approximately 305 meters north of East Morgan Avenue. The alignment is positioned in a railway/utility corridor that runs through a residential area of St. Francis, and is approximately 1000 meters west of Lake Michigan (Figure 2). Construction of this portion of the Lake Parkway alignment required the rerouting of an existing rail line, an existing street, and an overhead high voltage electrical line. Since the depressed roadway would be as much as 9 meters below grade, buried high pressure sludge, sewer and water lines were also of major concern. One of these lines was a 2100 mm water line that carries most of the City of Milwaukee's drinking water. Such utilities pass below the roadway parallel or perpendicular to the alignment. These utilities required special consideration during construction by either avoiding them if they crossed below the lowest points of the cutoff walls, or grouting around them to cut off potential leakage paths where they passed through the cutoff walls.

SUBSURFACE CONDITIONS

The soil borings drilled for the subject alignment revealed a highly layered profile of fill, clay, silt and sand underlain by very stiff clay till. These glacial sediments show no definite pattern of layering. The clay till below the sediments is highly overconsolidated, and is present within 6 to 12 meters in of the ground surface on the south end of the alignment, and as deep as 15 to 18 meters on the north end of the alignment. Limestone bedrock is present at a depth of 30 meters or more below the existing grade.

Groundwater was measured at depths of 0.6 to 3.7 meters in deep and shallow wells along the alignment as part of the original geotechnical investigation. Shallower perched water was also anticipated in areas with sand and silt layers.

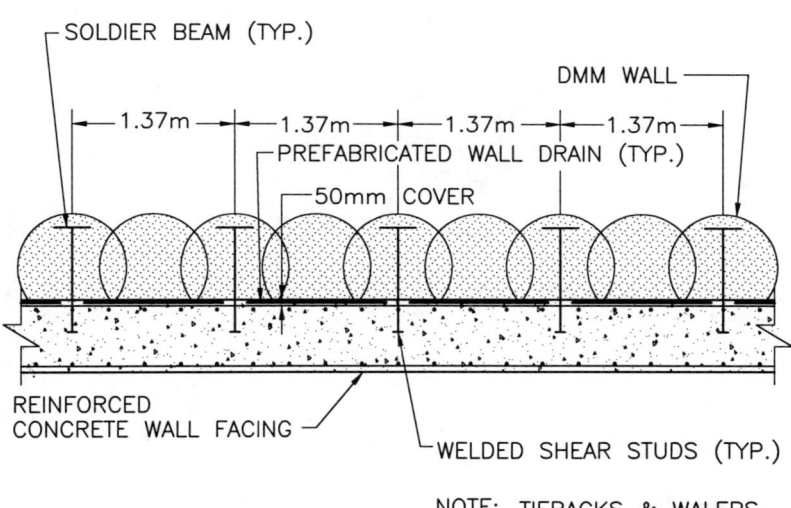

Figure 1 - Typical Wall Plan

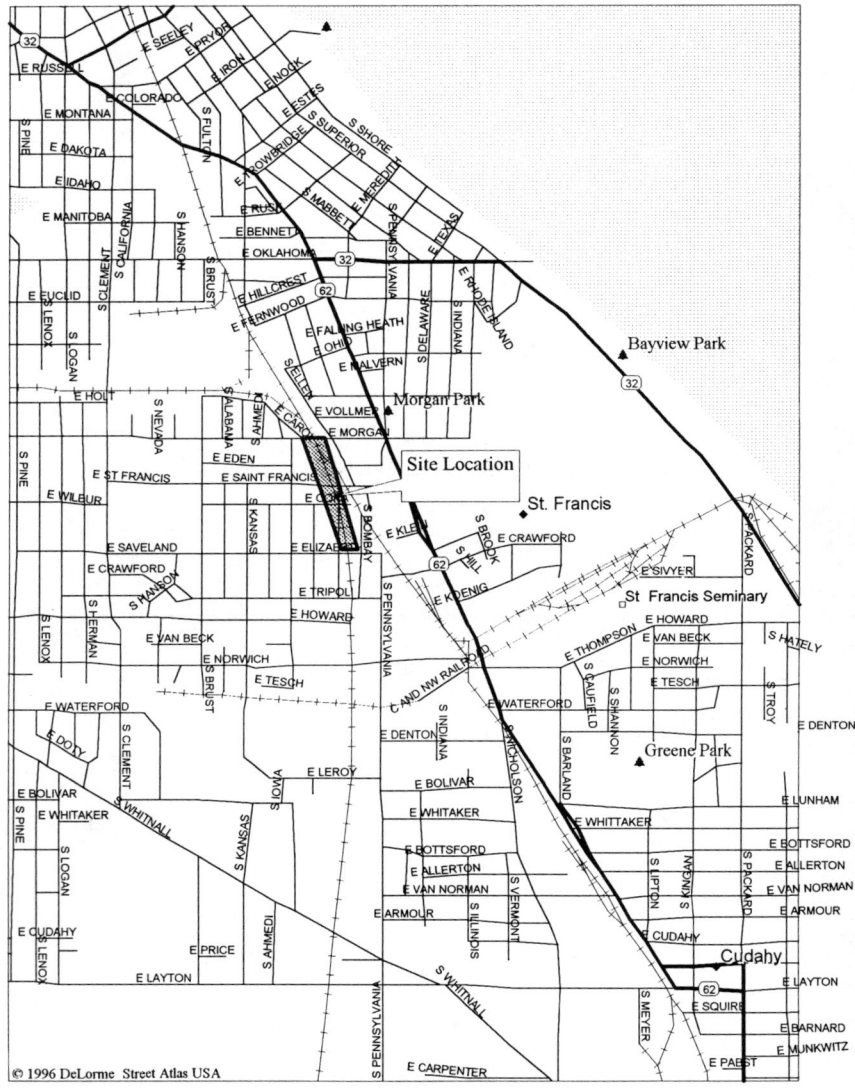

Figure 2 - Vicinity Map

CUTOFF WALL DESIGN/DESIGN CRITERIA

The design and construction of the DMM cutoff wall was the responsibility of Geo-Con and Woodward-Clyde, while the tiebacks and facing walls were designed by Schnabel Foundation Company. The design/build documents prepared by the WDOT identified the following criteria for the cutoff wall/retention system design:

- Minimum design life of 75 years
- A maximum groundwater infiltration rate of 6200 liters/day per meter (500 gpd/lf)
- A maximum groundwater table drop of 152 mm at a distance of 15 meters behind the cutoff walls.
- Maximum lateral wall movements not exceeding 25 mm.
- A minimum facing wall thickness of 610 mm at the base of the wall.

Driven sheeting was deemed an unacceptable scheme due to the potential for leakage through the interlocks, and between sheets driven out of interlock.

DMM WALL DESIGN

The geology of the site consists of highly layered deposits of sand, silt, silty clay, and clayey silt. These layered deposits are underlain by very stiff clay till which gradually slopes down to the north. To effectively create a cutoff around the perimeter of the depressed roadway section, a combination of DMM structural walls and non-structural cutoff walls were built. The termination depths of the cutoff walls were determined using the finite element program SEEP/W and hand calculations. A wall permeability of 1×10^{-9} to 1×10^{-7} m/s was assumed in the analyses. These analyses indicated that the design criteria identified above could be met by either keying the cutoff walls into the underlying very stiff clay till layer, or where the depth to the till layer was greater, penetrating intermediate layers of stiff clay, silt and clayey silt. In many cases, the thickness of these layers was substantial. The results of the SEEP/W analyses generally resulted in expected groundwater inflows of 2.48 to 30 lpd/m (0.2 to 2.4 gpd/lf), and drawdowns on the order of 25 mm for wall permeabilties of 10^{-9} to 10^{-8} m/s. These inflow rates increased to 62 to 186 lpd/m (5 to 15 gallons gpd/lf) and drawdowns of 76 to 229 mm when a wall permeability of 10^{-7} m/s was assumed. Accordingly, it was decided to key the cutoff walls into the till layer south of St. Francis Avenue where it was relatively shallow, and penetrate the shallower low permeability sediments with cutoff walls constructed north of St. Francis Avenue.

JET GROUT CUTOFFS

Jet grouting was used as the sole means of groundwater cutoff around underground utilities. The diameter of these utilities ranged from 300 to 3100 mm. The cutoffs were created by drilling vertical and angled holes to create soil-cement collars around the subject utilities that were of equivalent strength and permeability to the adjacent DSM wall. Where necessary, potential flow paths below the soil-cement collar and along the line of the pipes were lengthened by extending the collar laterally along the pipe behind the plane of the DMM wall.

CONSTRUCTION PROCEDURES

The structural wall was constructed in two main steps. The first step consisted of drilling and in-situ mixing of the soils with cement-bentonite grout. This was followed by installation of steel soldier beams on 1.37 meter centers within the freshly-mixed (i.e.; prior to curing) columns.

The grout was produced on-site using two high-shear lightning batch plants. Bentonite slurry was prepared by mixing dry bentonite with water. The slurry was pre-hydrated in the first batch plant, and was then pumped to a second plant for cement addition and conveyance to the soil-mixing rig. Each batch plant was equipped with a silo for dry storage and delivery (See Figure 3). The silos were mounted over each batch plant. Rotary vane feeders were installed at the bottom of each silo to facilitate ingredient metering and delivery to each plant in accordance with a predetermined cement-bentonite design mix.

Upon grout preparation, the cement-bentonite mixture was conveyed to the soil-mixing rig. The Deep Mixing system consists of a 136 metric ton crane supporting a set of leads that guides four, 0.9 m diameter mixing shafts/augers (Figure 4). As the shafts/augers were advanced vertically into the in-situ soils, the cement-bentonite grout was injected through the hollow stems of the mixing shafts and discharged at the auger heads. By combining auger flights and mixing paddles along the shafts, the soil is lifted and blended with the grout in a pugmill-like fashion. Once the wall design depth was reached, the mixing shaft rotation was reversed and the mixing process was repeated as the shafts were retracted, leaving homogeneously-mixed soilcrete columns. Wall continuity was assured in two ways: First, by the overlapping configuration of the auger flights and mixing blades of adjacent shafts. Second, by the overlap created between previously constructed columns and new columns (See Figure 5). Furthermore, the grout mixture was injected as the augers were advanced allow mixing during advancement and retraction of the augers.

After mixing, the steel soldier beams were set in the freshly mixed columns using a small vibratory hammer mounted on a 91 metric ton crane. The beams were lowered through steel templates that were surveyed in-place and anchored over the freshly mixed wall columns. The templates were used to assure that the specified spacing between the soldier beams was maintained, and that the plumbness/verticality of the beams was consistent with the specified tolerances. The templates were aligned with an established project baseline. The required alignment tolerances specified in the project drawings were as follows:

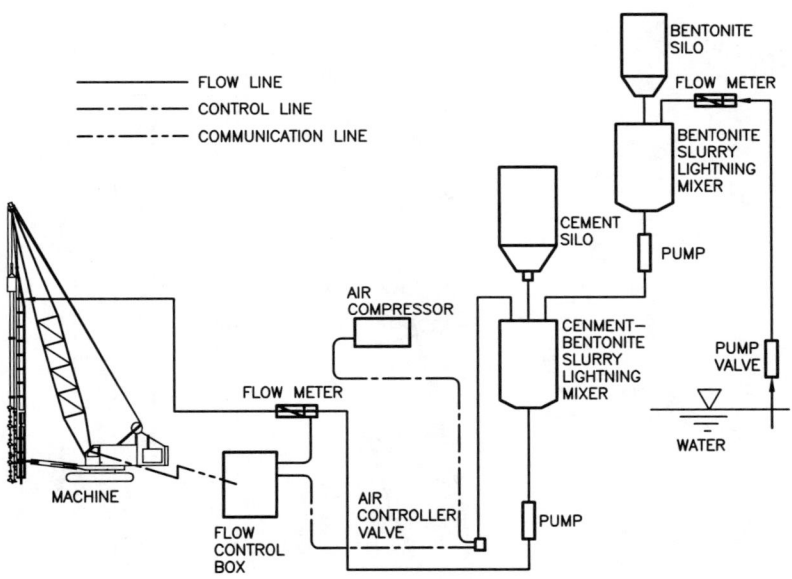

Figure 3 - Schematic Batch Plant

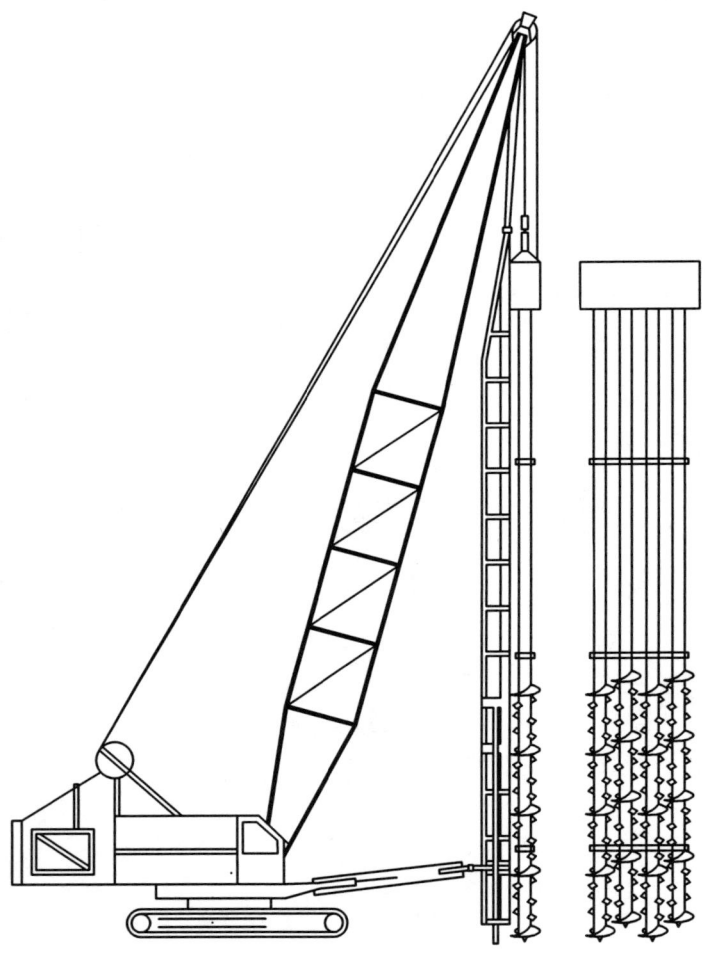

Figure 4 - Schematic of Four Shaft DMM Rig

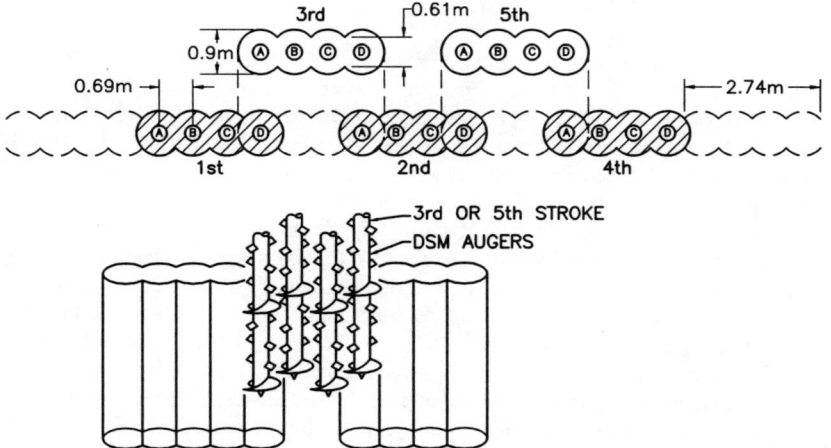

Figure 5 - DMM Column Overlap Pattern

- Plumbness: $\pm 1\%$
- Horizontal Tolerance of Top Location: ± 50 mm.

Overall, the soldier beams were readily installed within these criteria. Beams installed outside of these tolerances were removed and re-driven with a vibratory hammer when the mix was still plastic.

Other work associated with the DMM retaining wall consisted of installation of tieback anchors, wales, and a cast-in-place concrete facing by Schnabel Foundation Company. This was done upon DMM wall curing and mass soil excavation along the faces of the walls.

CONSTRUCTION MONITORING

The Deep Soil Mixed wall was monitored during construction to ensure that the specified design parameters were achieved. This consisted of obtaining "wet", remolded soilcrete samples for permeability and unconfined compressive strength testing. The samples were obtained at various locations and depths along the DMM wall alignment. Permeability testing was performed on collected samples after a 14-day cure time. Unconfined compressive strength was performed on collected samples after 3, 7, 14, and 28 days of curing. The design unconfined compressive strength of the wall was 0.5 MPa. The results of three day tests completed on the test cylinders were typically 2 to 4 times the design strength. Hydraulic conductivity was measured in the laboratory using flexible wall permeability tests, and measurements typically ranged from 1×10^{-8} and 9×10^{-10} m/s, below the 10^{-7} m/s target. "Wet" soilcrete samples were collected from desired locations using the sampling tool shown in Figure 6. This tool consisted of a steel tube that suspended a sampling bucket along its bottom. The bucket was hydraulically opened and closed as needed. In areas where sampling was required, the steel tube was suspended by a hydraulic excavator and lowered through the freshly mixed soil-cement-bentonite column to desired depth. At this point the sampling bucket was hydraulically opened, thus filling itself with freshly mixed soil-cement-bentonite material. The bucket was then closed and the tube retrieved. Grout molds were used for collecting grout samples. The samples were placed in air-tight containers with wet paper towels for storage until the testing time was reached. Unit weight samples of the fresh grout were taken at the batch plant for each batch mixed to keep the water-grout mix within the unit weight tolerances for the batch before injection.

Post-construction monitoring of the effectiveness DMM wall consisted of groundwater monitoring outside the wall. This was done through the use of piezometers. Deflection was also monitored through the installation of inclinometers at 3 wall locations, and survey monitoring points installed along the walls.

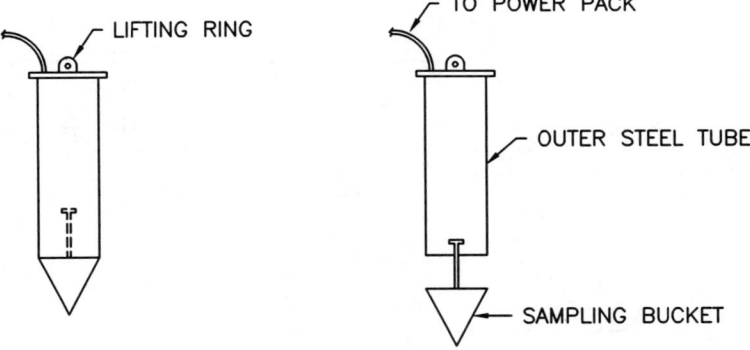

Figure 6 - DMM Wall Sampling Tool

Table 1a
Lake Parkway - Groundwater Level Monitoring
Wells South of St. Francis Avenue

Date	Water Level								
	MW-1	MW-1R	MW-2	MW-2R	MW-3	MW-3R	MW-4	MW-5	MW-6
4/17/97	Dry		8.93		7.00		8.55	9.39	12.98
4/23/97	Dry		9.07		6.87		8.72	9.62	13.32
5/7/97	16.79		8.12		5.68		8.90	9.56	14.00
5/15/97	16.90		8.40		6.43			9.94	
5/22/97	16.90				7.02		8.98	10.32	14.53
6/3/97	16.91				6.68		9.17	10.54	
6/13/97	16.91				7.21		9.34	10.54	14.93
6/20/97	16.46				5.80		9.13	10.05	14.12
6/30/97	13.11				3.90		8.32	9.85	13.24
7/3/97	11.00				4.07		7.76	9.17	11.12
7/30/97	13.19				6.76		8.74	10.33	14.28
8/5/97	13.56				7.27		8.89	10.42	14.51
11/25/97		12.48		6.50		11.11	10.60	11.94	16.04
12/10/97		12.16		5.94		10.80	10.70	11.92	16.06
2/2/98		11.95		3.63		9.04		10.49	14.95
2/5/98		11.98		3.95		9.12		10.30	14.71
4/27/98		11.50		3.94		8.45		9.88	13.62
5/6/98		11.57		3.99		8.57	8.72	9.93	13.69

Notes:
1. Monitoring Wells MW-1 and MW-3 were installed in Spring, 1997 and were knocked out by ditch grade operations during Fall, 1997.
2. Monitoring Wells MW-2 was installed in Spring, 1997 and was knocked out about 1 month later by excavators.
3. Monitoring Wells MW-1, MW-2 and MW-3 were replaced by MW-1R, MW-2R, and MW-3R in Fall 1997.

Table 1b
Lake Parkway - Groundwater Level Monitoring
Wells North of St. Francis Avenue

Date	Water Levels						
	MW-7	MW-7R	MW-9	MW-10	MW-11	MW-12	MW-13
Range	0.00	2.97	5.67	9.77	5.40	2.49	6.53
4/17/97							
4/23/97							
5/7/97							
5/15/97							
5/22/97	Dry			19.55			
6/3/97	Dry						
6/13/97	Dry		11.92	20.58			
6/20/97	Dry						
6/30/97	Dry		7.50	17.15			
7/3/97	25.45		7.43	16.42			
7/30/97			10.77	17.89			
8/5/97			11.17	18.53			
11/25/97		23.08	13.00	23.80	20.93	9.46	9.22
12/10/97		23.32	13.10	24.10	21.08	9.36	9.38
2/2/98		23.21	11.47	23.48	19.91	8.29	15.14
2/5/98		23.08	10.74	23.32	19.38	8.08	14.22
4/27/98		20.53	9.49	14.33	15.68	6.97	15.66
5/6/98		20.35	9.83	15.10	15.88	7.03	15.75

Notes:
1. MW-7 was installed Spring, 1997 and knocked out Late Summer, 1997.
2. MW-7 was replaced by MW-7R.
3. MW-8 was not installed due to lack of access.

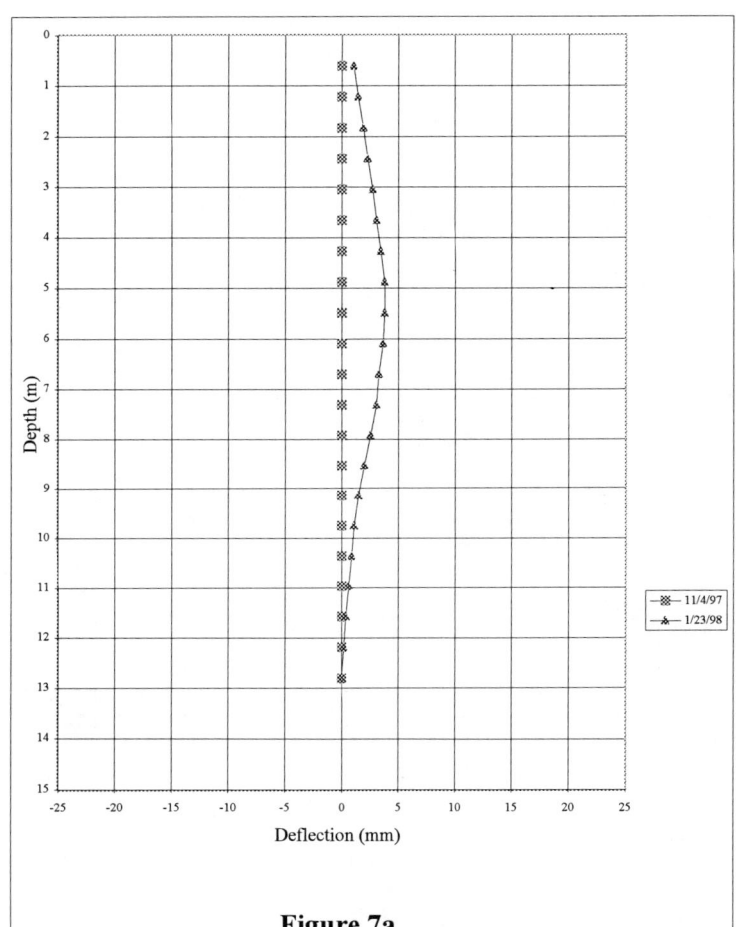

**Figure 7a
Lake Parkway: Inclinometer No. 1
Cumulative Displacement**

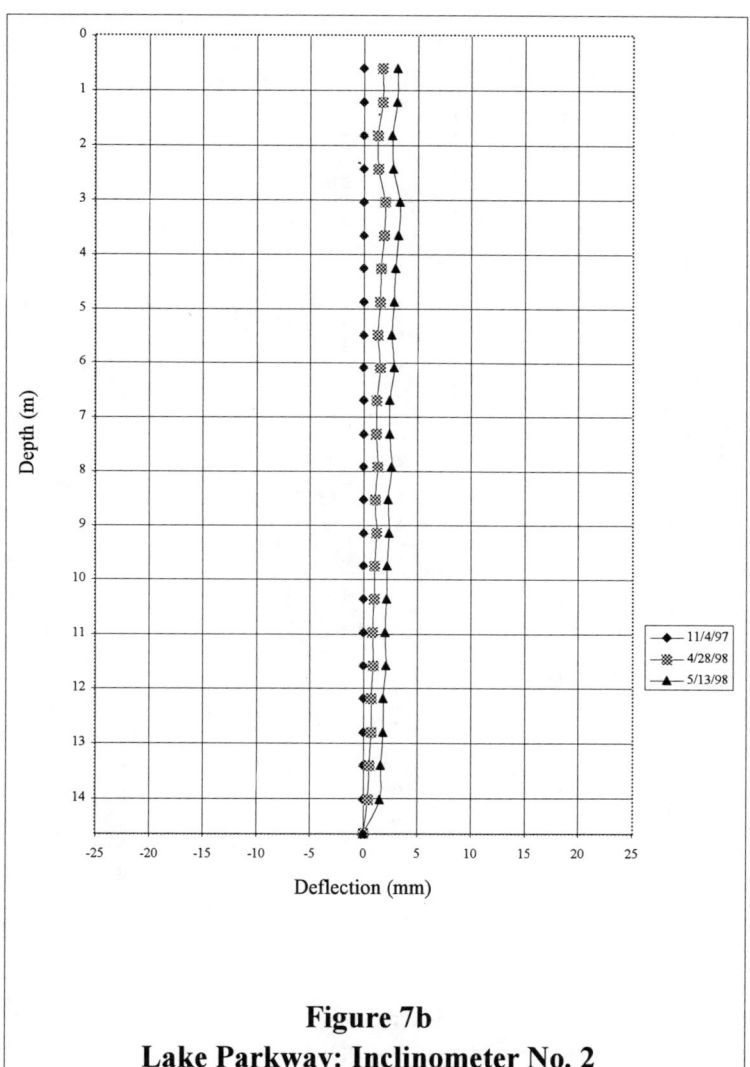

**Figure 7b
Lake Parkway: Inclinometer No. 2
Cumulative Displacement**

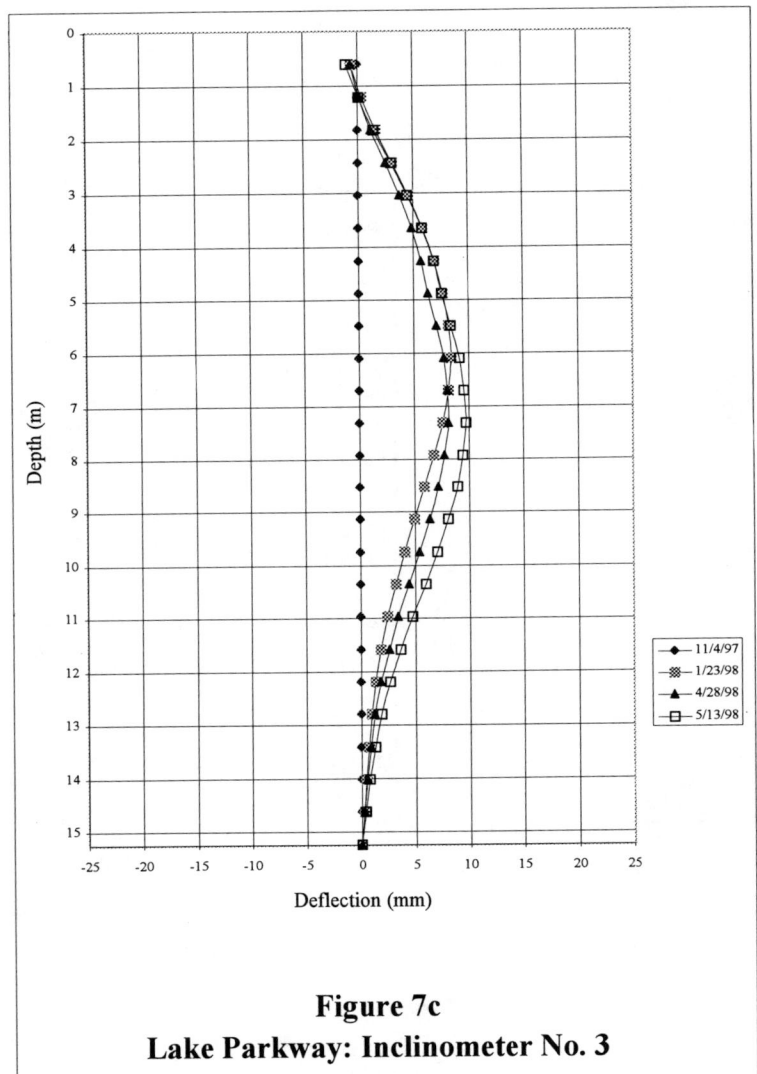

**Figure 7c
Lake Parkway: Inclinometer No. 3
Cumulative Displacement**

SOIL MIX WALL PERFORMANCE

Tables 1a and 1b present the monitoring well levels measured through May 6, 1998. In instances where the wells were installed at the beginning of construction, water levels were measured regularly. Measurements were discontinued in cases where the wells were damaged. As indicated in the tables, the water level ranges in the monitoring walls varied substantially from location to location. Given the highly layered profile and based on local experience, the variations are due to seasonal water table fluctuations, and localized drawdown through ports opened in the DMM wall during tieback installation. A review of the of ground water level readings shows a slow decrease in water table depth as tieback installation was completed and the ports through the wall were sealed.

Inclinometer readings were taken on August 8, 1997, November 4, 1997, and January 18, 1998. Inclinometer 1 is located in Wall RW-40-185 at Station 171 'LA' + 40; Inclinometer 2 is located in Wall RW-40-180 at Station 167 'LA'+70; Inclinometer 3 is located in Wall RW-40-179 at Station 162 'LA' + 80. The measurements show that the lateral movement measured by the inclonometers is well below the 25 mm maximum lateral movement limit. Optical survey data is generally consistent with the inclinometer data. Graphical results of inclinometer readings are presented in Figures 7a, 7b and 7c.

Excavation of the wall and construction of the permanent facing proceeded in the fall and winter of 1997 to 1998. Exposed portions of the DMM wall were exposed to repeated freezing and thawing, attributed to the unusually mild Wisconsin winter. As such, frost expansion and associated surficial crumbling of the exposed DMM wall was observed in areas. However, this phenomenon was considerably less pronounced where the walls were protected. At one bridge abutment location where deterioration was the most pronounced, a protective shotcrete layer was placed in advance of bridge abutment pile driving.

CONCLUSIONS

The Lake Parkway project provides an excellent example of the DMM earth retention technique. The approach resulted in the construction of a structurally sound, watertight and aesthetically pleasing finished wall. DMM provided a notable cost savings over comparable diaphragm wall systems, and resulted in a shorter construction time schedule.

ACKNOWLEDGMENTS

The authors respectfully acknowledge D/B team member Schnabel Foundation Company and their professionalism on this interesting and challenging project. The design support provided by Mr. Richard Tocher and Mr. David Manka of Woodward-Clyde Denver, and the exceptional efforts of field engineer Mr. Jeremy

Craven of Woodward-Clyde Milwaukee and DSM specialist Mr. John Thall of Geo-Con are also acknowledged.

Soil-Cement Pile/Column - A System of Deep Mixing

Osamu Taki[1], M. ASCE and Roy A. Bell[2], M. ASCE

Abstract

As an unique single drill stem deep mixing system, soil-cement pile/columns have been installed within the past 8 years on projects located in the San Francisco Bay Area. This system is capable of producing uniform soil-cement mixtures in clayey and/or sandy soils. The superior quality and predictable shape of the pile/column provides reliable foundation support and is also used in other applications.

In July, 1997 the Japan Building Center published the "Guideline of Design and Quality Control of Ground Improvement for Buildings - Deep Mixing System by Cement Reagent" (Sugimura Y. 1997). The guidelines were based on the evaluation of about 9,000 soil-cement pile projects in Japan, many with load tests, that were completed within the past 15 years.

In this paper we introduce the soil-cement pile/column installation system and mixing procedure, summarize design concepts developed by the Japan Building Center and their application in U.S. projects, and describe various applications and recent U.S. case histories.

Introduction

Present terminology used for some of the soil improvement methods and systems which introduce cement into the soil with various types of drilling and grout injection methods do not adequately distinguish one system from another. The engineering properties of the soil-cement products produced by the various methods are vastly different. Conclusions drawn from a project by one method will not apply to a project that is installed with a different method.

---

[1]President, SCC Technology, Inc., P.O. Box 1297, Belmont, CA 94002
[2]Senior Project Manager, GEI Consultants, Inc., 2201 Broadway, Suite 321, Oakland, CA 94612

After almost ten years from the first demonstration of the soil-cement mixed wall (SMW) system in Hayward, California, this wall system has become recognized by many geotechnical engineers in the U.S. The system was described as one of the deep mixing methods in the Geotechnical Special Publication No. 69, (Schaefer, V.R. 1997). The SMW system is also referred to in Japan as "continuous column wall" and "soil mixing wall" (Tanifuji, S. 1986) (see Photo 1).

Deep soil mixing (DSM) was the terminology used in the U.S. for the double and triple shaft systems used at Jackson Lake Dam project in Wyoming in 1987(Pujol-Rius, et. al. 1989) for the installation of a seepage cutoff system (Jasperse & Ryan 1987). Following that project, "deep soil mixing" was the terminology applied to the soil-cement mixed wall (SMW) system in the U.S.

Deep mixing was developed as a soft ground soil improvement method in Japan during the 1970's. In the Japanese Geotechnical Engineering Terminology Dictionary (JSSMFE 1985), "deep mixing method of soil stabilization" was described as a "generic term for soil improvement involving mixing by force together with chemical stabilizers such as lime or cement within the deep ground on site, but not including compaction. This generic method involves two primary functions: (1) mechanical mixing by force with mixing blades and (2) jetting with high pressure water or air with the chemical stabilizer." In Japan, the use of water-based cement grout is known as deep cement mixing (DCM) while the use of compressed air is known as dry jet mixing (DJM) (Taki and Bell 1997) (see Photo 2).

Originally the SMW system was developed as a means of installing earth retaining walls and to cut off ground water seepage for open excavation work in Japan. This system of overlapped columns can be installed in most soil conditions including gravel and cobbles. The SMW/DSM systems were not presented in any paper in the IS-TOKYO '96 International Conference on Deep Mixing (Yonekura, etc. 1996), inferring that the SMW system is not "deep mixing."

We believe the SMW/DSM system is a drilling system that injects a cement-bentonite slurry and that it is not a deep mixing system. Although the SMW drilling shafts have some paddles that stir and lift the cuttings, very little mixing is accomplished, particularly in clayey soils. Use of the terms "deep soil mixing" gives the impression that it is a mixing system, but it is primarily a drilling and replacement grouting operation, particularly in clayey soils.

Photo 1. Soil mixed wall   Photo 2. Deep mixing, double shafts

Soil-cement pile/column system

The soil-cement pile/column system is composed of: 1) single hollow-stem drilling equipment with a swivel hydraulic coupling at the top of the drill shaft and fitted with a special drill/mixing tool, 2) grout mixing plant, and 3) grout pump with a hose to the swivel at the top of the drill shaft. The system is illustrated in Figure 1.

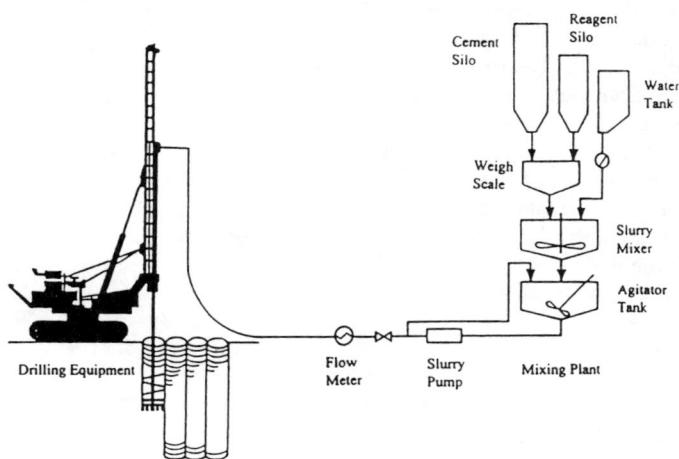

Figure 1. Soil-cement pile/column system

The installation process shown in Figure 2 consists of the following steps:
1. Set the drill shaft at the center of a pile for start of drilling.
2. Drill down to the cut-off depth without grout injection.
3. Drill with injection of cement grout below cut-off depth.
4. After reaching the bottom, continue rotation and grout injection for about 1 minute of bottom mixing.
5. Withdraw while continuing in counter-rotation without grout injection.

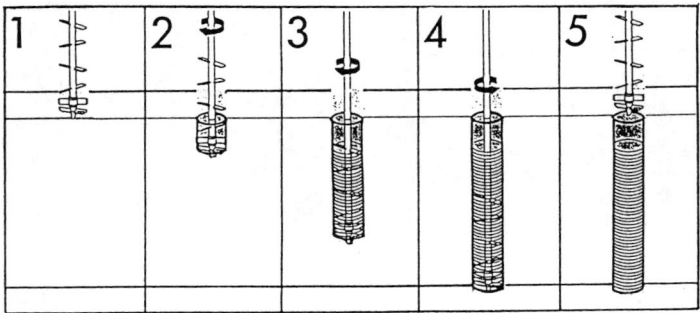

Figure 2. Installation process

Drilling/mixing tool: The key element of the drilling/mixing tool is the share blade, which is a relatively new invention. The share blade is not fixed to the drill shaft, but is free from the rotation of the shaft. The share blade is slightly longer than the drill diameter so that it cuts into the side of the in-situ soils. The share blade has a slight inclination from the axis of the drill shaft.[3]

The cutting-mixing blades are inclined slightly from the horizontal so that they tend to lift the soil-cement mixture when the drill is rotating in the downward penetration direction. The cutting-mixing blades press the soil-cement mixture into the pile/column during withdrawal when in the counter-rotation direction.

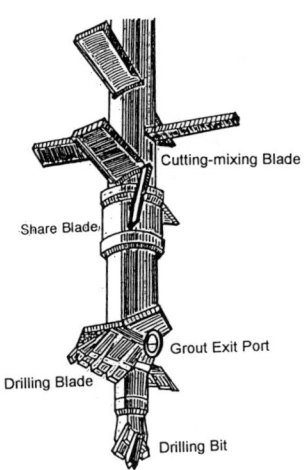

Figure 3. Drill/mixing tool

Mixing process:

1. Cement grout is ejected from the grout exit port located just behind the drill blade where

---

[3]The system and mixing tool are proprietary to SCC Technology, Inc.

the grout mixes with the cuttings as the drill penetrates into the in-situ soils. The soil cuttings and grout rotate with the drill blade and are only partially mixing together.
2. As the drill penetrates deeper, the grout-soil mixture encounters the share blade which is essentially not rotating, causing shearing action which thoroughly blends the soil-grout mixture (the share blade actually rotates about 40 degrees per minute during drill penetration due to its slight inclination).
3. As the drill continues to penetrate deeper, the soil-cement mixture is further blended by the cutting-mixing blades.
4. When the drill reaches the planned penetration, rotation continues for about 1 minute with continued flow of grout and then the grout flow is shut off.
5. It is necessary to perform a bottom mixing routine after completion of planned penetration to thoroughly mix and compact the bottom portion of the pile.
6. After bottom mixing, the tools are withdrawn with counter-rotation of the drill shaft. The slight angle of the cutting-mixing blades compresses the soil-cement into the pile/column during withdrawal of the drill/mixing tool.

Uniform soil-cement mixing: Uniform soil-cement mixing is achieved by the mixing process resulting in no clods of soil or clods of cement, a nearly uniform moisture content and uniform distribution of cement throughout the pile/column. The most important requirements for the successful installation of a soil-cement pile/column are: 1) thorough and uniform mixing of soil and cement grout, 2) proper water-cement ratio of grout, and 3) proper grout injection ratio (Taki and Bell 1997).

The mixing tool has proven to be effective in developing thorough mixing in both sandy and clayey soils. The typical rotation speed of the drill stem is 30 to 60 rpm. At this speed, the rate of penetration is typically about 1m/min. This results in about 1.5 to 3.0 cm of penetration for one rotation of the drill shaft, and therefore, since the blades are in pairs, one on either side of the drill stem, they cut only half of the penetration per rotation, thus producing very small-size cuttings. The rate of drill penetration and share blade rotation is illustrated in Figure 4.

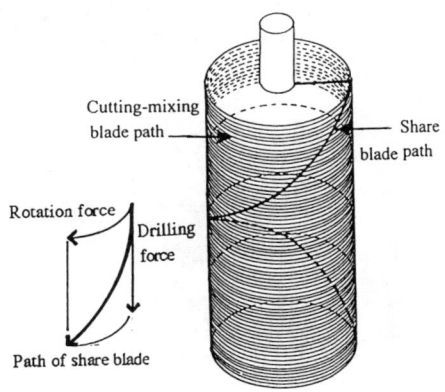

Figure 4. Drill/share blade penetration

It is necessary to use a low water-cement ratio to develop a high-strength soil-cement mixture. It is also beneficial to use a low injection ratio to avoid excessive use of cement and to minimize spoil. By minimizing spoil, the amount of wasted cement is also minimized. When all these advantages are achieved, the soil-cement mixture is relatively dry and stiff. The special mixing tool with share blade is needed to adequately develop a uniform soil-cement mixture with the relatively low water content.

Strength of soil-cement

The main factors that influence the strength of soil-cement are: soil type, cement content, water content of cement grout, curing time and thoroughness of mixing. Generally the strength of soil-cement in sandy soil is higher than in clayey soil. Strength is proportional to cement content and is inversely proportional to the water-cement ratio of the cement grout. The primary prerequisite for achieving the design strength for production piles is to create a uniform soil-cement mixture.

Compressive strength: The relationship of soil-cement strength to soil type and cement content is illustrated in Figure 5 (Hibino 1989). The data in this figure were developed using grout with a water-cement ratio that varied from 0.6 to 1.2 and with a grout injection ratio that varied from 0.23 to 0.35 by volume.

The design strength of soil-cement is achieved by considering the results of laboratory tests, field tests and mix design data from past projects. Strength is evaluated from the results of unconfined compression tests. Ideally, pile strength could be determined by digging up and testing segments of full-size piles. This is seldom practical and therefore unconfined compressive strength of soil-cement is usually evaluated by testing core samples obtained by core drilling or by testing molded samples taken just after pile installation with a special sampler.

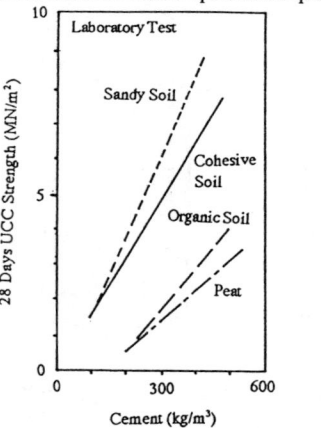
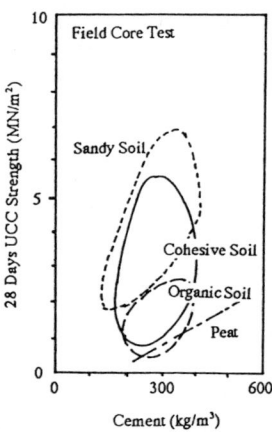

Figure 5. Relationship of ucc strength, cement-content and soil type

Sugimura, 1997, describes the following typical soil-cement strength characteristics that we believe can be useful in design studies:

1. Design strength: Design strength (F_c) of soil-cement (Figure 6) is determined from the average of unconfined compressive strength (q_{uf}) of core samples.

   $F_c = q_{uf} - 1.3\sigma$
   $f_c = 1/3 \, F_c$
   where F_c: Design strength
   f_c: Required strength under normal static loading
   σ: standard deviation of strength test results

   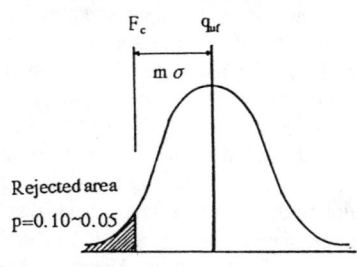

   Figure 6. Design strength

2. Shear strength: Shear strength (F_τ) of
   $F_\tau = \tau_{so} + \sigma_n \cdot \tan \varphi$
   where τ_{so}: shear strength at $\sigma_n = 0$
   σ_n: normal stress

   Sugimura, 1997 (p.p. 43) concludes from the Japanese testing data that the internal friction angle φ of soil-cement is about 30° for most soil types.

   From the direct shear tests, $F_{\tau so} = 0.29 \, F_c$

   From past research with ucc strength (q_u) and tensile strength, $F_{\tau so} = 0.337 \, F_c$

   For the typical condition where $F_c \leq 2 \text{MN/m}^2$, $F_{\tau so} = 0.3 \, F_c$, where $F_{\tau so}$: design shear strength

3. Creep characteristics: Long term design load should be lower than 1/4 of consolidation yield stress (P_c) which is defined as shown in Figure 7. Consolidation testing is recommended for evaluating creep characteristics of organic soil-cement:

   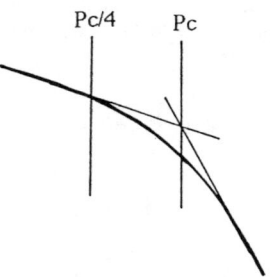

4. Modulus of deformation for design:
   $F_{E50} \fallingdotseq 180 \, F_c$
   F_{E50}: Modulus of deformation at 50% strain for design
   F_C: Design strength

   Figure 7. Interpretation of yield stress, e-log P relation in consolidation test of soil-cement

5. Poisson's ratio μ: From the results of triaxial compression tests and ucc tests, μ varies from about 0.19 to 0.30. A value of $\mu = 0.26$ is typically used for design analyses.

Quality Control/Quality Assurance

Quality control (QC) is required during the installation of soil-cement pile/column to achieve the desired uniformity of the soil-cement mixture and to obtain the design strength. Quality assurance (QA) is obtained from the installation records and from the results of strength tests on representative specimens of the soil-cement.

The first step of QC is to develop the mix design. This is usually accomplished by first preparing and testing soil-cement mixtures in the laboratory and then confirming results with a pilot drilling/mixing test program at the project site. The laboratory program should not be expected to produce the same mixing conditions or strength results as the field program. We recommend the emphasis be put on the pilot test in the field.

Core drilling is often used to obtain test specimens for QA. Piles to be cored can be randomly selected, but additional core drilling and testing should be performed when questionable soil conditions or mix conditions are observed during installation.

Core samples should be taken from near the middle of the pile radius as shown in Figure 8. Uniformity is dependent upon the degree of mixing and is best with an absence of clods. Uniformity is evaluated from core samples by measurement of the amount of soil-clods in Figure 9. The depths of core samples for strength testing should be measured from the top of pile and should include intervals containing the weakest soil layers. The top of pile is often the most severely loaded and therefore should usually be included in the testing program. Core samples should be visually examined for continuity and uniformity of the soil-cement mixture, and for strength. Continuity is defined as the percentage of continuous unbroken, full diameter core. Continuity of core should be more than 95% in sandy soil and more than 90% in cohesive soil.

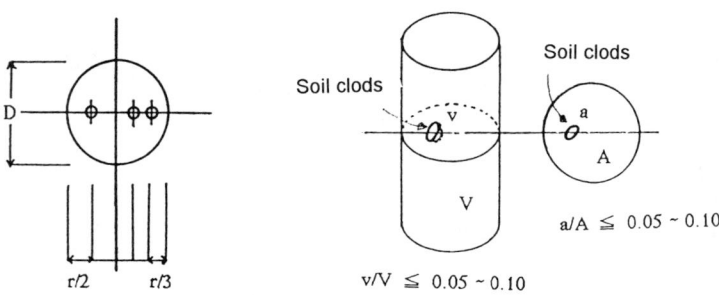

Figure 8. Coring locations   Figure 9. Evaluation of uniformity

Table 1 is a QC/QA testing items for soil-cement pile work in the field.

Table 1. QC/QA Testing Items for Soil-Cement Pile Columns

	Subject	Test Items	Instrument	Frequency
QC	Usage of cement	Total weight	Delivery record	Daily
	Cement grout	Specific gravity	Mudbalance	Each batch
	Injection of	volume	Flow meter	Each pile
	Mix condition	Drilling speed	Record	Each pile
		Rotation speed	Record	Each pile
		Wet sampling	Trap door	1 ~ 2 time/Day
	Pile length	Shaft length	Drill stem	Each pile
	Pile diameter		Tool diameter	Daily
QA	Continuity	Core drilling	Visual	Random sampling
	Uniformity	Core drilling	Visual	Random sampling
	Strength	Core	UCC test	Random sampling

Design concepts of soil-cement pile

Design concepts for soil-cement pile foundations are shown in Figure 10. Based on the Japanese Guidelines (Sugimura 1997) the vertical load capacity of the pile should consider both skin friction and end bearing. As indicated in Figure 10, lateral capacity is the sum of the base friction of the top of the pile and the surrounding soil. However, lateral capacity is normally based on the friction between the bottom of the building foundation and the gravel leveling layer at the top of the soil-cement piles. Lateral capacity (Q_u) is therefore based on the following:

$Q_u = P\mu$ where P = vertical load, μ = tan φ (Coefficient of friction), φ = Angle of internal friction of gravel (35°)

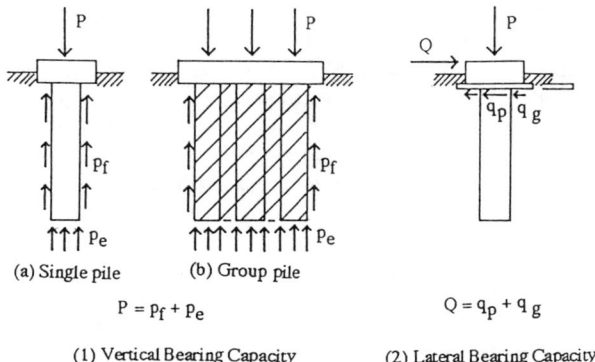

Figure 10. Design concept of soil-cement pile foundation

For the design level earthquake, piles loads should not exceed yield capacity and settlements should be small enough to maintain structural integrity of both the foundation and superstructure. For the maximum credible earthquake, pile loads can approach soil-pile interaction yield capacity or pile structural capacity provided the supported structure does not overturn. Modes of fatal failure are illustrated in Figure 11.

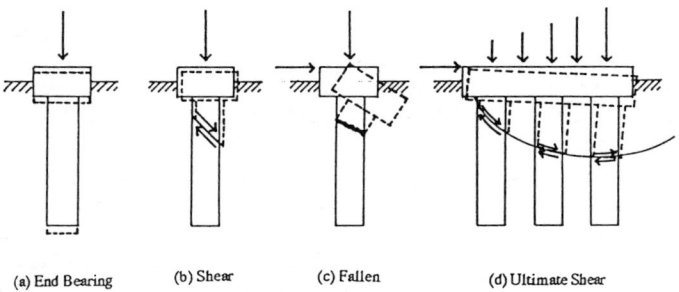

(a) End Bearing   (b) Shear   (c) Fallen   (d) Ultimate Shear

Figure 11. Fatal failure of pile foundation

Applications

Soil-cement pile/columns can be used for ground stabilization, load bearing piles, earth retaining systems and seepage cutoff walls. Especially in the case of buildings located on soft clays and loose saturated sands and silts in the Kobe area, soil-cement piles provided economical foundations that performed well during and following the 1996 strong earthquake shaking (Mitchel, et. al. 1995).

Soil-cement pile/column foundations can provide:
1. Settlement prevention when installed as end bearing piles, through shallow soft ground or into an inclined bearing stratum or through liquefaction susceptible soils.
2. Settlement reduction when installed as friction pile in deep soft ground.
3. Ground stabilization when installed as columns within liquefaction susceptible soils or within weak or compressible strata to prevent excessive settlements, bottom of excavation heaving or lateral displacements of excavation shoring.
4. Ground reinforcement as shear pile or relieving platforms for slope stabilization.
5. Shoring the sides of excavations and for seepage cutoff.

Case histories
1 Building foundation in loose sand
   This has been a primary application in Kobe, Japan. This building (B1 ~ 3F) is in an area that experienced strong shaking during the 1996 Kobe earthquake. The closely spaced soil-cement piles were installed below the isolated foundations through loose, liquefaction sensitive silty sands (Figure 12). Shallow spread footings were supported on the tops of unreinforced soil-cement piles (compared to pile cap construction for pile supported buildings).

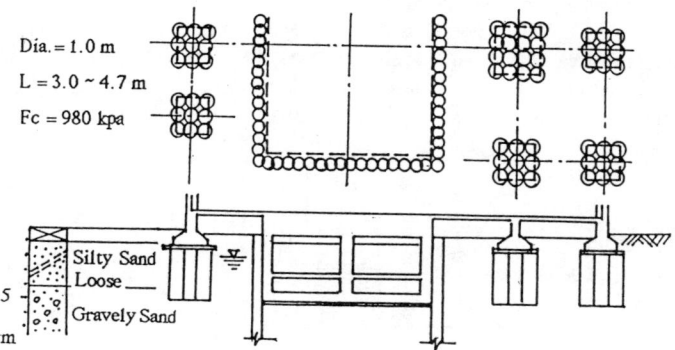

Figure 12. Building foundation in loose sand (Tenocolumn Association 1995)

2 Tank foundation

Sites where the supporting soils/bedrock were overlain with contaminated soils were developed by installing soil-cement piles in and through the contaminated materials, thus eliminating the high cost and safety hazards of removal and disposal of contaminated materials. Where the supporting stratum was at variable depths below a portion of the tank shell, soil-cement piles were used to transfer ringwall foundation loads to the supporting stratum (Figure 13). Some of the soil-cement piles were installed through caving and soil squeezing conditions that would be difficult to avoid with other installation methods. See reference (Bell, Hancock and Taki 1995) for additional information. The tank ringwall foundation was designed and constructed as a shallow footing bearing on the soil-cement piles. Soil-cement piles were also used below a portion of the tank floor to avoid excessive differential floor settlement.

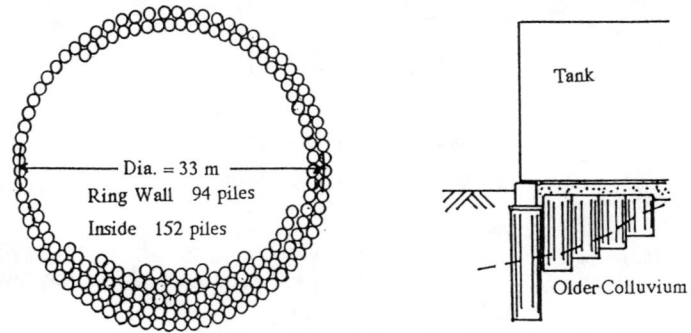

Figure 13. Foundations for large storage tanks (Bell, Hancock, Taki 1995)

3 Seepage cut-off walls

Excavation for the underpass road construction (Figure 14) in an area with shallow ground water was designed as a permanent cutoff wall. About 900 soil-cement columns up to 16m long, 0.75m diameter and 0.6m center to center spacing were installed surrounding the area. Specified Ucc strength was 1380 kpa and hydraulic conductivity could not exceed 1×10^{-6} cm/sec. Soil-cement columns were installed adjacent to a busy railroad track and next to an existing building also functioned as temporary excavation support.

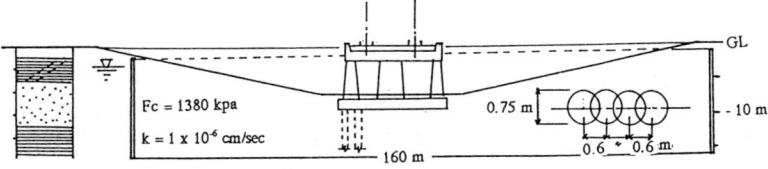

Figure 14. Seepage cutoff walls for a railroad underpass

4 Slope stabilization

A marine laboratory building was planned to be moved from its waterfront site to the top of a nearby dune sand hill. Soil-cement piles were installed to increase the stability of the hillside (Figure 15).

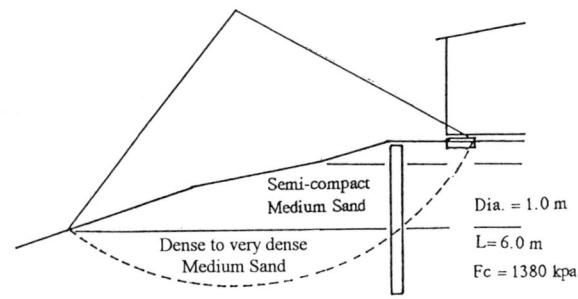

Figure 15. Slope stabilization

Conclusions

1. The unique soil-cement pile/column is installed by a deep mixing system that produces thoroughly and uniformly mixed soil-cement with a relatively low water-cement ratio.
2. The soil-cement pile/column system produces a well-defined pile/column shape with intimate bond at the sides and tip of each pile/column installation.
3. The shearing and mixing by the drill and share blade allows the installation of soil-cement piles as short as 1 meter with a completely uniform soil-cement mixture.
4. The share blade is the key to uniform and thorough mixing of cement grout with the

soils that are penetrated by the special drilling/mixing tool.
Soil-cement pile/column strength is related to soil type, cement grout, injection ratio, water-cement ratio and thoroughness of mixing.
5. The soil-cement pile/columns can be installed in loose saturated soils with no caving, squeezing or loss of adjacent ground. The in situ soils are always supported during installation by the soil-cement mixture.
6. Soil-cement piles can be designed for end bearing and/or side friction support.
7. Soil-cement pile/column strength is related to soil type, cement grout, injection ratio, water-cement ratio and thoroughness of mixing.
8. Soil-cement pile foundations performed well during the 1996 Kobe earthquake.
9. Soil-cement pile/column can be used for load support, earth retention, ground water cut-off, slope stabilization and mitigation of liquefaction effect.

References

Bell, R.A., Hancock, T.M. and Taki, O. (1995), "Hillside Tank on Mix-in-Place Soil-Cement Piles", X Pan-Am Conference on Soil Mechanics and Foundation Engineering, Guadalajara, Mexico, vol. 2 pp. 783 ~ 792
Hibino, S. (1989), "Summary and Histories of Tenocolumn Method", Foundation Engineering & Equipment (Kiso-Ko), pp. 90 ~ 96
JSSMFE Committee (1985), "Geotechnical Engineering Term Dictionary", JSSMFE
Jasperse, B.H., Ryan, C.R. (1987.10), "Geotech Import: Deep Soil Mixing", ASCE Civil Engineering, pp. 66 ~ 68
Mitchell, J.K., Baster, C.D.P. and Munson, T.C. (1995), "Performance of Improved Ground During Earthquakes", ASCE Geotechnical Special Publication No. 49, pp. 1 ~ 36
Pujol-Rius, A., Griffin, P. and Taki, O. (1989) "Foundation Stabilization of Jackson Lake Dam", 12th International Conference of Soil Mechanics and Foundation Engineering, Rio de Janeiro, Brazil, Vol. 2, pp. 1403 ~ 1406
Schaefer, V.R. (1997), "Ground Improvement, Ground Reinforcement, Ground Treatment, Developments 1987-1997", ASCE Geotechnical Special Publication No. 69, pp. 130 ~ 150
Sugimura, Y. (1997), "Guideline of Design and Quality Control of Ground Improvement for Buildings - Deep Mixing system by Cement Reagent", Japan Building Center
Taki, O. and Bell, R.A. (1997), "Soil-Cement Pile/Column", SCC Technology, Inc.
Tanifuji, S. (1986), "Geotechnical Manual", Civil Engineering Research Center
Tenocolumn Association (1995) "Reconnaissance Report of Buildings in Kobe by Tenocolumn Foundation, 1995 Kobe Earthquake-Prompt Report"
Yonekura, R., Terashi, M., Shibasaki, M. (1996), "Grouting and Deep Mixing", Proceedings of IS-Tokyo '96, Volume 1, A.A. Balkema

EXPERIENCES WITH GEOJET PILES IN SANDY CLAYS

by W. M. Isenhower,[1] J. A. Arrellaga,[2] S. T. Wang,[1] and J. O. Johnson,[3] Members, ASCE

ABSTRACT

This paper reports the authors' recent experiences with the construction and load testing of GeoJet piles for the West Field Improvements of San Francisco International Airport. The piles were being tested to prove their capability for use under an underground flood retention facility.

GeoJet piles are a foundation system constructed using a combined jet grouting and drilling process to form a column of soil cement in the ground. A structural member is placed into the cylindrical zone of fresh soil-cement, which has an initial shear strength low enough for the insert to fall into position under self weight or by adding a small amount of energy. Once the soil-cement sets, loads on the structural member are transferred through the soil-cement into the surrounding soil.

The subsurface conditions at the test site appear to be consistent with the typical stratigraphy found in much of the Bay area in San Francisco. Under 0.9 to 1.5 m of top fill material, soft Bay Mud was found and extended to the depth of 5.2 to 6.1 m below the existing surface. The soft Bay Mud is underlain by interbedded layers of silty sand and sandy clay. Old Bay Mud was encountered between depths of 16.8 to 30.5 m. An interbedded dense sand layer occurs at depths of 24.4 to 27.4 m, within the Old Bay strata.

Three test piles and three reaction piles were installed in a triangular pattern with a 3.81 m center-to-center spacing. The test piles had lengths of 12.2, 13.7, and 15.2 m. Different lengths of the test piles were specified so that the optimum length of production piles could be determined. The nominal diameters of all piles were 0.76 m. The full-length structural insert used in the piles were 0.41-m.-OD diameter pipes with a wall thickness of 9.5 mm. The structural inserts were instrumented with telltales at the pile tips and full strain gauge bridges on the structural inserts to measure axial load along the length of the piles. Two test piles were loaded in compression and in tension.

Axial load tests were performed on the 12.2-meter and 13.7-meter test piles. Prior to testing, the axial capacities and load-settlement curves in compression and tension were computed. Good agreement was found for axial capacity in compression and in tension

---

[1] Project Manager, Lymon C. Reese & Associates, Austin, Texas, (512) 458-1128, ensoft@ensoftinc.com.
[2] Associate Engineer, Lymon C. Reese & Associates, Austin, Texas, (512)458-1128.
[3] Vice-President, Condon-Johnson & Associates, Oakland, California, (510) 534-3400.

when the axial stiffness of the GeoJet piles was computed on the basis of the ACI equation for low-strain Young's modulus. Good agreement was also found for the compression load vs. settlement curves. The tension load vs. uplift curves differed from the predicted curves due to the locked-in stresses remaining from the compression tests that were performed beforehand.

After testing was completed, both test piles and one reaction pile were cored to recover samples of the soil cement. Several core samples were under-strength and others were over-strength. The cause for the variation in compressive strengths in the soil cement is unknown, but is not due to either insufficient cement content or incomplete mixing of the soil cement. However, the variation in strength apparently did not affect axial capacities because the strength of the soil cement exceeded the undrained strength of the soft clays at the site.

INTRODUCTION

A GeoJet™ pile is a deep foundation distinguished by its method of construction and materials. A GeoJet pile is cylindrical zone of soil cement formed in-situ into which a structural insert is placed. After the soil cement cures, foundation loads applied to the structural insert are transferred through the soil cement zone to the supporting soil (Spear, et al., 1994).

A typical GeoJet pile is constructed by advancing a rotating soil-processor into the soil at a controlled rate of rotation and penetration. Simultaneously, a computer-controlled grout of Portland cement under high pressure impacts the cuttings through jet-grouting ports in the soil processor. The rates of rotation and penetration and the jet-grouting pressure are varied depending on the type of soil and soil properties. A cylindrical zone of soil-cement is created as the soil processor progresses downward. On reaching the desired depth for design, the tool is removed slowly while grout is discharged to keep the jet ports from being plugged.

A structural insert, typically an open-ended pipe or an H-section, is placed into the cylindrical zone of fresh soil-cement. The soil cement column has an initial shear strength low enough that the insert can fall into position under self weight or by adding a small amount of energy. Once the soil-cement sets, loads applied to the structural insert are transferred through the soil-cement to the supporting soil.

The testing program discussed in this paper was conducted as part of a project for a proposed underground-retention structure for part of the West Field Improvements at the San Francisco International Airport, San Francisco, California (Lymon C. Reese & Associates, 1997). Three GeoJet piles of different embedment depths were constructed at the site on September 9, 1997, by Condon-Johnson & Associates of Oakland, California. These tests were made for the purpose of providing information about their capacities under axial loading in compression and tension.

SUBSURFACE CONDITIONS

Field exploration was conducted by Treadwell & Rollo (1996). The maximum depth of the soil borings was approximately 30 m. The subsurface conditions at the test site appear to be consistent with the typical stratigraphy found at much of the San Francisco Bay area. Under 0.9 to 1.5-m of top fill material, soft Bay Mud was found and extended to the depth of 5.2 to 6.1-m below the existing surface. The soft Bay Mud is underlain by interbedded layers of silty sand and sandy clay. Old Bay Mud was encountered between depths of 16.8 to 30.5-m. An interbedded dense sand layer occurs at depths of 24.4 to 27.4-m, within the Old Bay strata.

A high content of fine particles appears in some of the descriptions of the clayey sand; therefore, internal friction angles based on the Standard Penetration Test are questionable. Therefore, the available data on shear strength for strata that have a complex structure were evaluated by use of engineering judgment. The upper 4.6-m of the soil profile were excavated prior to construction of the test piles. The results of the analysis of the data obtained from the soil investigation, omitting the upper 4.6-m of soil, are presented in Table 1, including values used for strength parameters in some of the interbedded sand/clay layers.

Table 1. Soil Profile at Test Site After Excavation

Depth, m	Soil Description	Strength, kPa	SPT Blowcount
0 - 0.6	Soft Bay Mud (CH)	19.2	-
0.6 - 2.1	Soft clay, w/organic smell	33.5	-
2.1 - 3.6	Med. to dense clayey sand (SC)	-	17 bpf
3.6 - 9.1	Dense silty sand (SM)	-	18 to 34 bpf
9.1 - 10.6	Sandy clay (CH)	100.5	-
10.6 - 12.2	Very dense silty sand (SM)	-	46 bpf
12.2 - 24	Stiff Old Bay Mud (CH)	76.6	-

DETAILS OF TEST PILES, INSTRUMENTATION, AND LOADING

The arrangement of the testing and reaction piles is shown in Fig. 1. The six GeoJet piles form a triangular pattern, with the reaction piles located at the corners and the test piles located on the sides as indicated. The spacing between each test pile and the nearest reaction is 3.8-m center-to-center. GeoJet units of 0.76-m diameter were used for both test and reaction piles. Steel pipes of 406-mm outside diameter and 15.2-m length were used as inserts in all reaction piles. The wall thickness of all steel pipes is 9.5 mm. The test piles have three different penetrations, namely 12.2 m, 13.7 m, and 15.2 m. Each reaction pile had a 15.2-m penetration which was judged as necessary to provide a sufficient capacity in uplift.

The objective of the testing program was to identify the optimum length for production piles. Initial computations indicated that the 13.7-m penetration was optimum for design. Thus, the 13.7-m pile was tested first. It was planned that if the 13.7-m pile could support the design load then the 12.2-m pile would be tested second, otherwise the 15.2-m pile would be tested. The load testing program found that both the 12.2-m and 13.7-m GeoJet piles could satisfy the design requirements. Unfortunately, the 15.2-m pile could not be tested because construction operations were progressing into the area of the test piles.

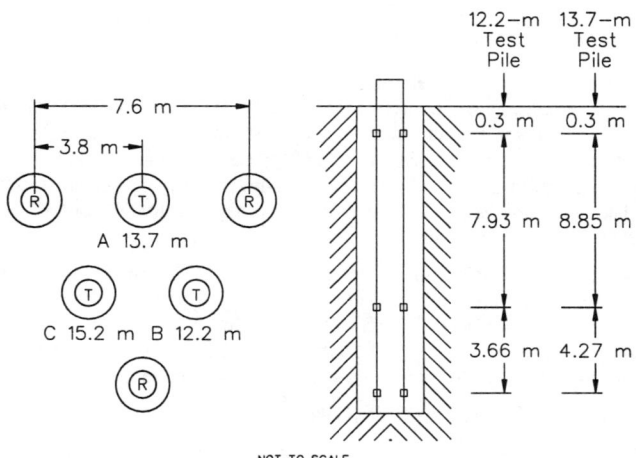

Fig. 1. Layout of Test and Reaction Piles and Location of Strain Gauges on Test Piles

Strain gauges were placed on the steel inserts in the approximate locations shown in Fig. 1. The transfer of axial load to the supporting soil along the length of the GeoJet units is obtained from the strain gauges. Mobilized skin friction can be measured by the difference in strain gauge readings along the upper portion of the GeoJet units and tip resistance can be measured by data measured by the strain gauges located near the tip.

Because of a tight construction schedule and site conditions, only four sets of strain gauges were installed on the 12.2-m GeoJet pile, three sets of strain gauges on the 13.7-m GeoJet pile, and two sets of strain gauges on the 15.2-m GeoJet pile. Several gauges were found to be malfunctioning during testing, perhaps due to insufficient time for curing the water-proof coating of the strain gauges prior to construction.

Each level of gauges shown in Fig. 1 consists of a pair of electrical-resistance, strain gauges on each side of the pipe, composing a full Wheatstone bridge sensitive to strain in the vertical direction, with compensation for the Poisson's effect. This bridge configuration was chosen to eliminate the effects of bending, if present, from the measurements. The gauge circuits were completed and checked over a two-day period before placement into the soil-cement mixture. A portable, digital, strain indicator with a switch-and-balance unit was used for measurements.

DETAILS OF LOADING ARRANGEMENT AND INSTRUMENTATION FOR VERTICAL MOVEMENTS

The loading and reaction systems were constructed by Condon-Johnson & Associates, Inc. Three GeoJet units were used for testing and three additional ones for reaction. In each test, the loading system consisted of two reaction piles, a reaction beam, a bearing block, a calibrated hydraulic ram with a capacity of 2,670 kN, and a calibrated load cell of similar capacity. The load cell was placed between the reaction beam and the hydraulic ram, so the load applied directly to the test pile could be measured precisely. The calibrated hydraulic ram was used as a backup to the load cell. The pressure in the hydraulic ram was obtained with a pressure gauge and a calibration curve provided values of axial load as a function of hydraulic pressure.

Vertical movements at the top of the foundation were measured by four LVDT gauges which were mounted on a stationary reference frame. The LVDT gauges were placed 90 degrees apart around the top of the steel insert of the test pile. Four LVDT gauges were used in order to obtain an average vertical movement in case tilting of the pile head occurred as load was applied. An automatic-level surveying instrument was also used to obtain back-up readings of the vertical movement of the test pile and to check for any movement of the reference frame.

Each test pile contained one telltale, which was terminated one foot from the bottom of the pile. The telltale consisted of a 13 mm diameter pipe, inside of which an unstressed 9.5 mm steel rod was placed. Each telltale was used to measure the overall elastic compression or elongation of the test shaft under load by measuring with a LVDT the relative movement between the upper end of the steel rod and the top of the test shaft.

LOADING SCHEME FOR FIELD TESTING

It was decided to test the 13.7-m GeoJet pile first, using the option to test the pile with 15.2-m of embedment if the capacity measured in the first load test did not meet the design criteria for the project. Alternatively, if the test pile with 13.7-m penetration met the design criteria, the one with 12.2-m penetration would be tested next to assess the proper length for production piles. The latter procedure was adopted.

For each test pile, the compressive-loading test was conducted first. The uplift test was conducted on the same pile after the compressive test was completed. In general, it took an average of about 4 hours to rearrange the loading system for applying tensile loads for the uplift tests.

Testing of the 13.7-m and 12.2-m GeoJet units was completed between September 24 and September 27, 1997. The method used to apply the axial loads to the top of test units was the "quick-load" test procedure, which is specified in the American Society for Testing and Materials (ASTM) D-1143, "Quick Load Test Method for Individual Piles."

This method is commonly used and has received wide acceptance in industry. The quick-load test is performed by adding set increments of load for prescribed increments of time and recording settlement as a function of time for each loading increment. For this particular application, the load was added in 178 kN increments until the specified capacity (1,960 kN) of the GeoJet pile was reached or until the total movement at the top of the pile was over 13 mm, whichever occurred first. The increment of time selected to apply each load increment was 5 minutes.

After the applied load reached about 890 kN, the load was removed in decrements. A second cycle of loading was then applied. For the second cycle of loading, the load was added in increments of 220 kN until the previous maximum load was reached, after which the test pile was again loaded in 178-kN increments until reaching two times the design load (1,960 kN). The total test load was then removed in four equal decrements to zero with readings taken at each decrement.

Readings were taken at 0 minute, 2.5 minutes, and 5.0 minutes after an increment of load was applied to the test pile. The total applied load, measured by the load cell, was immediately reported as the testing progressed. The vertical movements measured by the four dial gauges were averaged and a load-versus-settlement curve was plotted in the field to help control the loading sequence. The readings for the reaction piles were monitored during the testing using conventional surveying equipment and techniques to check for any unusual upward movements of the reaction systems.

The readings collected during each incremental period included readings from the strain gauges, the load cell, the jack pressure, and readings from four LVDT gauges. The LVDT gauges gave the movement of the top of the test pile and one LVDT gauge gave data for the telltale.

TEST RESULTS

The summary of the test results, along with the interpreted load-distribution curves, are presented below for discussion.

12.2-m GeoJet Pile

As planned, the test pile was first loaded in compression, then tension. The maximum compressive load of 1,960 kN was reached at the total movement of about 11.4 mm. The load-versus-settlement curve under the compressive load for the 12.2-m GeoJet pile is presented in Fig. 2 along with the predicted load-settlement curve computed using SHAFT 3.0 (Reese and Wang, 1995). After the maximum load was reached, loads were removed in approximately equal decrements of 445 kN. The rebound movement at the top of the GeoJet pile after the load was completely removed was about 5.3 mm (11.4 mm minus 6.1 mm). The permanent set under the two loading sequences was 6.1 mm.

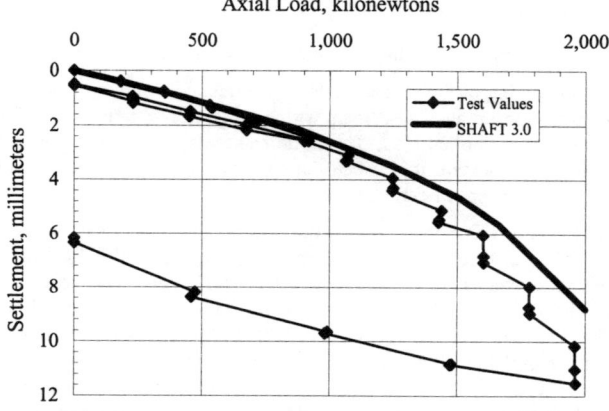

Fig. 2. Load-Settlement Curves for 12.2-m GeoJet Pile

The modulus of elasticity of the GeoJet pile is required for prediction of the load-settlement curve using SHAFT 3.0 and the subsequent analysis of load-test data to compute the distribution of axial load along the length of the pile. The ACI Section 8.5 equation for concrete modulus was used for this purpose, using an estimate of the average compressive strength of 2,100 kPa (300 psi) for the soil cement (ACI, 1995). Subsequent coring and testing of the soil cement found an average compressive strength value of 2,150 kPa (308 psi) for the 12.2-m GeoJet pile.

The load-distribution curves for compressive loading the 12.2-m GeoJet pile are presented in Fig. 3. A linear interpretation was used for portions of the curves because the strain gauge installed at the depth of 4.6-m did not function during the test. However, the curves clearly indicate that the tip resistance was either fully undeveloped or marginally developed at the maximum applied load. The movement of the telltale installed near the pile tip showed hardly any significant movement until the last two loading increments. The conclusion is that 95% of the maximum applied load (1,960 kN) was provided by side friction.

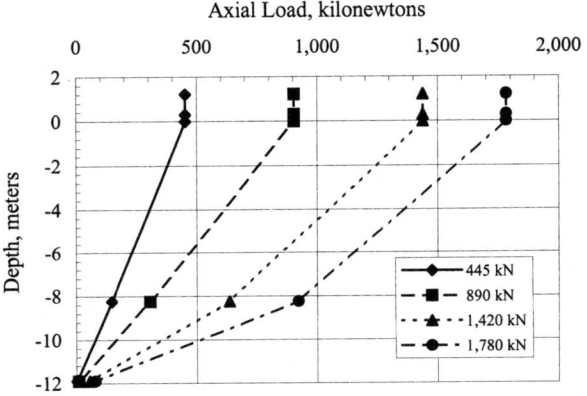

Fig. 3. Compressive Load-Distribution Curves for 12.2-m GeoJet Pile

Elastic shortening of the GeoJet units was computed assuming a composite section consisting of the steel insert and soil-cement with a compressive strength of 1,390-kPa (200-psi). For the maximum applied load, the elastic shortening was computed to be 4.6 mm. Therefore, the relative movement between the soil and the GeoJet pile is about 7.6 mm, which satisfies the design specification.

The uplift test was conducted on the 12.2-m test pile about 4 hours after the compression test had been completed. The maximum applied uplift load of 890 kN was reached at the total movement of about 8.4 mm. The load-versus-movement curve under the uplift loads on the 12.2-m GeoJet pile is presented in Fig. 4. After the maximum load was reached, loads were removed in decrements of 220 kN. The elastic rebound movement at the top of the GeoJet pile after the load was completed removed was about 3.3 mm, resulting in a permanent set of 5.1 mm. Less capacity in uplift was found than in compression because the soil was remolded to some extent during the loading in compression. It was expected that more permanent movement would be developed for uplift tests because the soil was remolded after the compressive test. The stress-strain relationship for remolded soft clay is much softer than for the undisturbed soft clay.

At the maximum applied load, the elastic shortening was computed to be 2.0 mm. Therefore, the relative movement between the soil and the GeoJet pile was about 6.4 mm at maximum applied load, which is less than the allowable value of 7.6 mm established in the specifications. The summary of the elastic compression and elongation and the relative movement between the pile and soil under the compressive and uplift loading is presented in Table 2.

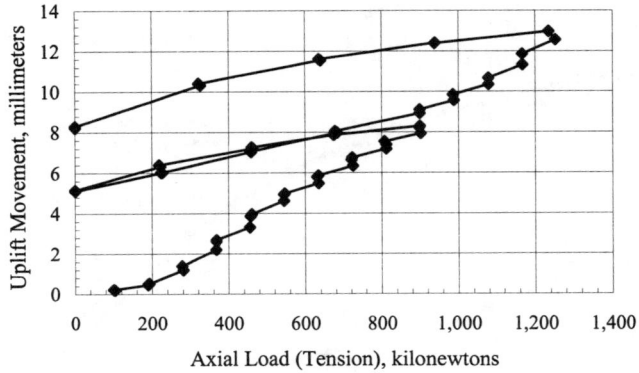

Fig. 4. Tension Load-Uplift for 12.2-m GeoJet Pile

Table 2. Summary of elastic and permanent movements for 12.2-m GeoJet Pile

Length of GeoJet Pile	Direction and Magnitude of Load	Total Movement (mm)	Elastic Movement (mm)	Compression or Uplift (mm)
12.2-m	Compression, 1,960 kN	12.2	4.6	7.6
12.2-m	Tension, 890 kN	8.4	2.0	6.35

13.7-m GeoJet Pile

The same loading procedures were employed for compression and uplift tests on the 13.7-m GeoJet pile as for the 12.2-m GeoJet pile. A maximum compressive load of 1,960 kN was reached at a total movement of about 11.2 mm. The load-versus-settlement curve under the compressive loads on the 13.7-m GeoJet units is presented in Fig. 5. A predicted load-settlement curve computed using SHAFT 3.0 is also shown in Fig. 5. After the maximum load was reached, loads were removed in decrements of 445 kN. The rebound movement at the top of the GeoJet pile after the load was completely removed was about 5.1 mm (11.2 mm minus 6.1 mm). The permanent set for the two loading sequences was 6.1 mm.

The load-distribution curves for the first compressive loading sequence for the 13.7-m GeoJet pile are presented in Fig. 6. Similar to results found for the 12.2-m GeoJet pile, the tip resistance was either undeveloped or only slightly developed under the maximum applied load. The movement of the telltale installed near the pile tip showed no significant movement. It can be concluded that 98% of the maximum applied load (1,960 kN) was provided by the side friction for the 13.7-m test pile.

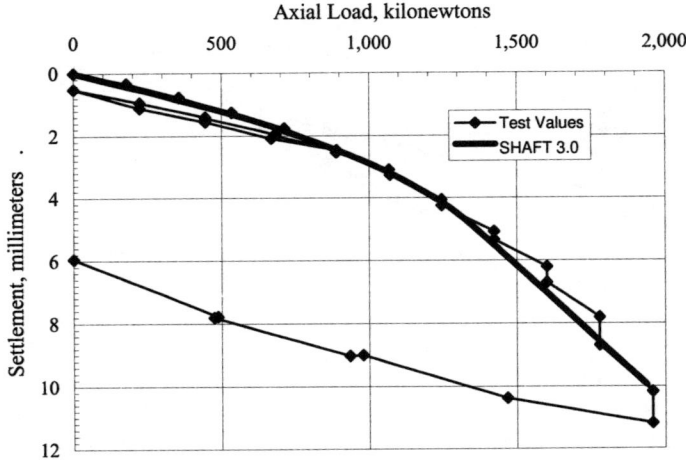

Fig. 5. Load-Settlement Curve for 13.7-m GeoJet Pile

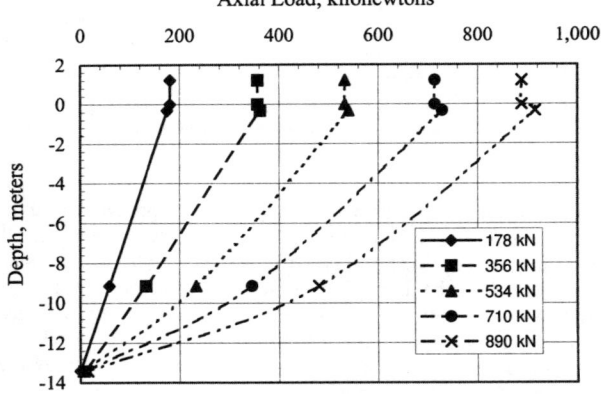

Fig. 6. Compressive Load-Distribution Curve for 13.7-m GeoJet Pile

Elastic shortening of the GeoJet units was computed assuming a composite section consisting of the steel insert and 1,380-kPa (200-psi) soil-cement. For the maximum applied load, the elastic shortening was computed to be 4.8 mm. Therefore, the relative movement between the soil and the GeoJet pile is about 6.4 mm, which satisfies the design specification of 7.6 mm.

The uplift test was conducted on the 13.7-m test pile about 4 hours after completion of the compression test. The load-versus-movement curve under the uplift loads on the 13.7-m GeoJet pile is presented in Fig. 7. Two applications of load were made; 890 kN and 1,020 kN. The first application of 890 kN resulted in a pile head movement of 8.4 mm. The load-uplift curve exhibited a change of curvature due to the presence of locked-in stresses from the compression test performed earlier. The pile was then unloaded to zero in decrements of 220 kN resulting in a rebound movement of 2.8 mm (8.4 mm minus 5.6 mm). The maximum uplift load of 1,020 kN was reached at a total movement of about 9.65 mm. After the maximum load was reached, loads were removed. The rebound movement at the top of the GeoJet pile after all load was completed removed was about 3.3 mm (9.65 mm minus 6.35 mm), resulting in a permanent set of 0.75 mm under the last application of load and a permanent set of 6.35 mm since the start of testing.

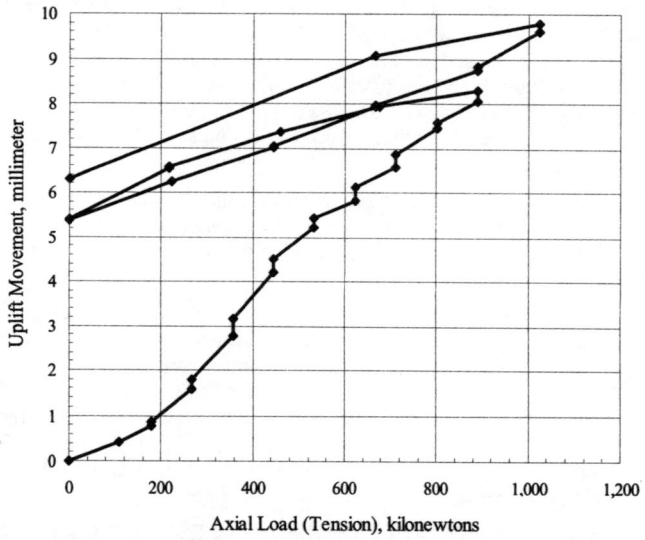

Fig. 7. Tension Load-Uplift Curve for 13.7-m GeoJet Pile

For the maximum applied load, the elastic shortening was computed to be 2.5 mm. Therefore, the relative movement between the soil and the GeoJet pile was about 7.1 mm, which is less than the allowable value of 7.6 mm established in the construction specifications.

The summary of the elastic compression, elongation, and the relative movement under the compressive and uplift loading is presented in Table 3.

Table 3. Summary of elastic and permanent movements for 13.7-m GeoJet Pile

Length of GeoJet Pile	Direction and Magnitude of Load	Total Movement (mm)	Elastic Movement (mm)	Compression or Uplift (mm)
13.7 m	Compression, 1,960. kN	11.2	4.83	6.35
13.7 m	Tension, 1,020. kN	9.65	2.5	7.11

COMPRESSIVE STRENGTH OF SOIL-CEMENT

In general, field experience has indicated that for the GeoJet process the soil-cement mixing in sand is quite efficient and results in a uniform mixture. With regard to clayey soils, experience has shown that the soil-cement may contain some clusters of clay. Clayey soils have generally required higher cement content and more mixing effort to achieve satisfactory results.

A common understanding about soil-cement is that compressive strength depends on the properties of the soil being mixed with the cement grout. Typically, soil cement made from soils with high sand content may have unconfined-compressive strength in the ranges of 3,450 to 6,900 KPa (500 to 1,000 psi), while soil cement made from soft clay may have strength in the ranges of 550 to 1,030 kPa (80 to 150 psi), dependent on the amount of cement in the final mixture. Hydration of cement can be slowed or reduced in situations where organic compounds are present.

The volume of cement grout mixed with soil is usually recorded by computer in the GeoJet process. Unfortunately, the records of these volumes were not recorded due to computer malfunction. This prevents determination of the amount of cement contained in the soil cement.

The quality of soil-cement was investigated by coring both 12.2-m and 13.7-m GeoJet test piles and one reaction pile. The unconfined-compressive strength of the soil-cement varied from 7,670 KPa (1,113 psi) for the sample cored at depths of 6.1 to 6.7-m from one of the GeoJet reaction piles to as low as 152 KPa (22 psi) for the sample cored at the depth of 11.6-m from the 12.2-m GeoJet pile. The average compressive strengths were 2,150 kPa 312 psi) for the 12.2-m GeoJet pile and 2,560 kPa (371 psi) for the 13.7-m GeoJet pile. It is concluded that the zones of low compressive strength may have resulted from organic material in the soft clays retarding hydration of the cement.

ANALYSIS OF BOND RESISTANCE OF GEOJET PILES TESTED AT SITE

Although the compressive strength of soil-cement in the test units is irregular, the factor of greatest importance is that the bonding strength between the steel pipe and the soil-cement is sufficient for the specified maximum axial load. Previous experience with GeoJet piles has found that bonding strength between the soil-cement and the surface of the steel insert is approximately 10% of the unconfined compressive strength of the soil cement. Computations of bond strength were made incrementally along the length of the 12.2-m and the 13.7-m test pile. The computations indicate that the bond resistance for both 12.2-m and 13.7-m GeoJet piles leads to computed loads that are greater than two times the design load. Further, a computation of the axial capacity of each of the units on the basis of load transfer at the interface of the soil-cement and the natural sustain, not shown here due to page length limitations, found that both of the units would sustain loads larger than the design load. The computed axial capacity of the two units was verified during the loading tests because no slip failure between the steel pipe and the soil-cement was detected, even after several cycles of loading and unloading.

CONCLUSIONS

This paper summarizes a load test series on GeoJet piles in which the piles were loaded in both compression and in tension. Both the 12.-m and 13.7-m GeoJet piles were found to meet the design requirements for this project.

The possibility exists at the project site that the hydration of the soil-cement may have been impeded or delayed by the presence of organic materials in the Bay Mud. The irregularity in the strength of the soil-cement samples, cored from the test piles, indicates that compressive strength of some zones of the soil cement was less than originally anticipated.

The results of this load testing program confirm that settlement predictions for GeoJet piles can be computed using the same design procedures commonly used to compute settlement of drilled shafts, provided that the compressive strength of the soil cement is used to compute elastic modulus in the ACI equations. Ultimate axial capacity of GeoJet piles can be computed using the same procedures as used for drilled shafts, provided that the bond between the structural insert and soil cement is not broken before the axial capacity in skin friction is fully developed. As for other types of foundations, the axial capacity used for design is governed by either the allowable bearing capacity or by the allowable settlement under loading.

ACKNOWLEDGEMENTS

The authors wish to acknowledge the support and contribution of the following individuals. The site investigation and soil characterization were performed by Treadwell & Rollo, Environmental and Geotechnical Consultants of San Francisco, California. The GeoJet process was developed by Mr. Lonnie Schellhorn of Gualala, California. The data acquisition system and much of the instrumentation for the loading tests were provided by Mr. John Lemke of Delta Geotechnical Services, Sacramento, California. Lastly, we thank the San Francisco International Airport, Facilities, Operations, and Maintenance Division for considering and selecting GeoJet piles for use on this project.

REFERENCES

American Concrete Institute, *Building Code Requirements for Structural Concrete (ACI 318-95) and Commentary (ACI 318R-95)*, ACI Committee 318, American Concrete Institute, Detroit, 1995, 369 pp.

American Society for Testing and Materials, Standard Test Specification D-1143-81 (Re-approved 1987), Standard Test Method for Piles Under Static Axial Compressive Load, Section 5.6 "Quick Load Test Method for Individual Piles," 11p.

Lymon C. Reese & Associates (1997), "Testing of GeoJet Units at Site of Proposed Detention-Pond Structure, West Field Improvements, San Francisco International Airport," Report Submitted to Condon-Johnson & Associates, Oakland, California, November 13, 1997, 55 p.

Reese, L. C., and Wang, S.-T. (1995), *Computer Program SHAFT, Version 3.0 for Windows*, Drilled Shafts under Axial Loading, User's Manual, Ensoft, Inc., 227 p.

Spear, D., Reese, L. C., Reavis, G. T., and Wang, S.-T. (1994), "Testing of GeoJet Units Under Lateral Loading," Proceedings, International Conference on the Design and Construction of Deep Foundations, Orlando, Florida, Vol. 2, pp. 969-979.

Treadwell & Rollo (1996), "Geotechnical Investigation, West Field Improvements, San Francisco International Airport, San Francisco, California," Report to the City and County of San Francisco, California, Contract 3494, 110 p.

THE APPLICATION OF DEEP MIXING PILE WALLS FOR RETAINING STRUCTURES IN EXCAVATIONS

Yong Shao[1], S. M. ASCE,
Chunming Zhang[2],
Emir Jose Macari[3], M. ASCE

ABSTRACT: Compared to other types of retaining structures used in excavations, the deep mixing method (DMM) pile has many advantages, such as low cost, low vibrations and noise, no pollution, simple and light equipment, and no need for bracing nor dewatering measures. Deep soil mixing pile techniques have been successfully used in many excavation projects in many parts of the world. This paper is based on a case study of the Sunlight Park Hotel in Shanghai, China and describes in detail the design principles and the construction techniques for the DMM piles as retaining structures for the excavation. The paper also presents the results of centrifuge tests used to simulate the DMM pile system, focusing on the deformation mechanisms of retaining structures and the change of the earth pressures. Finally, the paper presents the results of a finite element analysis that is compared with the centrifuge results and the performance of the actual DMM pile system.

INTRODUCTION

Deep soil mixing (DMM) piles, originally developed as a method to improve the soil conditions in soft soil areas, have been used as retaining structures for excavations in Shanghai, China since 1987. Because of the very thick layer of saturated soft clay in the coast areas of China, such as in Shanghai, most excavation systems are supported by steel sheet piles, contiguous bored piles, or diaphragm walls. The reasons for introducing the DMM pile system as a potential retaining structure in excavation are listed below:

- Horizontal struts and other bracing are not needed because the DMM piles are generally designed as self-supported gravity retaining structures; therefore, the large open space allows the excavation to proceed more efficiently.
- No wellpoint systems outside the pit are required for dewatering during excavation because the DMM piles intersect each other to form a continuous soil mixing pile wall, which is good in preventing ground water from seeping inside the excavation pit. Meanwhile, this also avoids problems associated with

---

[1] Ph.D. Graduate Student, Georgia Institute of Technology, Atlanta, Georgia, USA
[2] Senior Engineer, Shanghai Railway Institute, Shanghai, China
[3] Associate, Professor, Georgia Institute of Technology, Atlanta, Georgia, USA

adjacent buried pipelines or other infrastructure because the displacements caused by dewatering is commonly a large portion of the total displacement.

These reasons and several others such as lower cost, low vibration and noise, no pollution, and the light equipment requirement, have resulted in the tremendous popularity of the DMM piles for retaining structures in Shanghai in recent years. Because this remains a relatively new technique, many issues remain to be resolved and studied before one may have a clear understanding of their effective use in practice. The most common questions that may arise are:

- What is the basic principle and analysis method for designing DMM piles as retaining structures?
- How to evaluate the deformation and pore pressure during the excavation?
- What should the proper configuration be for the walls?

This paper attempts to answer some of these questions based on the designing experiences and field instrumentation for an excavation with DMM piles as a retaining structure and some results from the centrifuge test for the same project.

PROPERTIES OF DMM PILES

Unlike concrete piles, DMM piles are formed by stirring the soil in-situ with slurry containing a cement-based agent. The piles gain their strength as a result of a series of physical and chemical reactions between the soil and the cement. Therefore, the properties of DMM piles depend not only on the type of soil and cement but also on the amount of cement added into the soil. In order to make full use of the advantages of DMM piles in designing retaining structures for excavation, it is very important that one clearly understands the reactions that occur within the soil-cement mixture and the resulting mechanical properties.

From laboratory blending tests and in-situ tests, the properties of DMM piles have been shown to be dependent on a number of factors, such as:

- Type of soil and cement
- Ratio of cement to soil
- Water content of cement-soil mixture
- Age of the cement-soil mixture
- Use of other additives

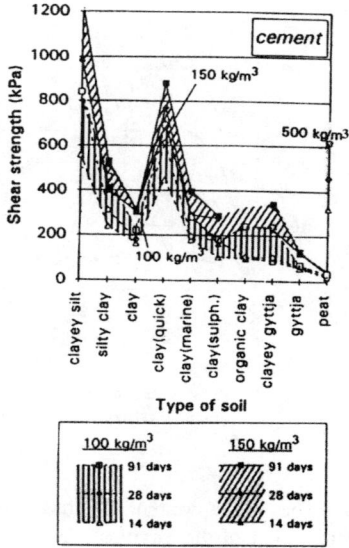

FIG. 1. Shear Strength of Soil-Cement Mixture with Different Types of Soil [Åhnberg, 1996]

Comparisons of the results from unconfined compression tests on different types of soft soil-cement mixtures have shown that considerably varying strengths and stiffnesses can be obtained, see Fig. 1.

One of the most important properties for the soil-cement material is the unconfined compressive strength, which may range from 300 to 4000 kPa depending on the amount of cement that is added into the soil (Fig. 2). The shape of the stress-strain curves is very different for the different specimens. As the stress approaches the maximum strength, for the soil-cement mixture with strength greater than 2000 kPa, the failure occurs in a brittle fashion suddenly without plastic deformation and the residual strength is very low (as A_{20}, A_{25} in Fig. 2). On the other hand, for the soil-cement mixture with strength less than 2000 kPa, the stress-strain response exhibits large plastic deformations before failure (as A_5, A_{10}, A_{15} in Fig. 2).

The results of these tests leads us to conclude that it is desirable to add cement into soil with the cement-soil ratio a_w between 8% ~ 12% for DMM piles in excavation projects for the reasons that follow. In addition, the cement-soil ratio should not be less than 5% because the strength of that mixture will not be sufficient to act as a retaining structure. Finally, it is not recommended that the cement-soil ratio be larger than 15% because the brittle nature of such mixture may result in sudden collapse without warning in the case of overloading.

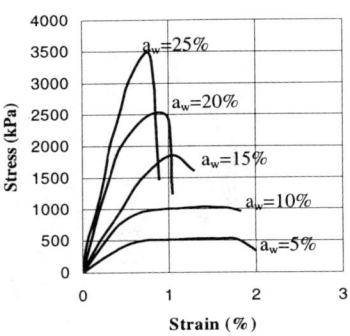

FIG. 2. Stress-strain curves of soil-cement mixture (After Zhou, 1985)

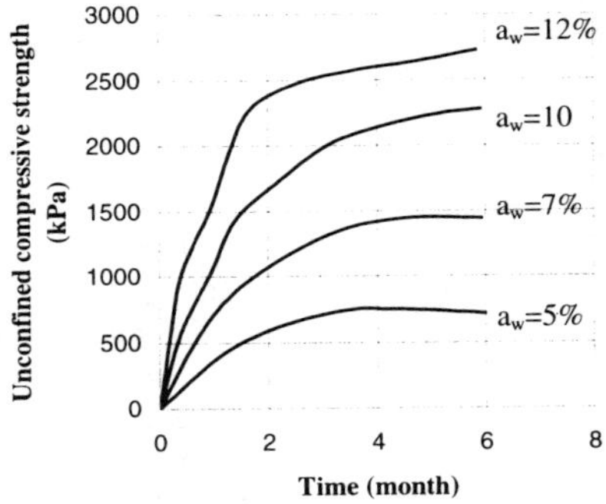

FIG. 3. Change of soil-cement strength with time (After Zhou, 1981)

The water content, w, has great influence on the unconfined compressive strength q_u of the mixture. As indicated in Table 1, q_u increases from 260 to 2320 kPa as w decreases from 157% to 47%. Therefore, it is very important to control the water content prior to the hardening of the soil-cement mixture.

TABLE 1. Relationship of Water Content to Unconfined Compressive Strength [Wang, 1979]

Water content (%)	Natural soil	47	62	86	108	125	157
	Soil-cement mixture*	44	59	76	91	100	126
Unconfined compressive strength**, q_u (kPa)		2320	2120	1340	730	470	260

* $a_w = 10\%$, ** Strength at 28 days

The strength of soil-cement mixture increases with time, however, unlike concrete, the strength continues to increase after 28 days of curing (Fig. 3). For example, the strength at 120th day is about 2.03 times as that at 28th day for a_w = 7%. One suggestion is to use the strength at 3 months as the design strength.

It is shown that some additives, such as fly ash, may be added to soil in order to increase the strength of the mixture. Tests have shown that the strength of the mixture with fly ash may increase on the order of 10% [Zhou, 1988].

Permeability tests have shown that the permeability of the soil-cement mixture decreases as a_w increases. For a cement-soil ratio, a_w, of 10%, the permeability of the mixture is on the order of 10^{-7} cm/sec, which makes the retaining structure relatively impermeable and not needing dewatering operations in the excavation.

DESIGN CONSIDERATIONS

Configuration of DMM pile walls

From laboratory tests, bending and tensile strengths of DMM piles have been found to be much lower than compressive strengths. Therefore, one should make full use of compressive strength in designing the retaining walls for an excavation. The best design choice for these conditions is the self-supported gravity retaining walls. In practice, each individual DMM pile intersects into adjacent ones to form a long continuous wall (Fig. 4).

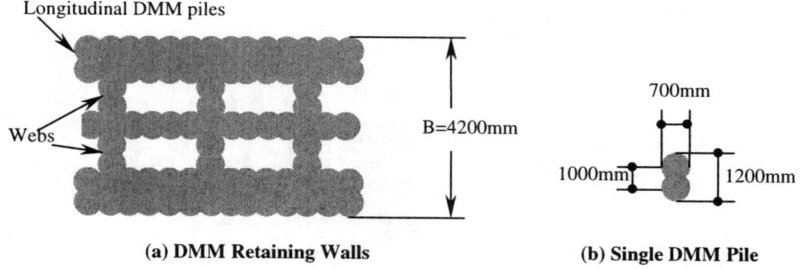

(a) DMM Retaining Walls   (b) Single DMM Pile

FIG. 4. Cross Section of DMM Retaining Structure

However, for economic reasons, it may suffice to solidify the soil throughout the entire thickness (B) of the DMM piles with intermediate webs at a specified spacing. It is very important, however, to make sure of a good connection between webs and longitudinal DMM piles since those webs play an important role in obtaining the required overall stiffness of the retaining walls.

Design Procedures

There are several calculation methods that may be used for the design of DMM pile walls as retaining structures for excavations. Generally, DMM walls may be treated as either a rigid structure or a flexible structure depending on its dimensions. Although it resembles a gravity retaining walls, the DMM walls is commonly thinner and is embedded deeper, and the DMM walls itself is much more flexible than a concrete walls. Therefore, there have been several controversial issues in the design community such as the determination of lateral earth pressures on the walls, how the retaining walls deforms, and the relationship between lateral earth pressure and pore water pressure behind the walls. From recent engineering experience and the use of field instrumentation, it has been concluded that it is reasonable to apply gravity retaining walls design methods to DMM walls to check for stability but not for displacement. Deformation patterns should be evaluated using beam theory or finite element techniques.

Figure 5 presents a proposed simplified earth pressure diagram where the lateral earth pressures are determined based on Rankine Earth Pressure theories. It is assumed that active earth pressures below the excavation depth are uniformly distributed, which has been verified by field monitoring (see case study).

Factors of safety against failure are checked in the same fashion as with the design procedures for gravity retaining walls, therefore, one must check the overall stability, anti-overturning, anti-sliding and anti-seepage. In addition, one must also to check anti-heaving and anti-piping for the excavation. The appropriate factors of safety are:

- FOS (Overall) = 1.1
- FOS (Overturning) = 1.4
- FOS (Sliding) = 1.3
- FOS (Seepage) = 1.3
- FOS (Basal heaving) = 1.5
- FOS (Piping) = 1.5

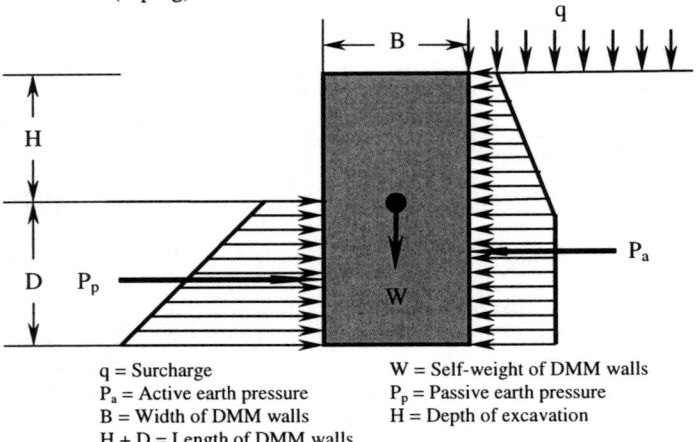

q = Surcharge
P_a = Active earth pressure
B = Width of DMM walls
H + D = Length of DMM walls

W = Self-weight of DMM walls
P_p = Passive earth pressure
H = Depth of excavation

FIG. 5. Earth Pressures on DMM Walls

Similar to the design of other types of retaining structures, the tentative geometry of DMM walls is first assumed based on previous experience. Following the design procedures, checks of all appropriate FOS are performed. If any of the stability factors are not satisfied, one must reselect the geometry and perform the checks again.

CASE STUDY

Project Descriptions

The Sunlight Park Hotel excavation project is one of the successful examples where DMM piles were used as retaining structures. The excavation pit was approximately a 94m by 63m rectangle in the plan view (Fig. 6), excavated to a depth of 6.75m below the ground surface. The representative soil properties of the site are shown in Table 2.

Table 2. Soil Properties in Sunlight Park Hotel

Layer	(1)	(2)	(3)	(4)
Soil type	Misc. fill	Soft clay	Very soft clay to soft clay	Soft clay with sand
Color	Variegated	Brown	Gray	Gray
Thickness (m)	0.3 ~ 2.4	0.5 ~ 1.8	2.1 ~ 3.6	8.0 ~ 15.3
Water content (%)	--	34.3	36.7	30.0
Unit weight (kN/m^3)	18.2	18.6	17.8	18.9
Void ratio	--	1.08	1.21	0.90
Cohesion, c_{cu} (kPa)	--	29	20	75
Friction angle, ϕ_{cu}	--	18	21	27
Unconfined comp. Strength (kPa)	--	65	48	57

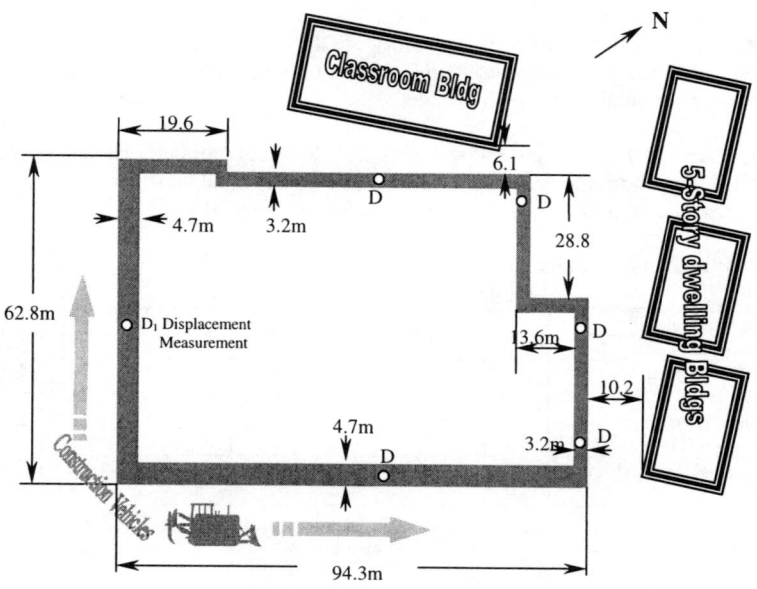

FIG. 6. Excavation site in a Plan View

In the vicinity of the excavation pit were several buildings that needed to be monitored and considered in the design. A five-story building complex was located at approximately 10m from the north side of the excavation pit and an elementary school building was located 6m from the western side. One of the main reasons for considering DMM piles was because of the low vibrations and noise that would be generated during the excavation program. The displacement of those buildings was strictly monitored and controlled during the excavation.

Choice of Schemes for Retaining Structures

The use of traditional steel sheet piles with a dewatering system for the excavation posed the following problems at this site:
- Because of the large excavated area and the complexity of main structures, the excavation pit had to remain exposed for about one year. The cost of renting steel sheet piles for such a long period was considerable.
- Owing to the large excavated area and depth, several rows of deep wellpoints would be needed to provide dry working conditions in the pit. A large scale dewatering system, such as that required for this project, would inevitably cause large settlements in the surrounding ground, which would pose a great danger to the dwelling buildings and classroom buildings.
- Due to the large excavated area, it would be difficult to install the required horizontal struts for the bracing support system.

After considering the above issues and after thorough technologic and economic comparison with other alternatives, a DMM pile retaining walls was finally selected for the following reasons:
- Because of the very low permeability of DMM walls, dewatering inside the pit would have very little impact on the ground water levels outside the pit.
- No horizontal struts were needed since the DMM walls are self-supported structures.
- Large construction space would allow for the use of more efficient equipment.
- Compared to the steel sheet pile scheme, the total cost for the retaining structure system was estimated to be lower by about 30%.

DMM Pile Descriptions

Table 3. Properties of DMM Pile

No. of sample		(1)	(2)	(3)	(4)	(5)	(6)	(7)	(8)	(9)
Water content(%)		33.4	34.6	36.0	23.9	27.9	35.7	35.9	34.0	23.8
Unit weight(kN/m^3)		17.7	18.3	17.3	18.8	19.1	18.4	18.7	17.6	18.6
Void ratio		1.05	1.01	1.14	0.79	0.82	1.01	0.98	1.47	0.81
Unconfined compressive strength (MPa)		1.13	1.36	1.26	1.52	--	1.21	1.27	1.52	2.08
Tri-axial test	c(MPa)	--	--	--	--	0.34	--	--	0.14	0.06
	ϕ	--	--	--	--	42.2	--	--	42.5	0.06
Permeability(10^{-7}cm/s)		--	--	--	3.4	--	--	0.77	2.67	--

Note: Normal Portland cement (Grade 425) was used for these tests.

Following the procedures described earlier for checking all the relevant FOS, a final configuration and dimensions for the DMM walls was chosen. The width of the DMM retaining walls in the west and north sides was 3.2m made of 4 rows of piles, and a 4.7m width with of 5 rows of piles in the east and south side because the construction vehicles would travel on the latter sides. The piles were generally embedded to 10m except the middle of row, which were 3m deeper in order to increase the anti-piping ability. The ratio of cement to solidified soil was chosen to be 10%. Table 3 presents the test results of mechanical properties for the DMM piles. Test samples were obtained from pile cores.

Field Instrumentation

Since the principles and calculation methods for designing DMM walls for retaining structures remain not fully understood, this project was viewed as a research study, hence, additional field instrumentation was used. Measurements of the lateral earth pressure distribution on the walls and variations in pore water pressures during excavation were made in order to verify the conventional design assumptions.

Lateral Earth Pressures

As shown in Fig. 7, before the excavation, the measured earth pressure is linearly distributed with depth, displaying a coefficient of earth pressure at rest (K_0) of 0.66. Once the excavation begins, the walls moves towards the excavation pit, therefore, the earth pressure becomes an active earth pressure behind the walls and a passive earth pressure in front of the walls. The measured earth pressures in the active zone and the upper part of the passive zone, are very similar to the assumed values. However, the measured earth pressure in the lower part of the passive zone is much less than the assumed value. This is because a relatively small displacement has taken place at level, hence not fully mobilizing the passive zone. This difference can be overcome if the interaction between soil and walls is considered. Finite element analysis is one way to consider this interaction.

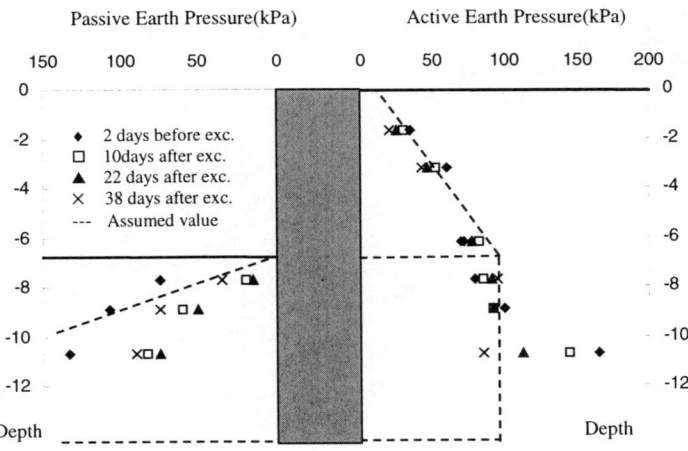

FIG. 7. Lateral Earth Pressure Profile

Pore Water Pressures

Piezometers were embedded behind the walls and the measured values of the pore water pressure are shown in Fig. 8. One may see from this figure that the measured values are close to the hydrostatic pressure during the entire excavation process, in other words, the dewatering inside the pit does not result in much variation of the ground water level outside of the excavation. This means that the DMM walls have very good anti-seepage ability.

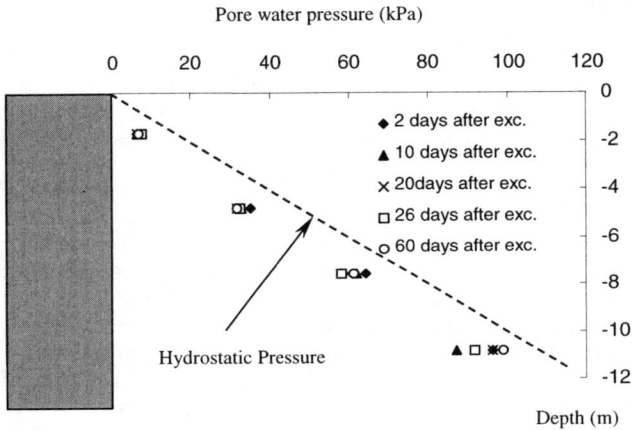

FIG. 8. Variation of Pore Water Pressure

Displacement of the Walls

In general, horizontal displacements on the top of the walls increase as the excavation progresses deeper. The maximum measured displacement at the top was 18.2cm or 1.52% of the wall height. Figure 9 presents additional details on the displacement of the DMM walls.

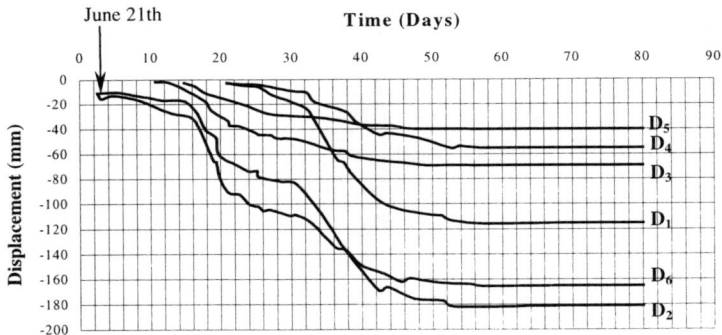

FIG. 9. Horizontal displacement at the top of the DMM walls

CENTRIFUGE TEST

The geotechnical centrifuge has been a device used for several decades to allow engineers to better understand the load-deformation mechanisms in earth structures. This study also used the centrifuge to analyze the deformation mechanisms associated with DMM pile walls in deep excavations (Zhang, 1995). A detailed description of the centrifuge testing program is out of the scope of this paper, however, some remarks regarding the testing procedures and observed results follow.

Preparation for the Model DMM Pile Walls

Based on the dimensions of centrifuge test box, a scale factor of 100 was used in the test (i.e. the centrifuge model is spun at 100 times the acceleration due to gravity.) Similarly as with the proposed prototype configuration, the model DMM pile walls were composed of 4 rows or 5 rows of DMM piles, with the ratio of cement to soil 10%. The unconfined compressive strength of the model piles was between 1.1 ~ 1.3 MPa.

The soil used in the model was the same clay from the site investigation. The clay was thoroughly remolded and mixed with de-aired water to a moisture content of about 90%. The soil was one dimensionally consolidated in the soil container in several layers.

Test Results

The tests were performed in a strong specimen box with one side made out of "lucite" in order to be able to observe the deformation mechanisms. Pictures were taken using a high-speed synchronous data camera and a real time TV system to observe the deformation of soil and walls at different acceleration values. Several miniature earth pressure cells and piezometers were installed in the soil model to measure the earth pressure behind the walls and pore water pressure variation during the test. A summary of some of the test results is shown in Table 4.

FINITE ELEMENT ANALYSIS

Because conventional design methods assume the lateral earth pressure from the Rankine Earth Pressure theory, soil-walls interactions were not considered, which, as was stated earlier, leads to an improper passive earth pressure assumption in the lower part of passive zone. A better analysis approach for such a problem is the use of the finite element method, especially for the deformation analysis.

By taking into account the elastic rebound and residual stress during the excavation, FEM analysis may reflect the influence of a particular stress path. The basic assumptions are:
- Plane strain problem.
- Initially, the DMM walls and soil are at rest (i.e. "K_0" state).
- Since no dewatering is used outside the pit and the DMM walls was rather impermeable, the excavation can be reasonably assumed to be undrained and hence Poisson's ratio is taken as 0.495.

- The material models for both the DMM walls and soil is elasto-plastic following the Mohr-Coulomb strength criteria.
- Staged excavation is simulated by FEM in order to capture the maximum value of force and deformation.

A summary of the FEM results is shown in Table 4.

Table 4. Results from centrifuge and FEM calculation

Methods	Maximum Displacement (cm)			
	Toe of Walls	Top of Walls	Surface Settlement	Basal Heave
Centrifuge Test	4.8	15.3	7.4	2.0
FEM	7.9	11.3	8.3	1.5
Field measurement	10.3	18.2	6.2	2.7

The values from the three different methods are relatively close to each other except for the horizontal displacement at toe of the walls. This is because values of FEM results are the maximum value calculated from first stage to the end of excavation, no long term effect are considered. While the field measurement values are the maximum values measured during the entire project, the displacement may increase even after excavating to the final depth. The centrifuge test, on the other hand, was performed mainly to study the failure mechanism of DMM, hence some test conditions did not strictly match the in-situ conditions.

FUTURE WORK

Because of the advantages of DMM piles previously discussed, this type of retaining structure for excavation projects has become widely accepted. However, practice has also shown some of its shortcomings, such as the very low tensile strength and brittleness of the soil-cement mixture that forms the retaining walls. Hence, as with any other developing technology, there are several issues that warrant further research. This section outlines some of the recommendations that have resulted from the use of DMM pile walls in actual engineering projects.

Increase the strength of DMM Pile and its anti-bending ability

In order to increase the tensile strength of DMM piles, H-type steel beams or steel sheet piles should be placed in the middle of the DMM piles. The resulting composite DMM pile system should be considered as a bending structure in the design stages and not as a gravity walls. In Japan, this kind of DMM pile has been referred to as the SMW method. In China, the steel piles are often replaced with bamboo sticks that are inserted into the middle of DMM piles. The idea is that the elastic modulus of bamboo is much lower than that of steel and it works more compatibly with the cement mixture than steel, in addition to the fact that the cost of bamboo is much lower.

Reduce the Active Earth Pressure

When the excavation depth is relatively large, the lateral earth pressures at the bottom of DMM walls might be too large to be sustained by the DMM walls. This issue may be resolved by adding one row of continguous bored concrete piles to be placed behind the DMM walls and extending them into a firm layer. The bored piles serve as a curtain to reduce the lateral earth pressure on the DMM walls. The amount of stress reduction depends on the spacing between the bored piles and DMM walls, the spacing between each bored pile, and their diameter.

Use Different Configurations

The DMM piles arranged in a continuous arch will favorably change the loading condition that they will experience. Commonly, bored piles are constructed at the toes of the arch, then the earth pressure exerted on the DMM walls will be transferred to the bored piles which are supported by horizontal struts.

CONCLUSIONS

This project proved that the DMM piles are a good choice as retaining structures for deep excavations in very soft soil areas, owing to such advantages as lower cost, reduced vibration and noise, lower pollution, the need of simple and light equipment, and short period of construction.

DMM retaining structures may be designed with the help of the theories used for conventional gravity retaining walls. The assumed lateral earth pressures are similar to the measured values except within the lower part of the passive zone where the assumed earth pressure is much larger than measured values.

A more precise analysis can be performed if one accounts for the interaction between the DMM walls and soil. This is especially true for the deformation analysis. The finite element method is a useful technique for the analysis of these types of structures.

In order to use DMM retaining structures for deeper excavations, measures must be taken to increase the strength of DMM piles and anti-bending strength.

These are some of the most popular topics that should be studied in the future.

REFERENCES

Åhnberg, H. (1996). "Stress dependent parameters of cement and lime stabilized soils." *Grouting and Deep Mixing*, Vol. 1, A. A. Balkema, Rotterdam, 387-392

Wang, Y. L. (19790. "Improve the strength of soft clay by soil-cement mixing technology." *report*, Shanghai

Zhang, S. D. et al. (1991). "Laboratory study and engineering practice for deep soil mixing piles as retaining structure." *Proceeding of the workshop for deep excavation construction technology*, Shanghai Association of Civil Engineering, 114-121

Zhou, G. J. et al. (1981). "Deep mixing method for reinforcing soft clay." *report*, Shanghai

Zhou, G. J. et al. (1985). "Deep Mixing Method." *Soil improvement handbook*, Chinese Architecture and Industry Press, 405-410

Zhang, C.M. et al. (1995) "Centrifuge test for Sunlight Park Hotel deep excavation project", *Test report*, Shanghai

DRY JET MIXING FOR STABILIZATION OF VERY SOFT SOILS AND ORGANIC SOILS

David S.Yang[1], Jack N.Yagihashi[2], and Steve S.Yoshizawa[3]

ABSTRACT

Dry Jet Mixing (DJM), a type of deep mixing technology, uses mixing blades to mix dry reagents such as cement powder or lime with in situ soils to increase the strength and reduce the compressibility of the very soft ground. DJM was developed by the Japanese Ministry of Construction together with private construction organizations. The dry cement or lime is supplied from the plant by air and injected through the ports located at the lower portion of the mixing shaft. After depositing the reagent, the air is released at the ground surface. Due to the feature of using dry powders, DJM provides the flexibility of reagent selection for treatment of various soils that slurry type deep mixing technologies are inefficient or uneconomical to treat. This paper introduces the DJM equipment and construction procedures and presents the engineering properties of the improved soils produced by DJM. Two case histories are used to illustrate the application of DJM for the stabilization of a very soft ground and the treatment of peat and organic soils.

INTRODUCTION

DJM is a soil treatment and improvement technology which pneumatically delivers powdered reagent into the ground and mixes it with in situ soils to form a soil-reagent column. The chemical reaction of the soil and stabilizing reagent increases the strength of the soil reagent column. Columns are installed according to predetermined patterns such as column group, wall, grid, or block for various ground stabilization or reinforcement purposes. DJM was developed and researched during 1977 and 1979 by the Civil Engineering Research Institute of the Japanese Ministry of Construction in conjunction with the Japanese Construction Machine Research Institute. By the end of 1994, 1,820 DJM projects, with a total volume of 12.6 million m^3 treated soil, had been completed (DJM, 1995).

[1,2,3] RAITO, INC., 1818 Gilbreth Road, Suite 145, Burlingame, CA 94010.
Phone: (650) 259-1210 E-mail: dsyang@jaccess.com

The selection of reagent for use in the DJM method is based on the properties of soils to be treated and the purposes of the soil treatment. Dry powders or particles with size less than 5 mm can be delivered and mixed with in situ soils. Currently, cement, cement based reagent, slag cement, and lime are most frequently used, followed by slag, fly ash, and gypsum. The strength gain of the soil-reagent mixture is derived from the hydration and pozzolanic reactions of the soils with the reagent. According to the soil type and the required engineering properties of the treated soil, the reagent dosage can be adjusted to different soil strata. Due to the use of dry reagent, the volume of reagent injection is smaller than other slurry type deep mixing technologies. Consequently, the quantity of spoil and the impact on the surrounding ground are lower.

a) DJM Rig

b) DJM Mixing Blade

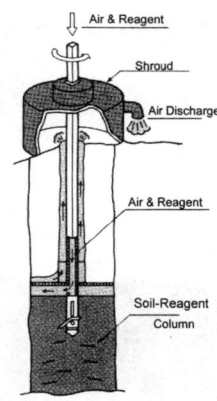

c) Reagent Delivery and Mixing (after DJM 1993)

Figure 1. DJM Equipment

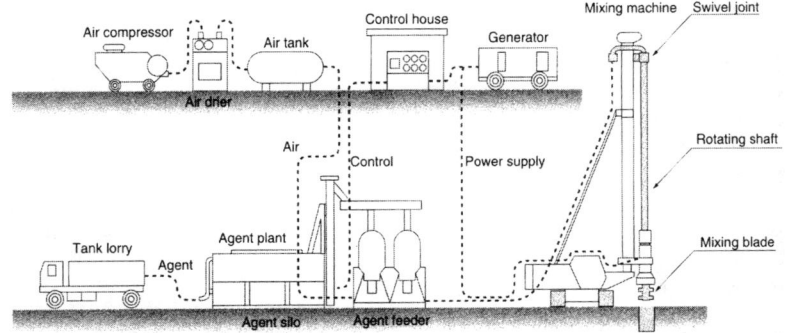

Figure 2. Reagent Supply System

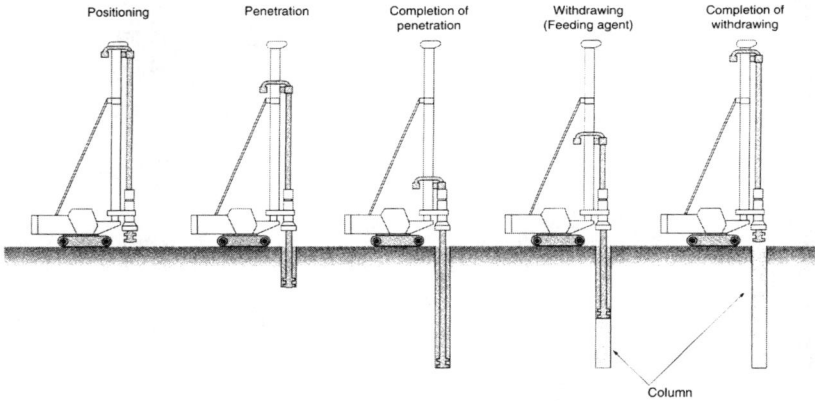

Figure 3. Installation Procedure

Figure 4. DJM Columns

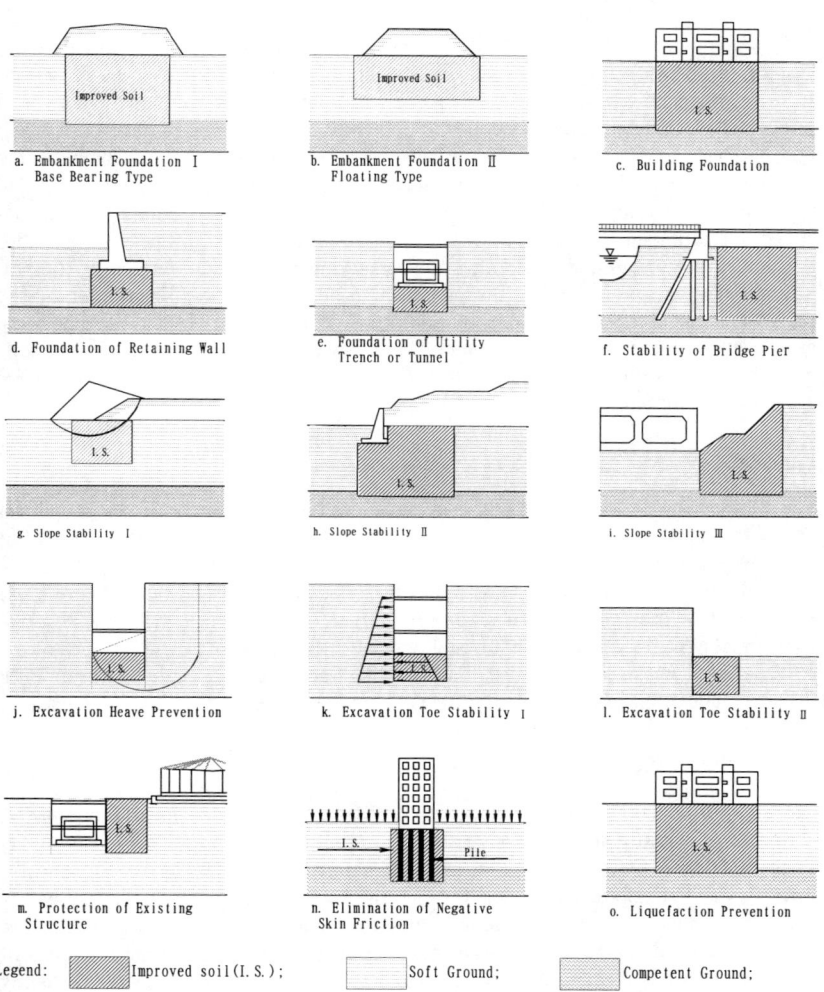

Figure 5. Typical Applications of DJM (after DJM 1993)

EQUIPMENT AND INSTALLATION PROCEDURE

DJM equipment consists of a soil-reagent mixing system and a reagent delivery system, as shown in Figures 1 and 2. The reagent is transported by compressed air from the reagent silo through the reagent feeder and the hollow stem of the soil mixing shaft and then deposited in the soil void at the lower portion of the soil mixing shaft. The soil mixing system consists of soil mixing blades, which cut the soil and create voids for reagent injection and distribution during the soil mixing. The separation of air and reagent occurs in the soil voids, where the air pressure and speed are dramatically reduced after being discharged from the nozzle. After depositing the reagent, the air flows upward along the outside surface of the soil mixing shaft and is discharged through a valve on the particle collection shroud at the ground surface. Schematics of the mixing blade and the air/reagent passage route are shown in Figure 1c. The installation procedure is shown in Figure 3, and soil-reagent columns produced are shown in Figure 4. Typical applications of DJM are shown in Figure 5.

ENGINEERING PROPERTIES

The engineering properties of the treated soils are dependent upon the soil properties before treatment, reagent type, and reagent dosage. Lime is effective for treatment of clayey soils, especially marine clays. Ordinary Portland Cement and slag cement are generally effective for treating most types of soils. Cement based reagents are more effective than others for treating peat and organic soils. Since ordinary Portland Cement, slag cement, and cement based reagent constitute approximately 90% of the reagents used in the DJM process, the term "soil-cement" will be used to represent the soil-reagent mixture produced by DJM using these three cementitious reagents.

1. Strength

The strength of soil-cement produced by DJM can be obtained in a laboratory by performing unconfined compressive strength tests, triaxial compression tests, direct shear tests, and tensile tests. The most common type of test is the unconfined compressive strength test, and its results are used for design and construction quality control and quality assurance. The test samples include laboratory samples and core samples prepared before and after the construction, respectively. The strength of soil-cement is dependent upon the soil type and cement dosage. DJM is most frequently used to treat liquefiable sand, organic soils, peat, and very soft silt or clay with water content more than 100%. Pore water is the sole source of water for cement hydration, and therefore, DJM is generally used to treat soils below or adjacent to groundwater table. The cement dosage generally ranges from 100 to 400 kg per cubic meter of in situ sand and fine-grained soils, and ranges from 200 to 600 kg per cubic meter of peat or organic soils. The lime dosage generally ranges from 50 to 300 kg per cubic meter of marine clay soils.

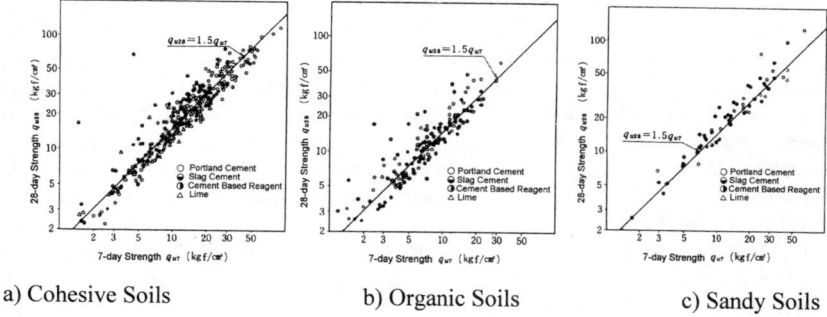

a) Cohesive Soils   b) Organic Soils   c) Sandy Soils

Figure 6. Relationship between 7-Day and 28-Day Strengths (after DJM 1993)

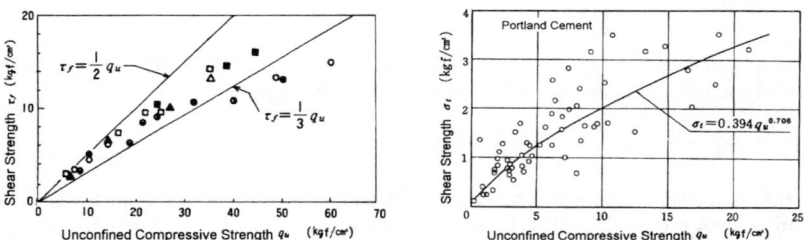

Figure 7. Relationship between τ_f and q_u (after DJM 1993)

Figure 8. Relationship between σ_t and q_u (after DJM 1993)

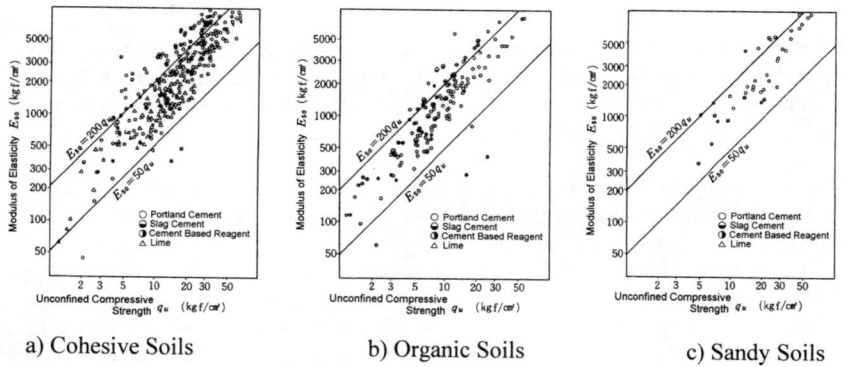

a) Cohesive Soils   b) Organic Soils   c) Sandy Soils

Figure 9. Relationship between E_{50} and q_u (after DJM 1993)

The strength gain of soil-cement produced by DJM is generally faster than those produced by slurry-type deep mixing. The relationship between 7-day and 28-day unconfined compressive strengths (q_u) of soil-cement and soil-lime mixture is shown in Figure 6. Regardless of soil type and different reagent used, the 28-day strength is approximately 50 percent higher than the 7-day strength. The correlation between unconfined compressive strength and shear strength is shown in Figure 7. The shear strength is obtained by direct shear test. The ratio of unconfined compressive strength to shear strength is close to 2 when the unconfined compressive strength is less than 10 kgf/cm^2. This ratio increases with the increase in compressive strength of soil-cement produced by higher cement dosage. The tensile strength obtained by splitting tensile strength test of soil-cement produced by DJM varies from 10 to 20 percent of the unconfined compressive strength. The ratio decreases with the increase of unconfined compressive strength. A correlation between the tensile strength and the unconfined compressive strength is shown in Figure 8.

2. Modulus of Elasticity

The modulus of elasticity (E_{50}) of soil-cement and soil-lime mixture is proportional to the unconfined compressive strength as shown in Figure 9.

3. Compressibility

The compressibility of soil-cement was represented by the consolidation test results. Soil-cement produced by DJM has a very clear consolidation yield pressure (P_y), a value similar to the undisturbed clay and soil-cement produced by slurry type deep mixing (Yang, 1997) as shown in Figure 10, and can be expressed by the following formula:

$$P_y = 1.27 \, q_u$$

When the compression pressure is lower than the consolidation yield pressure (P_y), the volume change is very small. When the compression pressure exceeds P_y, the volume change increases rapidly due to the high void ratio of the soil-cement created during the soil mixing process.

When the soil-cement columns are used to reduce the settlement of soft ground, the settlement is estimated by considering the soil-cement columns and the untreated soil together as composite ground. The load distribution ratio between soil-cement and the untreated soil is considered to be proportional to the ratio of the coefficients of volume change of soil-cement and untreated soil. Using the coefficient of volume change of the untreated soil, m_{vr}, and P_y, the dimensionless relationship between the coefficient of volume change of treated soils, m_v, and the average compression pressure on the composite ground, P, can be developed, as shown in Figure 11. The average compression pressure on the composite ground is generally substantially lower than the consolidation yield pressure. The consolidation settlement of the soil-cement and the composite ground are usually negligible.

4. Uniformity

Soil type is one of the major factors affecting the strength of soil-cement. It also affects the procedures for the evaluation of the soil-cement produced by deep mixing. Deep mixing is not designed to dissolve the in situ soils before blending it with the cement powder or grout. The cutting heads and mixing paddles are designed to break up the in situ soil and mix it with cement powder or grout. In the case of cohesionless soils with minor fines, sand and gravel particles become dispersed aggregates of the soil-cement mixture. When the fines content and its plasticity increase, the soil lumps increase. In the case of highly cohesive clays, part of the clay cannot be broken down during the soil mixing process and remains as lumps inside the soil-cement mixture. As long as the lumps disperse inside the soil-cement, the performance of the soil-cement will be satisfactory. Therefore, the uniformity of soil-cement should be evaluated from large-scale viewpoint. To focus heavily on the existence of clay lumps might mislead the evaluation on the mass performance of the soil-cement structures.

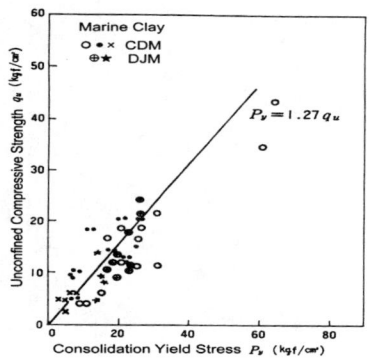
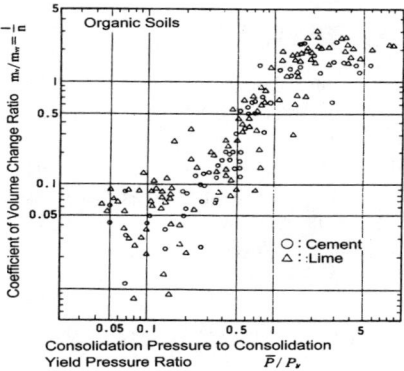

Figure 10. Relationship between q_u and P_y (after DJM 1993)

Figure 11. Relationship between m_v/m_{vr} and P/P_y (after DJM 1993)

CASE EXAMPLES

Case Example 1: Torishima Dike Reconstruction Project

Background

Torishima Dike is approximately 7m in height and is used for flood control to protect the heavily populated commercial and residential areas along Yodo River, a major river meandering through the city of Osaka. During the 1995 Kobe earthquake (Magnitude 7.2), liquefaction occurred in the sand underlying 2 km section of the dike located near the border between Kobe and Osaka, where the Yodo River discharges to the Seto Inland Sea. Due to the liquefaction of 10 to14m of foundation soils, the dike suffered severe damage as shown in Figure 12. Without the protection of the dike, the commercial and residential areas were exposed to high potential of flooding during the hurricane season. The dike had to be reconstructed before the arrival of the annual hurricane season in September 1995.

a) Damaged Dike

Subsurface Materials

Post-earthquake subsurface investigation revealed that the soils underlying the collapsed dike consisted of 10m interbedded layers of sands and silts with standard penetration values ranging from 3 to 10 blows per 30 cm. It was concluded that the liquefaction potential of the foundation soils was high and foundation treatment was needed before the reconstruction of the dike.

Selection of Soil Improvement Method

The soil improvement methods to be used at site had to meet the following requirements:

1. Short construction time to allow for dike reconstruction before the hurricane season.

b) Reconstruction

Figure 12. Torishima Dike

2. Low noise and no vibration to minimize the disturbance to the neighborhood.
3. Equipment that can work on soft ground without stability concerns, further settlement, or damage to the sensitive ground.
4. No need to import large quantity of construction materials such as sand or gravel to minimize traffic congestion and material shortage problems in the earthquake-damaged region where major reconstruction was ongoing.

Among the various soil improvement methods evaluated, the Deep Mixing method was selected. The relatively small space needed for the operation of deep mixing rigs would allow the limited working space to accommodate more rigs to work simultaneously and shorten construction time. Soils treated by deep mixing would gain strength in a short period of time and would allow for earlier embankment reconstruction. The powder type deep mixing-DJM and the slurry type deep mixing-CDM were compared and the former was selected due to its advantages of no need to transport large quantity of water for slurry mixing and less construction spoils for disposal off site.

Design of DJM Ground Treatment

Grid type treatment pattern was designed to 1) Reduce the shear strain and the build-up of excessive pore water pressure during the seismic-induced ground shaking, 2) Contain the local liquefied zones, should liquefaction occur, and 3) Reinforce the foundation and increase the safety factor against slope failure or lateral spreading. A typical cross-section of the dike and treatment zone is shown in Figure 13a. A modular grid is shown in Figure 13b. In two outside zones under the slopes and toes, only the 10m thick loose sand was treated. In the center zone, the treatment was extended an additional 4m into the clay layer. The field unconfined compressive strength of the soil-cement was designed as 5 kgf/cm^2 at 28-day curing age. Based on data from laboratory mix design, a cement dosage of 100 to150 kg per m^3 of in situ soil was needed to produce the required strength.

Construction of DJM

DJM work was commenced immediately after the removal of the damaged embankment. At the peak period, a total of 32 DJM rigs were deployed simultaneously with each rig treating an area approximately 50m by 60m as shown in Figure 12b. It was the DJM that made this large-scale operation possible in the 50m by 2000m strip where there was no water supply system for slurry preparation and no space for spoil handling and transportation.

Construction Evaluation

The quality control and quality assurance were based on the core samples retrieved from the hardened soil-cement columns. The average unconfined compressive

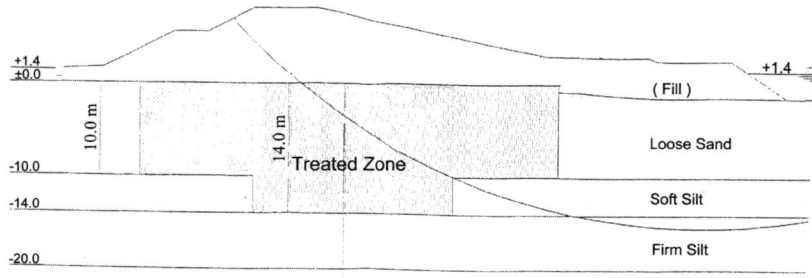

a) Dike Cross Section

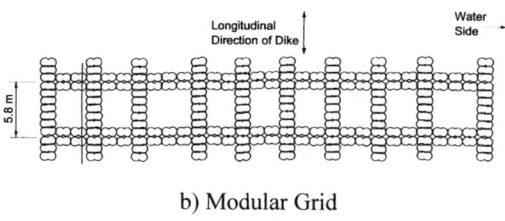

b) Modular Grid

Figure 13. Dike Cross Section and Modular Grid

strengths in the different work sections ranged from 6 to 7 kgf/cm^2, depending on the distribution of clay lenses within the sand stratum. The unconfined compressive strength generally reached 10 kgf/cm^2 in clean sand layer and was slightly above 5 kgf/cm^2 in the clay lenses. The unit weight of the soil-cement produced by DJM is generally lower than the in situ soils before treatment, due to the use of air for delivery of cement. At this project site, the lighter unit weight provided the advantage of reducing the total settlement after placement of the new embankment. The permeability of soil-cement produced by DJM is higher than those produced by slurry type soil mixing and was considered semi-permeable. This feature reduces the impact on the groundwater table and the natural groundwater flow.

Summary

Approximately 600,000 m^3 of liquefiable soils were treated for the reconstruction of a 2,000 m long section of dike that was severely damaged during the earthquake. Using up to 32 sets of DJM equipment and working 11hours per day, the soil improvement work was completed in three months, avoiding the probable flooding damage that might hinder the reconstruction of the region suffering severe earthquake damage. The dike reconstructed immediately after the soil improvement work had been performing satisfactorily.

Case Example 2: Tomei Highway Expansion Project

Background

Tomei Freeway which connects Tokyo and Nagoya has been the artery of Japanese culture and economy since the 17th century. Even with the addition of railroads and bullet train rails, the traffic volume along the Tomei Freeway continues to increase. Therefore, the expansion project was carried out to expand the four-lane freeway into a six lane freeway – three lanes in each direction. The section near Isebara, located along the foothill of the Tanzawa ridge, is underlain by consecutive sections of ridges and valleys. The ridges consist of weathered rock while the valley materials consist of organic clays with peat. Excessive total and differential settlement and embankment instability were expected, if the new embankments were placed without improving the strength and compressibility of the organic soils.

Selection of Soil Improvement Method

The soil improvement method to be used had to meet the following requirements:

1. Effective in treating organic soils and peat.
2. Capable of adjusting mix design to obtain desirable strengths in various soil types. A lower dosage of cement is required to maintain the strength of the treated embankment soil within the same range as the existing embankment while a higher dosage of cement is needed to improve the organic clays and peat under the embankment.
3. Sufficient equipment stability when operating on temporary working pad placed over existing embankment slope.
4. Flexible in the adjustment of area treatment ratio to cope with various loading conditions.

DJM was selected. DJM has been used in numerous projects for treating organic soils and peat with satisfactory results. The cement dosage at different depths can be controlled reliably without affecting the construction speed. Instead of moving up and down along the lead as in the slurry type soil mixing equipment, the motor and gear box of DJM are located near the bottom of the rig, which lowers the center of gravity of the rig and improves the stability when working at less level ground. The space between the soil mixing shafts is adjustable between 80 cm and 150 cm, which enables the adjustment of column spacing and, consequently, area treatment ratio. In addition, DJM shortens construction time, since there is no need for installation or relocation of water supply systems for slurry preparation.

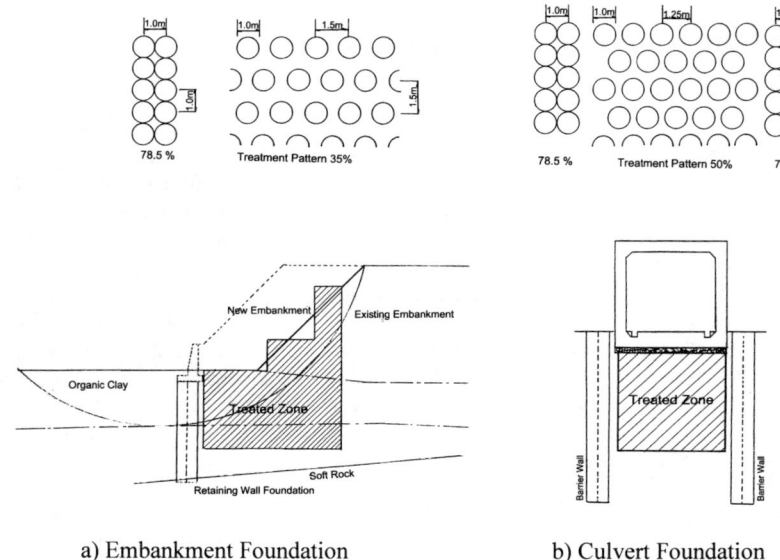

a) Embankment Foundation   b) Culvert Foundation

Figure 14. Cross Sections and Treatment Patterns

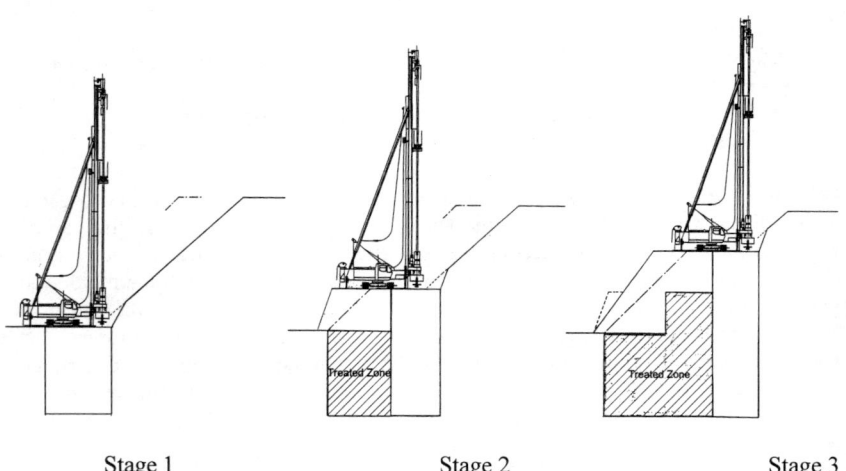

Stage 1   Stage 2   Stage 3

Figure 15. Staged Construction

Design of DJM Ground Treatment

Slope stability and settlement criteria were the basis for the design of DJM ground treatment. The required minimum factors of safety against slides were 1.25 and 1.1 for static and seismic conditions, respectively. Soil-cement columns are installed in the soft soil to form a composite ground. The average strength of the composite ground is then used for stability and settlement analyses. As a general practice in Japan, a minimum area treatment ratio of 35 percent is required to consider the treated ground as composite ground. With a treatment ratio of 35 percent, an unconfined compressive strength of the soil-cement of 7 kgf/cm^2 was determined for the zone under the new embankment. An allowable settlement of 1.5 cm was selected. This value is the average annual overlaying thickness of the pavement of the Tomei Freeway. With the treatment ratio and strength determined to satisfy the stability requirement, the total settlement of the new embankment is relatively small in comparison to those of the existing embankment, if the depth of treatment was extended to the competent soil layer below the highly compressible organic clay. To minimize the differential settlement between the new and existing embankment, the depth of soil treatment was designed to leave a few meters of soils untreated as shown in Figure 14. The layout of the soil-cement is also shown in Figure 14.

Construction of DJM

A level working pad is needed for the safe and efficient operation of DJM rig. Staged construction at 3 elevations was used for soil treatment as shown in Figure 15. Two mix designs were used. Lower cement injection rates, 100 kg to 120 kg cement per m^3 of target soil, were used in the existing embankment zone to maintain the strength of soil-cement at about 0.8 kgf/cm^2 – the average strength of the existing embankment soils. The higher cement injection rate, 170 kg to 230 kg cement per m^3 of target soil, was used in the organic clays to obtain a minimum unconfined compressive strength of 7 kgf/cm^2 after treatment.

DJM was also used to produce soil-cement foundations for supporting a retaining wall and box culvert, as shown in Figures 14a and 14b, and a soil-cement barrier wall to eliminate the transfer of embankment load to the new section of box culvert under the new embankment, as shown in Figure 14b. This soil-cement barrier wall reduces the differential settlement between the existing and new sections of box culvert. The area treatment ratios were 35%, 50%, 78.5%, and 78.5% for embankment foundation, box culvert foundation, retaining wall foundation, and barrier wall, respectively.

Summary

A total of 50,215 m^3 organic clays, peat, and fill were treated for use as foundations of the new embankment, retaining walls, and box culvert. Two sets of DJM rig were used. The deep mixing work was commenced in April 1994 and completed in November 1994 without interrupting the use of the four-lane freeway. Staged construction procedure was used to perform the soil treatment within a limited working space. In addition to the foundation treatment for the new embankment, DJM also improved the foundation for the retaining wall and the box culvert, and eliminated the mobilization of pile driving equipment to the congested work zone parallel to an existing freeway.

CONCLUDING REMARKS

DJM pneumatically delivers powdered reagent into the ground and mixes it with in situ soils to form soil-reagent columns for ground improvement. Currently, Portland Cement, slag cement, cement based reagent, and lime are most frequently used, followed by slag, fly ash and gypsum. Dry powders or particles with size less than 5 mm can be delivered which increases the flexibility of reagent selection for treatment of soils with unusual properties such as peat, organic soils, and contaminated soils.

REFERENCES

DJM (1993), "DJM Technical Manual," DJM Method Research Institute, Tokyo, Japan.

DJM (1995), "DJM Seminar '95," DJM Method Research Institute, Tokyo, Japan.

Yang (1997), "Deep Mixing," In Situ Ground Improvement, Reinforcement, and Treatment: A Twenty Year Update and a Vision for the 21st Century, Ground Reinforcement Subcommittee, American Society of Civil Engineers, Geo-Institute Conference, Logan, Utah, July 1997. pp. 130-150.

SWING METHOD FOR DEEP MIXING

David S. Yang[1], Jack N. Yagihashi[2], and Steve S. Yoshizawa[3]

ABSTRACT

Mechanical deep mixing methods treat the in situ soils from the ground surface to the designated depths using soil mixing augers or blades. Jet grouting treats the in situ soils in zones between designated depths. SWING (Spreadable Wing) is an innovative type of deep mixing technology that enlarges the soil mixing blade for soil treatment when the drilling reaches the designated depth where soil improvement is required. This procedure bypasses the upper soil layers that do not require treatment and subsequently saves cement cost and also reduces the quantity of spoils. SWING generally produces a 2m-diameter soil-cement column by mechanical mixing. By incorporating a high-pressure jet system like the jet grouting equipment into the mixing blade, the hybrid type of deep mixing equipment named SWING-JET, SWING-MJET, and SWING-HIJET are created. SWING-HIJET produces a 3.2m-diameter soil cement column using both mechanical mixing and jet mixing (SWING, 1997). The expandable characteristics of SWING enable the installation of soil cement column or wall under existing structures for various applications including underpinning. This paper introduces the SWING equipment and its construction procedures, and presents one case history. This case history illustrates the use of SWING to treat the soils within a designated zone near existing structures.

INTRODUCTION

Although the direction of the recent development of the deep mixing technology is diversified, the main trends target toward greater depth, larger diameter, higher quality, and the capacity to treat soils with difficult engineering properties or conditions while maintaining the economy of soil mixing work. SWING (Spreadable Wing) method was developed with these main trends as targets and contains two special features: 1) Spreadable mixing blade and 2) High-pressure jet at edge of the

---

[1,2,3] RAITO, INC., 1818 Gilbreth Road, suite 145, Burlingame, CA 94010
Phone: (650) 259-1210 E-mail: dsyang@jaccess.com

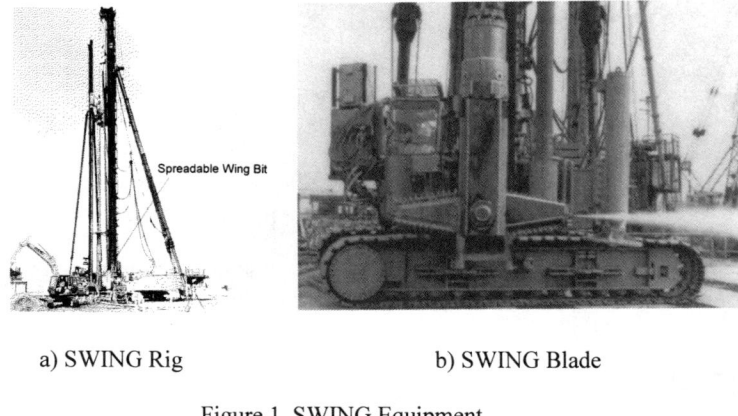

a) SWING Rig   b) SWING Blade

Figure 1. SWING Equipment

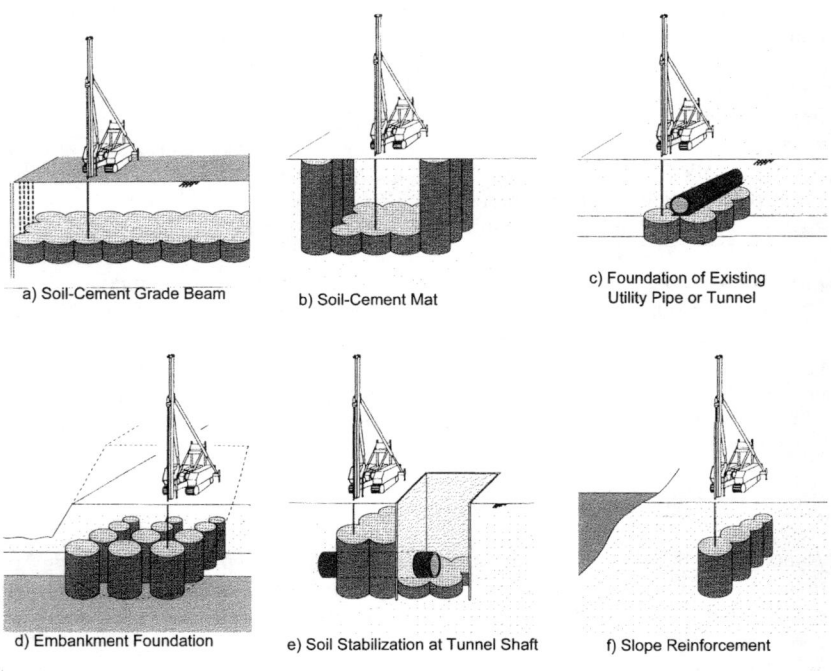

Figure 2. Application of SWING Method

mixing blade. The mixing blade can expand from 60 cm to 200 cm at any depth to produce larger diameter soil-cement column below a designated depth. With the high-pressure jet, SWING increases the diameter of the soil-cement column beyond 200 cm - the maximum mechanically possible diameter of soil-cement that could be produced by deep mixing with reliable quality. The SWING is guided by a vertical lead on a crawler base machine (Figure1-a) as other types of deep mixing equipment. Before expanding, the shaft drills a 60-cm diameter pilot hole through the upper soil strata that do not require treatment. Due to the smaller diameter, the drilling capacity of the shaft is high and can penetrate through harder strata to reach the deeper soft stratum efficiently. This makes the mechanical mixing of deep soft stratum possible without extensive mixing of the upper soil strata and makes the soil treatment of a thin soft layer cost efficient. The high-pressure jet at the edge of the mixing blade (Figure 1-b) performs soil mixing like the ordinary jet grouting equipment and increases the diameter of the soil-cement column to the range between 2.4m and 3.6m, depending on soil type and pressure used. In addition, jet mixing makes the soil-cement column in good contact with neighboring soil cement columns or existing structures. SWING is a hybrid type of soil mixing method combining the features of mechanical deep mixing and jet grouting.

SWING is generally used for soil treatment of limited zones at depths to produce soil-cement mats and soil-cement grade beams at bottom of excavation (Figure 2). These types of soil treatment are usually made by jet grouting, which is considered to be less reliable in terms of diameter and accuracy when compared with mechanical deep mixing. Due to its capacity of producing a large diameter column through a small pilot hole, SWING is also used to treat the soils below the existing structures. The major applications of SWING are shown in Figure 2.

SWING was developed in 1984. The 2m-diameter soil-cement column was produced solely by a mixing blade without the use of a high-pressure jet. Since 1986, a two-fluid high-pressure jet was added to the edge of the mixing blade and the diameter of the soil-column was increased to the range between 2.4m and 3.0m, depending on the soil type and the jet pressure used. SWING-JET with a jet pressure of 19.6 MPa (200 kgf/cm^2) and SWING-MJET with a jet pressure of 39.2 MPa (400 kgf/cm^2) are two models developed in1986 and 1988, respectively. In 1991, SWING-HIJET was developed, which incorporates three-fluid pressure jet into the system and extended the diameter of soil-cement to a range between 3.2m and 3.6m. The advantages of adding jet grouting to the mechanical deep mixing system are the capacity to produce larger diameter columns and the better contact to the neighboring columns and existing structures. However, caution is required while evaluating the portion of soil-cement produced by jet grouting since the additional diameter beyond the 2m diameter mechanical mixing zone will vary with soil type, jet models, jet pressure, and experience of the contractors.

Installation Procedure

1) Drilling of Pilot Hole
Drill the pilot hole with water to the bottom of soil layer to be treated.

2) Expanding the SWING Blade
Expand the SWING blade while rotating the shaft.

3) Pre-drilling the Treatment Zone
Pre-drill the zone to be treated upward using SWING blade and high-pressure jet. Lower the blade down to the bottom while keeping the blade at expanded position and shaft rotating. Grout may be injected during the downward drilling.

4) Injecting and Mixing
Inject the grout from the blade and high-pressure grout jet from the edge of the blade and perform both mechanical mixing and jet mixing.

5) Extract SWING Blade
Rotate the SWING blade back to the shaft after soil mixing.

6) Retrieving the Shaft
Withdraw the shaft to the ground surface.

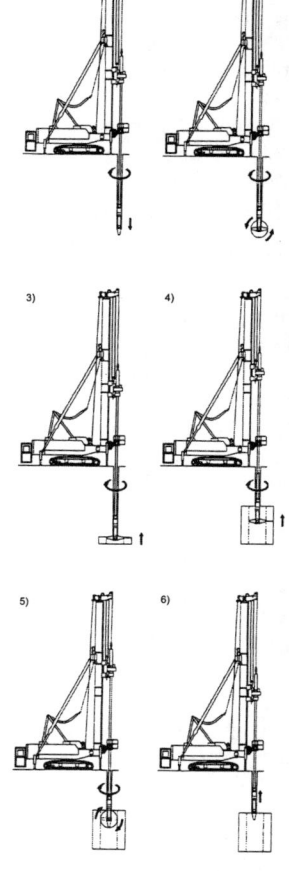

Figure 3. Installation Procedure

EQUIPMENT AND INSTALLATION PROCEDURE

Four models of SWING equipment were discussed above. The term "SWING" will be used as a generic term to represent the type that combines both mechanical mixing and jet mixing. The special features of SWING equipment are the expandable blade and the shaft that houses the multiple tubes for the operation of mechanical mixing and jet mixing. One inner rod is mechanically connected to the control unit of

 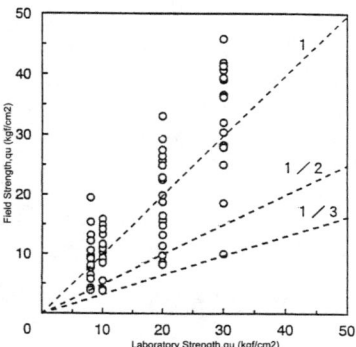

Figure 4. Soil-Cement Columns   Figure 5. Strength Data

the expandable blade and is operated by a hydraulic system inside the gearbox. Since the expansion of the blade is controlled mechanically, the position of the blade can be monitored at the ground surface. The high jet system is similar to the conventional jet grout equipment. Triple tube system is used when the jet pressure exceeds 39.2 MPa (400 kgf/cm^2). The installation procedures consist of six steps as shown in Figure 3. A view of the soil-cement columns produced by SWING is shown in Figure 4.

ENGINEERING PROPERTIES

1. Strength

The mix design of jet grouting is usually determined empirically. According to Japanese practice, several standard strengths have been established for three soil types -sandy soil, cohesive soil, and organic soils - to avoid performing a field test section for every project. Due to the higher variation in strength data of soil-cement produced by jet grouting, the standard strength values are set on the conservative side, which in turn causes high cost in using the jet grouting method. In the case of SWING, mechanical deep mixing produces the main portion of the soil-cement; therefore, laboratory testing can be performed to generate data for the selection of appropriate mix design for specific site conditions. The relationship between the unconfined compressive strengths (q_u) of soil cement samples produced in the laboratory and in the field by SWING is presented in Figure 5. In practice, the field strength to laboratory strength ratio is considered to range from 1/2 to 1. Based on previous field data, the unconfined compressive strength of the soil-cement within the 2m diameter mechanical mixing zone ranges from 392 to 4,410 kPa (4 to 45 kgf/cm^2). Within this range, an economical and reliable mix design can be selected from the laboratory trial mix data. Nevertheless, the strength of soil-cement produced by jet mixing outside the 2m-diameter zone is lower than that produced by mechanical mixing within the 2m-diameter zone using grout with the same water/cement ratio. For preliminary design purposes, the maximum unconfined compressive strength of 1,470 kPa (15 kgf/cm^2) in sandy soils and 1,176 kPa (12 kgf/cm^2) in cohesive soils are recommended for the soil-cement produced by jet mixing. In practical applications, the column produced by SWING is considered as composite column for the estimate of the average strength. In the case of a column group, the lower strength portions are considered as overlapping zones for load transfer within a composite ground.

2. Modulus of Elasticity

The grout injection ratio of jet mixing, defined as the ratio of grout volume to the volume of in situ soils, is generally higher than that of the mechanical deep mixing. As a result, more soil particles are discharged during the jet mixing process and the modulus of elasticity is generally lower than that of soil-cement produced by mechanical deep mixing. Soil-cement column produced by SWING contains two zones - the outside zone produced by jet mixing and the inner zone produced by mechanical mixing. The modulus of elasticity of the outside zone follows the relationship of $E_{50} = 100q_u$ for the soil-cement produced by jet mixing. The modulus of elasticity of the inner zone has lower modulus of elasticity than that produced by CDM (Cement Deep Mixing; E_{50} = 400 to 600q_u) and is closer to that produced by DJM (Dry Jet Mixing) and is represented as $E_{50} = 150q_u$.

3. Permeability

The majority of the applications of SWING has been the treatment of soft ground. Therefore, the strength gain has been more important than the reduction of permeability of the treated soils. The permeability of soil-cement produced by SWING is expected to be below 1x10^{-5} cm/sec and is considered to be satisfactory for most applications without testing. Therefore, there is a lack of data on the permeability of soil-cement produced by SWING. However, it is reasonable to adopt the permeability data of soil-cement produced by jet mixing and mechanical deep mixing and consider the coefficient of permeability to be in the order of 10^{-6} cm/sec.

CASE HISTORY - KAWASAKI UTILITY TUNNEL PROJECT

Background

During this decade, the large-scale projects of Eastern Japan were concentrated in the Tokyo Bay Area. The most noticeable one is the Trans Tokyo Bay Highway Project which involved the construction of man-made islands, undersea tunnels, and more than 2 million m^3 of soil-cement by various soil mixing methods. Accompanying the Trans Tokyo Bay Highway Project was the construction of the access road and utility tunnel systems to serve the Tokyo Bay Industrial Area. One of the major challenges to this so-called "Japan's last meg size project of this century" was the soft ground surrounding the Tokyo Bay. This case history presents the application of SWING for the construction of soil-cement mats or grade beams that controlled the movement of the excavation support walls during the construction of the utility tunnel extending from the bay shore to the inland industrial area.

Subsurface Conditions

The subsurface materials consist of 3m of granular fill and 10m of very loose silty fine sand underlain by 27m of soft to firm marine clay. The competent bearing stratum consisting of sandy gravels is located at a depth of approximately 50m below the existing ground surface. The standard penetration values (N-values) range from 2 to 5 blows/30cm in the silty fine sand and range from 3 to 5 blows/30cm in the upper 10m of marine clay deposits affected by the excavation. The marine clay has a high plasticity with cohesion ranging from 53.9 to 94.1 kPa (0.55 to 0.96 kgf/cm^2). The ground water table, affected by the tidal fluctuation, is located approximately 1m below the ground surface.

Selection of soil improvement Method

This marine clay has been notorious for the occurrence of large deformation and stability problems during excavations. For the construction of the utility tunnel, excavation to depths ranging from 10m to17m was required. To increase the passive resistance and to reduce the lateral deformation, soil improvement of 2m to 5m thick layers of marine clay located near the bottom of excavation was planned. The soil-cement produced would behave as grade beams or mats at the bottom of the excavation. Three soil improvement methods were evaluated:

1) Jet Grouting Method - This method is effective in treating thin layer of soil at greater depth. The effective diameter of soil-cement column produced by jet grouting in cohesive soils is affected by the N-value and cohesion of the cohesive soils and the withdrawal speed of the jet grouting rod. Based on the empirical data from the Jet Grout Technical Manual (JJGA, 1995), jet grouting is applicable for cohesive soils with N-values less than 9 blows/30cm and the effective diameter of the soil-cement columns suggested by the manual is not applicable when the cohesion of the clay exceeds 49 kPa (0.5 kgf/cm^2). In other words, Japanese practice bypasses the use of jet grouting if the cohesion of the soils is greater than 49 kPa (0.5 kgf/cm^2). The marine clay at this project site has a cohesion of 88.2 kPa (0.9 kgf/cm^2) which was considered to have adverse effects on the effective diameter and quality of the soil-cement column. Jet grouting method was unacceptable at this site.

2) Mechanical Deep Mixing Method - This method is effective in treating

cohesive soils. The effective diameter of the soil-cement column produced is reliable due to the use of mixing blades. However, the mixing blades can only rotate tangent to the pre-installed excavation support wall. The soil-cement column produced will not be in good contact with the wall. Reliable and sufficient contact of the soil-cement and the excavation support wall is crucial for the stability and control of wall movement. Jet grouting could be used to fill in the gap with soil-cement to provide sufficient contact for load transfer. However, the use of an ancillary method reduces the cost efficiency. In addition, the mechanical deep mixing method cannot use small diameter pilot holes like jet grouting and has to drill through 10 to 17m of upper soils before treating 2 to 5m thick layer of marine clay. This procedure incurs large volume of unnecessary drilling and also soften the upper soils which, in turn, reduces the bearing capacity and trafficability of the upper soils and increases the construction difficulties. Mechanical deep mixing was therefore considered to be unacceptable at this site.

3) SWING Method - This method can produce large diameter soil-cement column at depth through a small diameter pilot hole. The majority of the soil-cement is produced by mechanical deep mixing for better quality. The jet mixing in the outer zone creates a reliable contact with the excavation support wall and also provides a connection between neighboring soil-cement columns for load transfer. Since its special features meet the project requirement, SWING method was selected to treat the marine clay for this project.

Layout Design of SWING Soil Treatment

The purpose of soil improvement in this project was to produce a 2 to 5m thick layer of soil-cement to perform as a grade beam or mat near the bottom of the excavation as shown in Figure 6. The grade beam or mat would behave like an internal bracing for the support of two soil mix walls (SMW) on two sides of the excavation. The equipment selected was SWING-MJET which has a 200cm wide mixing blade and a 39.2 MPa (400 kgf/cm^2) high-pressure jet nozzle at the edge of the mixing blade. The jet mixing could produce an additional 40cm soil-cement ring surrounding the 200cm-diameter column produced by mechanical mixing. Based on the data obtained from the pre-construction test section, the quality of the soil-cement varies with the distance from the center of the column and is evaluated as follows:

1) Soil-cement within the diameter of 240 cm - High quality with minor variation in strength along the whole column length (Type 1). The unconfined compressive strength of soil-cement produced by mechanical mixing ranges from 1,960 to 2,940 kPa (20 to 30 kgf/cm^2) and those produced by jet mixing were approximately 980 kPa (10 kgf/cm^2).
2) Soil-cement within a ring between diameter 240 cm and 260 cm - More variation

SOIL IMPROVEMENT FOR BIG DIGS 119

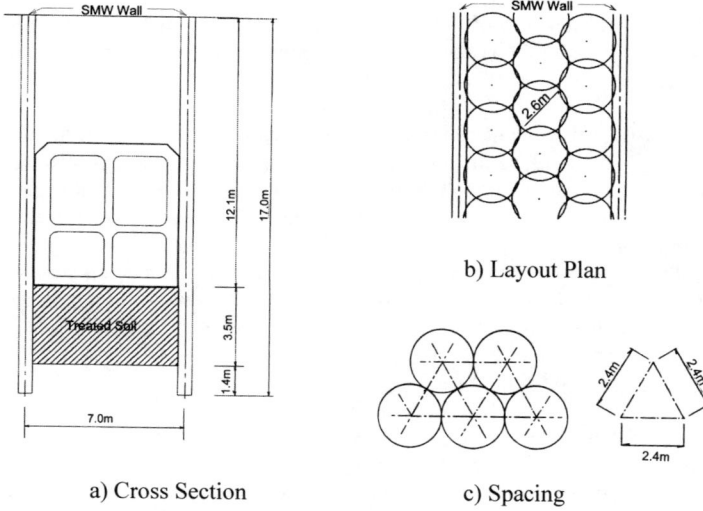

a) Cross Section

b) Layout Plan

c) Spacing

Figure 6. Cross-Section and Plan View of Utility Tunnel

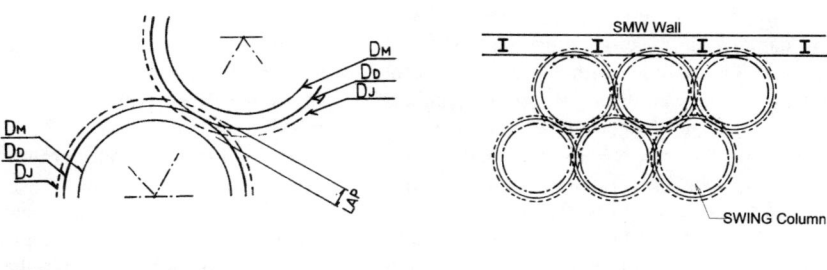

DM: Diameter of Blade (ϕ =2.0 m)
DD: Effective Diameter (ϕ =2.4 m)
DJ: Overlapping Zone (ϕ =2.4~2.6 m)

a) Overlapping Zone

b) Contact with SMW Wall

Figure 7. Layout Design

in strength. However, all strength test data exceed the design strength of 392 kPa (4 kgf/cm^2) (Type 2).
3) Soil-cement within a ring between diameter 260 cm and 280 cm - Higher variation in strength. The soil-cement contains greater number of larger clay lumps (Type 3).

The layout of the soil-cement columns was designed to obtain a series of subgrade beams or mat after soil treatment. The centers of the neighboring columns are arranged in an equilateral triangle pattern as shown in Figure 6c. This treatment pattern ensured that the majority of the grade beams or mat consisted of the Type1 high quality soil-cement. The Type 2 materials served as the overlapping zone for load transfer between columns as shown in Figure 7a. Type 3 material supplements Type 2 material near the SMW excavation support wall and provided a close contact between the wall and the soil-cement grade beams as shown in Figure 7b.

Construction of SWING Columns

Four sets of SWING-MJET equipment were mobilized for soil treatment. The utility tunnel was located along an existing road that was open to the traffic during the daytime. The contract requested that the soil treatment work be performed in the night and that the working area be cleaned for daytime traffic. To meet these difficult requirements, a special work plan was developed. The area between the two excavation support walls was excavated to a depth of 1.5m. A deck was then installed at grade for daytime traffic. During the night, the deck was removed for the SWING equipment to perform soil treatment work from a working pad at the first level of excavation. The spoils from the soil treatment were handled within this excavated area, and thus, eliminated the expensive daily cleanup of the roadway.

Soil improvement may cause excessive lateral pressures to incur ground movements or damage to existing structures. Although not as severe as grouting and the installation of lime columns, sand piles, or stone columns, the excessive lateral pressure generated during jet grouting and deep mixing should be controlled to avoid adverse effects to adjacent structures. In this project, SMW walls along two sides of the utility tunnel were installed before the soil treatment. Due to the soft ground conditions, the passive soil resistance behind the SMW wall was low and large lateral deformation of the wall might have occurred, which in turn might have caused vertical cracks and leaks of the SMW wall. Special attention was given during the soil mixing to ensure more soil particles were discharged as spoils to minimize the build up of internal pressure in the soil mixing zone and lateral pressure in the surrounding ground.

Evaluation of Soil-Cement

Core samples were retrieved after installation of the soil-cement columns. The unconfined compressive strength of the core samples ranges from 2,450 to 3,920 kPa (25 to 40 kgf/cm^2) with an average value of 3,430 kPa (35 kgf/cm^2). The strength of soil-cement produced by SWING exceeds those of soil-cement produced by other mechanical deep mixing methods with the same range of cement dosage. The modulus of elasticity ranges from 441 to 735 MPa (4500 to 7500 kgf/cm^2) with an average value of 568 MPa (5800 kgf/cm^2). The relationship between E_{50} and qu is close to the empirical expression of $E_{50} = 150q_u$.

Summary

A total of 10,200 m^3 of soil cement was installed near the bottom of the excavation along 8 km section of the 10 km long utility tunnel. Four sets of SWING-MJET equipment were mobilized. The soil mixing work was commenced in December 1994 and completed in April 1995 without interrupting the daytime traffic. Despite the deep excavation in the soft Tokyo Bay marine clay, no excessive lateral movement of the excavation support wall was observed. This project demonstrated the effectiveness of soil-cement grade beams or mats in the control of ground movement during deep excavation in soft ground and the efficiency of using SWING method for treating thin soil layers at greater depths.

CONCLUDING REMARKS

SWING is a hybrid type of soil mixing method combining the features of mechanical deep mixing and jet grouting. The mechanical mixing produces the main portion of the soil-cement column. The jet mixing enlarges the column diameter beyond the mechanical mixing portion and provides soil-cement connections between neighboring columns or contact zone with existing structures. The expandable mixing blade enables the treatment of a thin soft stratum at depth with minor disturbance to the upper strata. These features expedite the treatment of soft ground at depth and soft ground adjacent or under existing structures.

REFERENCES

JJGA (1995), "Jet Grout Technical Manual," Japanese Jet Grout Association, Tokyo, Japan.

SWING (1997), "SWING Method", SWING Association, Tokyo, Japan.

Pre-Construction Aspects of Deep Soil-Cement Mixing for CA/T Project

Prabir K. Das, P.E.[1], Justice J.G. Maswoswe, Ph.D., P.E.[2], Edward Y. P. Yin, P.E.[2]

Abstract

The Central Artery/Tunnel (CA/T) Project in Boston, MA is the largest and most complex urban highway project ever undertaken in the United States. Through a network of modern underground expressways and new viaducts, this multi-billion dollar project will replace the existing Central Artery (I-93), an elevated, forty-plus-year old expressway, and extend the Massachusetts Turnpike (I-90) to Logan International Airport. At the I-93/I-90 interchange, the underground portions are comprised of cut-and-cover, immersed tube, and jacked tunnel elements.

To support construction of the cut-and-cover portion of the I-90 extension at the interchange, large-scale ground improvement was required. The technology of deep soil-cement mixing, or Deep-Mix Method (DMM), is being utilized to construct soil-cement shear walls which will provide support-of-excavation during construction, and permanent lateral and vertical foundation support of the proposed cut-and-cover tunnels. This case study presents two particular construction staging challenges that needed to be addressed in order for production DMM installation to be viable at this site.

Introduction

An important aspect of the Central Artery/Tunnel (CA/T) Project has been the extension of the Massachusetts Turnpike (I-90) through the I-93/I-90 South Bay Interchange. The cut-and-cover tunnel portion of this extension, east of the interchange, is adjacent to and within the Fort Point Channel. It is also underlain by deep, soft Boston Blue clay with some of the lowest compressive strengths (40kPa) observed on the CA/T

---

[1] Chief Structural Engineer, Bechtel/Parsons Brinckerhoff, Central Artery/Tunnel Project, Boston, MA.

[2] Professional Associate, Bechtel/Parsons Brinckerhoff, Central Artery/Tunnel Project, Boston, MA.

Project. The site (refer to Figure 1) is bounded by 1) an active rail facility owned by the Massachusetts Bay Transit Authority (MBTA) and operated by Amtrak, 2) the USPS General Mail Facility, which is the primary mail distribution center for the northeastern U.S., and 3) Gillette, Inc., the largest private employer in Massachusetts. Disruptions to any of these abutters due to construction activities was deemed unacceptable. Hence, the impact of any construction-induced ground movements had to be maintained within acceptable limits. Furthermore, as shown in Figure 1, approximately a third of the Channel, south of the contract limits, had to be left unaltered to ensure continued flow.

The proposed alignments of the tunnels, traversing a land/marine interface and requiring excavations 12m to 18m deep and 60m wide in the thick deposit of soft clay beneath and adjacent to the Fort Point Channel, presented major challenges to constructing the tunnels. The siting of the tunnels presented major concerns regarding global stability and heave during construction; tunnel foundation capacity; and lateral ground movements during and post-construction. Several construction methods were evaluated during conceptual design including: 1) full and/or partial filling of the Channel, with conventional cut-and-cover tunnel construction; and 2) full and/or partial excavation within a marine cofferdam. These methods were ultimately deemed unacceptable due to adverse environmental concerns and prohibitive costs.

Deep soil-cement mixing, or Deep-Mix Method (DMM), was then considered and selected for its ability to create shear walls formed by contiguously installed soil-cement columns resulting in lateral and vertical load transfer capabilities; and its cost and schedule effectiveness. The shear walls were to be designed to address concerns regarding global stability, heave, and ground movements.

Deep soil-cement mixing was to be performed using multiple shaft auger equipment, capable of augering down 38m in a single stroke. During the downward (penetration) stroke, the soil would be precut and fluidized with water; and during the upward (withdrawal) stroke, grout would be injected and mixed uniformly with the pre-fluidized soil column. Advancement of the strokes would follow a primary/secondary sequence similar to slurry wall panel placements, allowing DMM secondary strokes to partially overlap primary strokes to create a continuous shear wall.

The Owner of the Project is the Massachusetts Highway Department, working in conjunction with the Federal Highway Administration. The joint venture of Bechtel/Parsons Brinckerhoff serves as Management Consultant and provided preliminary design. The joint venture of Maguire/Harris was the design consultant and prepared plans and specifications for the construction contract.

Subsurface Site Conditions

Soil stratigraphy for the site, from ground surface, is composed of man-made fills, organic silts, marine clay, and glacial till overlying bedrock. The bedrock typically consists of moderately to severely weathered and kaolinized argillite, sometimes extending 7m or more below top of rock. The thickness of the dense glacial till deposit varies from 1.5m to 6m. The most influential stratum is the deep deposit of marine clay, known locally as Boston Blue Clay. This deposit is approximately 23m thick, and varies in shear strength from 40kPa to 50kPa. The organic silt deposits in the Channel and on land have shear strengths of approximately 5kPa and 20kPa, respectively, and vary in thickness between 3m and 6m. Significant areas of the surface sediments consist of fill placed during colonial times to create land suitable for human habitation and development. The fill layer is approximately 8m thick. The Fort Point Channel traverses these fill areas.

Groundwater conditions at the site include a confined aquifer in the glacial till and an unconfined aquifer in the fill layer. There is no apparent connection between these two water sources, with piezometric levels in the deep glacial till deposits lower than the upper fill layer.

Site History

The Fort Point Channel has a history of development from a bygone shipping industry era. The shoreline progression, abutting structures, and wharves are shown in Figure 2. In the 1700s, the only area in this figure that was land is the southeast corner, below the dark line shown approximately parallel to the shaded Channel. Foundation remnants typically consist of granite structures constructed on timber cribbing bases supported on dense timber pile clusters. Most of the original moveable span bridges along the Channel have been replaced in recent years with fixed bridges, effectively eliminating a shipping industry revival. In the wake of shipping industry activity, abandoned bridge piers, deteriorating fender systems and hundreds of timber piles remain not only as an eyesore, but also as obstructions to ground improvement using auger-style methods. These obstructions presented the first challenge in preparing the site for DMM installation.

The second challenge was matching available DMM equipment to the site. The area where most of the ground stabilization is required is within the Channel; hence, this should ordinarily have been a water-based DMM operation. However, with an effective channel depth of 3m at low tide and 6m at high tide, Channel widths as narrow as 45m in some locations, and existing bridge structures/abandoned pile foundations/piers, a water-based DMM operation using large off-shore barges was not practical nor possible. Small barges could be used but they would not be schedule effective due to their inherently slower DMM production rate.

The following indicates how the Contract Documents required the Contractor to address these challenges.

Pre-Construction Challenge No. 1: Obstructions

DMM equipment, typically consisting of (multiple) hollow-stem augers, does not have the ability to cut through timber and granite obstructions. The concern was that auger flights and/or mixing paddles, which are welded to the drill stems, would shear off when these types of obstructions were encountered, or the augers would "bounce off" obstructions causing a soil-cement element to be installed excessively out of vertical tolerance and possibly compromising the continuity of the shear wall. Given these potential construction schedule-impacting problems, it was critical that obstructions be removed as expeditiously as possible prior to DMM installations. Figure 3 shows the locations of some of the known obstructions at the site.

Existing and/or abandoned piles and fender systems within the Channel were to be removed using equipment on barges. With the piles typically embedded 14m or more within the soft, cohesive clay, some for over a hundred years, the removal process was expected to be slow. This operation was meant to remove the predominant number of exposed piles. Exploratory dredging was to then be performed, to locate and remove piles that had been previously broken or been cut-off at or near the mudline.

To remove obstructions outside the Channel, a land-based operation of exploratory pre-trenching was to be performed. Where DMM elements were to be installed, a 1.5m deep general excavation was required to expose obstructions near the surface. When an obstruction was encountered within that zone, it was to be chased full-depth for removal. Based on historical research of former structures, it was anticipated that the predominant number of foundations would be encountered within 1.2m of existing grade (which corresponds to the depth of frost penetration in the local building code).

At locations that were not critical for installation of DMM shear walls, the dense pile clusters supporting abandoned piers could be left in-place. However, all remnants above the piles were to be removed and, if necessary, the pile clusters were to then be encapsulated with jet grout.

Pre-Construction Challenge No. 2: Equipment Restrictions

With a significant quantity of DMM to be installed in the Channel, water-based work should ideally have been performed using large, off-shore barge-mounted DMM equipment. In order to carry associated equipment and materials, these barges are generally the size of a football field. Given the effective Channel constraints on width and depth, a barge of this size would be unable to maneuver, and would sink into the mud at low tide due to insufficient draft. Figures 4 and 5 schematically represent the equipment types and sizes expected for land- and marine-based DMM operations. As can be seen, land-based equipment requires much less room in which to operate.

The solution to this problem was land-reclamation and to complete most of the work using a land-based DMM operation. A system of interlocking sheet piles was to be installed to enclose the proposed DMM limits within the water. The organic deposit would then be grout-stabilized using shallow-mixing methods to support a surcharge loading. Gravel fill would then be placed to an elevation serviceable by the adjacent land, so that DMM equipment had land access to the area to be treated. A typical section illustrating this land-reclamation method is shown in Figure 6.

Shallow pre-stabilization of the organic deposit was to be completed by in situ mixing grout into the organic deposit over its depth to develop 140kPa minimum unconfined compressive strength. This would produce a "relieving-platform" effect on the effective lateral earth pressure component on the sheetpile wall system and provide a stable work platform for equipment during installation of DMM elements. The perimeter sheetpile system would serve as backfill containment, limiting migration of spoils generated by the DMM process, and minimize the influence of tidal fluctuation of the groundwater level in the reclaimed zone.

This land reclamation scheme in the Channel was to be performed in three sequences; an east cell and a west cell simultaneously, followed by the middle cell. This sequence was necessary as the middle cell surrounds and protects an existing historic railroad bridge. Only after the existing bridge was bypassed with a temporary structure over the westernmost cell could demolition of the existing bridge occur, and the land reclamation method be completed in the middle cell.

On completion of the land reclamation to the limits circumscribed by the installed perimeter sheetpile containment systems, DMM installation using land-based equipment could then begin. Figure 7 illustrates a typical cross-section of the required extent/depth of DMM.

Note that primarily for environmental reasons, there were some relatively small areas east of the Dorchester Avenue Bridge (refer to Figure 1) where land reclamation was not viable. For those areas, DMM was to be installed using barge mounted equipment.

Project Status

At the time that this paper is being prepared, the contract is under construction. Existing piles and fender systems within the Channel, between the three bridges (Wye, Railroad and Dorchester Avenue), have been removed. Vibratory equipment on a barge-mounted crane was initially used to remove visible piles. Most of the timber piles came out smoothly and fairly intact. Thereafter, a barge-mounted clamshell bucket was used to locate and remove piles that were not visible, i.e., those that were at or below the mudline. The same procedure will be used to reclaim the area under the Railroad bridge when the bridge has been demolished.

The pile removal process in the Channel appears to have been successful because DMM was subsequently installed in the two areas between the bridges without significant problems with obstructions. The organics were shallow-mixed using a barge mounted, extended reach excavator with a 1.3m-wide mixing bucket. The bucket had rotating blades and nozzles through which the required grout was injected. The organics were mixed with grout by the rotating blades as the bucket moved through the organics from mudline to top of clay. Gravel was then placed on top of the treated organics to an elevation that was approximately 1.5m below high tide. Thereafter, land-based DMM equipment was used to perform the deep soil mixing, with groundwater being controlled by two sump pumps.

A temporary Dorchester Avenue bridge and roadway have been constructed on the newly installed DMM, just west of the existing bridge and are operating satisfactorily. Demolition of the existing Dorchester Avenue bridge is almost complete. A barge mounted DMM rig will then be used to treat the area once obstructions have been removed. Installation of the foundations (drilled shafts) for the temporary Railroad bridge (refer to Figure 1) is almost complete. The temporary bridge will then be constructed, the existing railroad tracks transferred on to it, the existing bridge demolished, and the area below it soil mixed..

However, removal of obstructions on land was not as simple. The initial removal of obstructions down to a general depth of approximately 1.5m proved to be inadequate, despite having chased full depth any obstructions that were encountered within the 1.5m-depth zone. Subsequent attempts to install DMM resulted in frequent encounters with obstructions that prevented further mixing. This appeared to indicate that the area had unknown obstructions that were located deeper than 1.5m.

Due to the sensitivity of adjacent railroad tracks and Dorchester Avenue, these obstructions could not be removed by an open-cut excavation. Consequently, the Contractor is currently in the process of installing excavation support systems to facilitate removal of these obstructions. Removal will be to general depths of about 6m and localized depths of as much as 9m. DSM will then resume once the obstructions have been removed.

Conclusion

This paper discussed two aspects of site preparation that were essential to the successful installation of DMM within an area with severe site constraints. Removal of obstructions prior to production DMM installation was necessary to provide for full-depth penetration and ensure longitudinal continuity of DMM elements. Reclamation of land was also necessary to maximize land-based installation of DMM given the depth and width limitations of the Channel, access restrictions due to the existing bridges, and the lower production rate of water-based DMM operations using smaller barges.

In areas between the existing three bridges, reclamation has been successfully accomplished. It is anticipated that reclamation of the area beneath the railroad bridge will be just as successful.

Obstruction removal has been satisfactory in the Channel but removal on land is still in progress. Observations during construction confirmed that advance removal of most obstructions was essential to the successful installation of deep soil-cement mix.

Acknowledgments

The authors wish to recognize the Massachusetts Highway Department, the Federal Highway Administration, and Bechtel/Parsons Brinckerhoff for their continual pursuit to advance construction technologies and to share their experiences with the engineering profession.

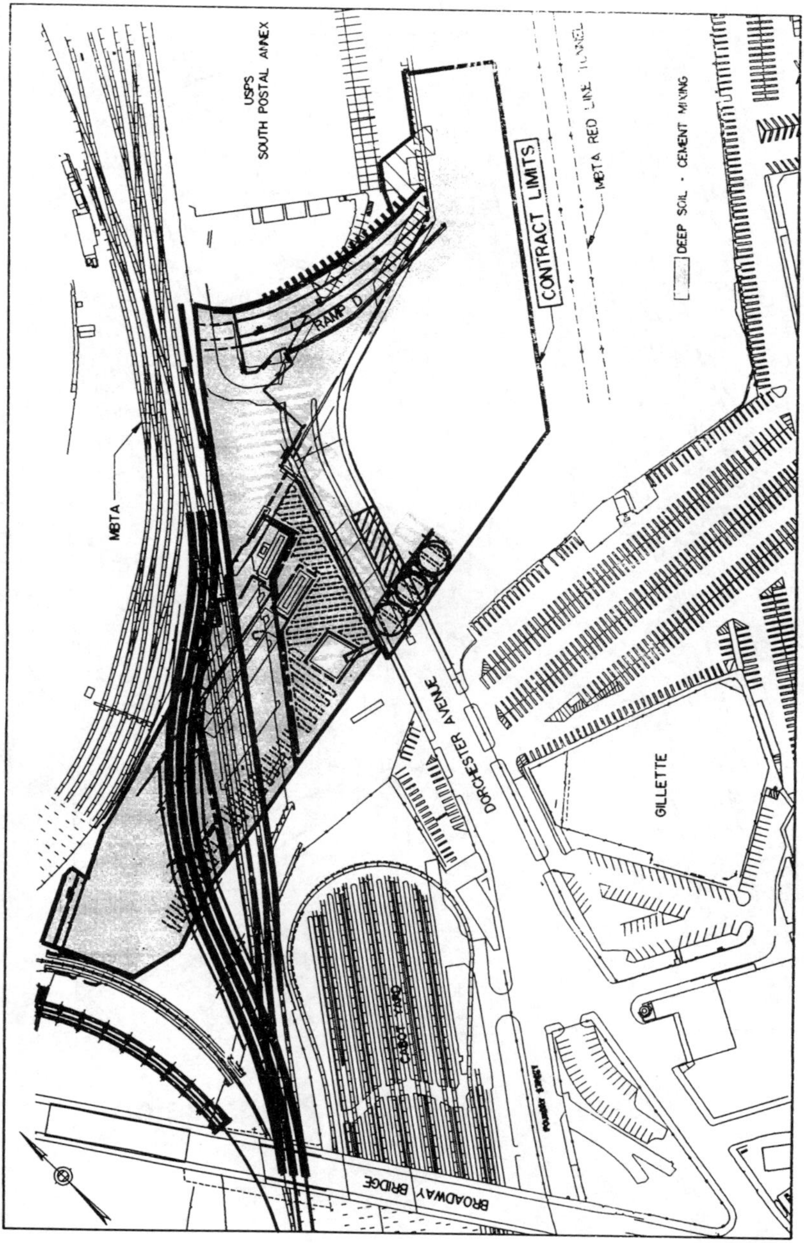

SOIL IMPROVEMENT FOR BIG DIGS 129

Figure 1 DMM Contract Limits

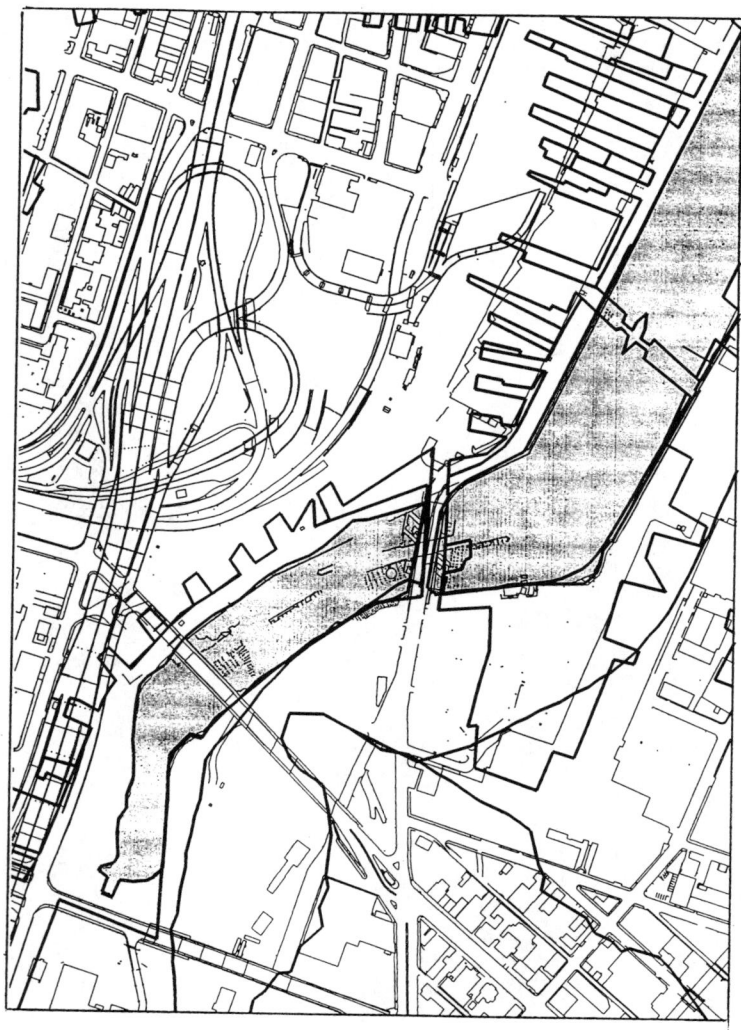

Figure 2   Shoreline Progression

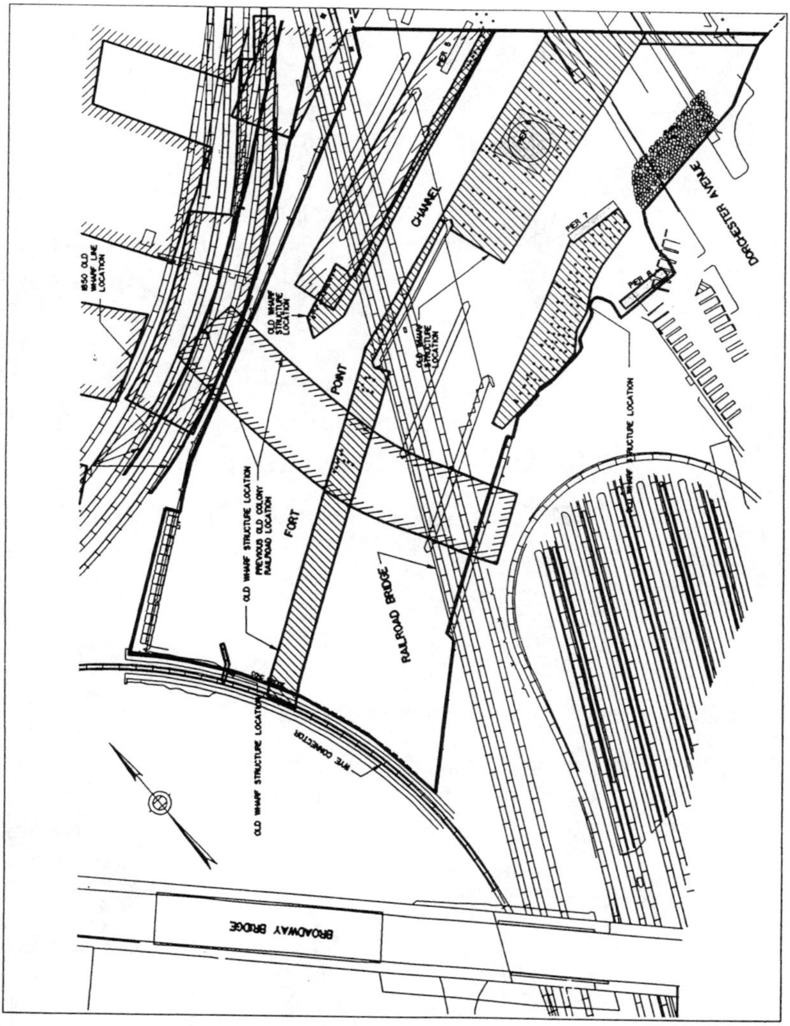

SOIL IMPROVEMENT FOR BIG DIGS 131

Figure 3 Obstruction Removal Plan

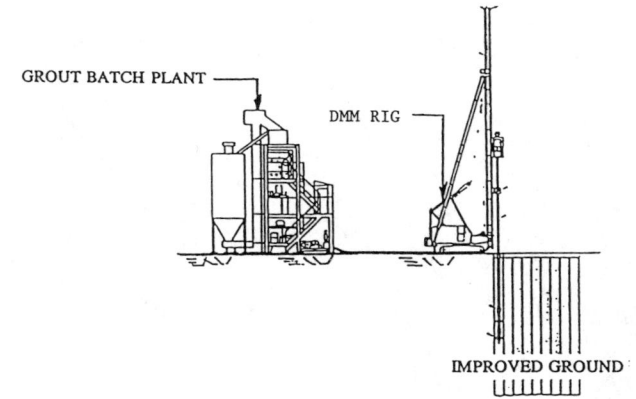

Figure 4   Schematic of Land-Based DMM Operation

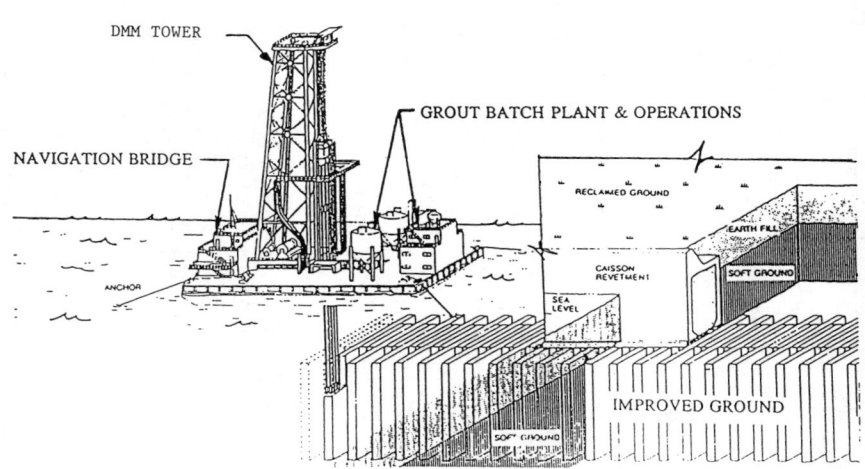

Figure 5   Schematic of Marine-Based DMM Operation

SOIL IMPROVEMENT FOR BIG DIGS

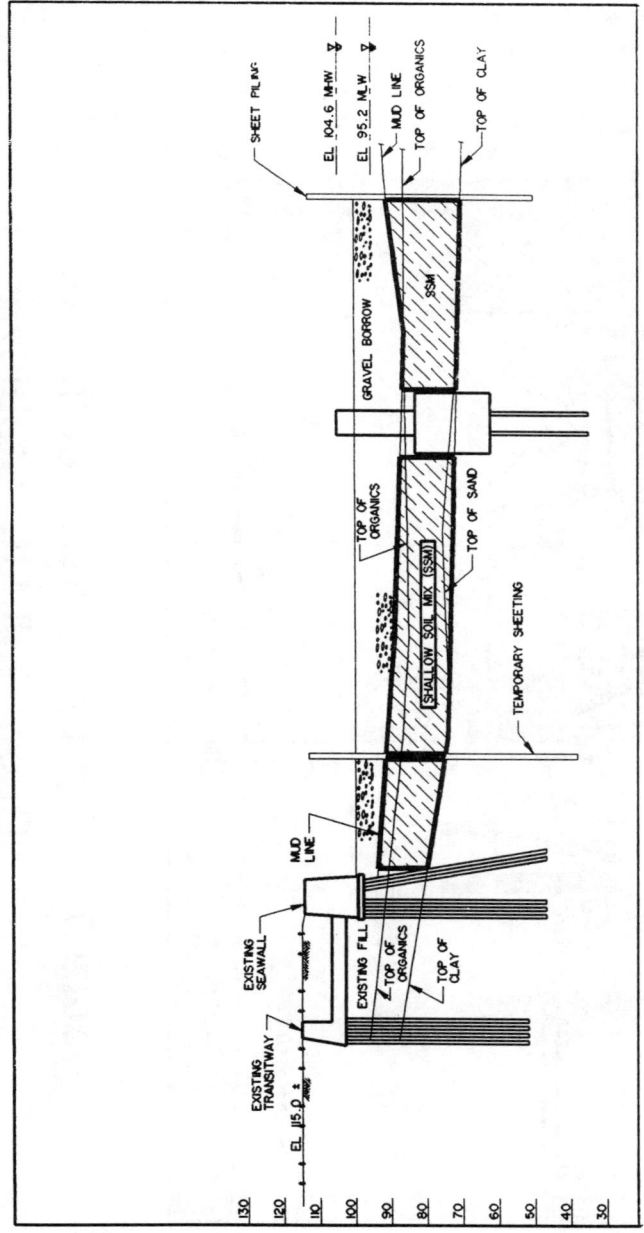

Figure 6 Typical Cross-Section of Shallow Soil Mix Stabilization

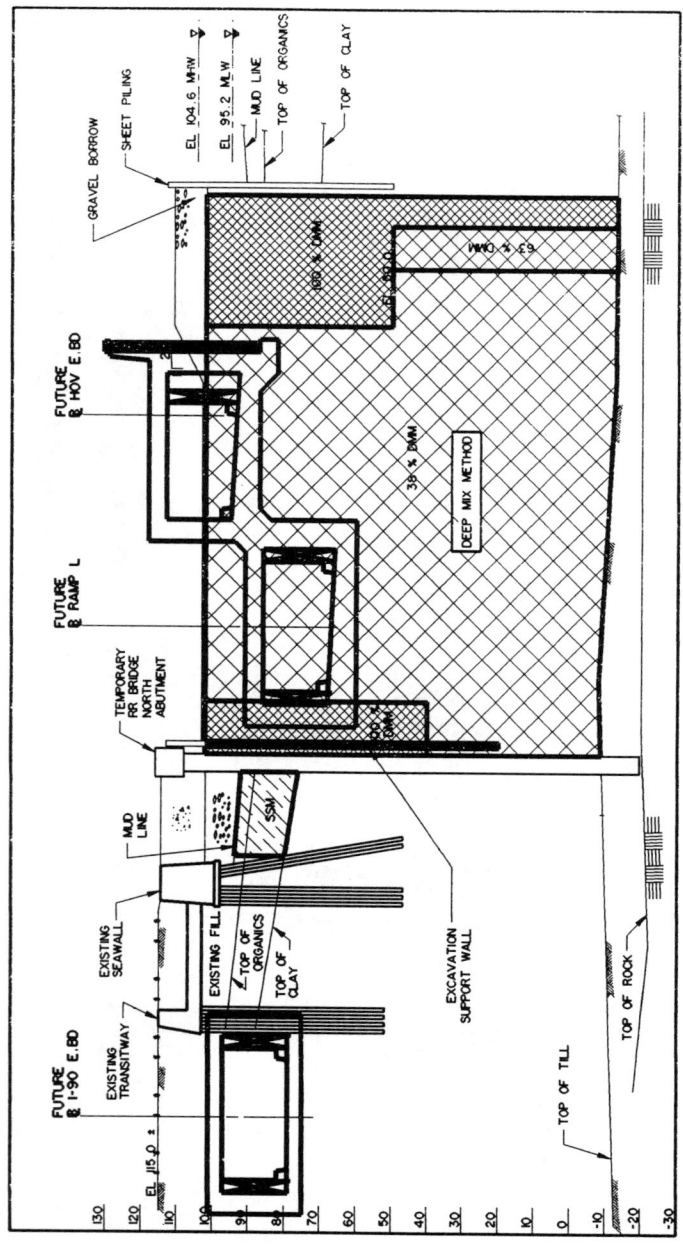

Figure 7 Typical Cross-Section of Deep Soil Mix Stabilization

STRENGTH GAIN OF ORGANIC GROUND WITH CEMENT-TYPE BINDERS

Melanie B. Hampton[1], Student Member, ASCE and Tuncer B. Edil[2], Member, ASCE

Abstract

The next frontier in soft ground improvement, particularly for soils containing a high percentage of humus, is deep in-situ mixing methods with cement-type binders. One of the difficulties in applying this technology to organic soils is that organic matter inhibits cementitious reactions. In fact, for all soils, inorganic and organic, the properties of the stabilized product are extremely difficult to predict due in part to the lack of understanding of the reactions between the soil, water, and binding agent. Current design methods rely heavily upon laboratory mixture tests in which the soil is mixed with different binders at different dosages, and specimens are cast and allowed to cure. The unconfined compressive strength of the specimens is then measured after a designated curing time. This paper presents a synthesis of mixture tests conducted in Delft, the Netherlands, and Madison, Wisconsin, on several peats and an organic clay. Although concrete mix designs are made using distinct relationships between water, cement, and aggregate contents, the same relations can not be extrapolated to stabilized soil. This paper will show that current experimental techniques are not sufficient to create a comprehensive model of strength gain in stabilized organic soil. Research is needed on the fundamental chemical reactions contributing to changes in the geotechnical properties of stabilized organic ground.

INTRODUCTION

The growing demand for space to accommodate new construction and the expansion of existing facilities has forced the construction industry to build upon soft ground, including areas of extensive peat deposits. Peat deposits are most often excavated and replaced with fill material. This solution is cost prohibitive if the peat

---

[1] Ph.D. Candidate, Department of Civil and Environmental Engineering, University of Wisconsin-Madison, Madison, WI 53706, bauerm@cae.wisc.edu
[2] Prof., Department of Civil and Environmental Engineering, University of Wisconsin-Madison, Madison, WI 53706, edil@engr.wisc.edu Tel: (608)-262-3225

deposit is extensive, or if satisfactory fill material is not readily available. Geosynthetics can be used to increase the stability of embankments on peat deposits, although this method does not address other construction complications due to the high water content and long term settlement potential of the peat. Long term settlement of peat deposits can be reduced by preloading; however, this option is notably time consuming.

In-situ improvement of peat and soft organic soils using deep mixing techniques offers tremendous economical and practical advantages over these other methods. Deep in-situ chemical stabilization using lime and cement has successfully improved the engineering properties of soft clays in Japan and Scandinavia for nearly 30 years (Rathmayer, 1996; Okumura, 1996). Extension of this technology to organic soils has been very slow since organic matter inhibits the cementitious reactions responsible for strength gain. In fact, for all soils, the properties of the stabilized product are extremely difficult to predict due in part to the lack of knowledge about what reaction products are formed when a particular binder is mixed with a soil.

The avenue of approach in previous research on stabilization of organic soils has been to select a matrix of different stabilizers and soils having different organic contents, degrees of decomposition, clay fractions, water contents and pH values. Mixture tests are conducted from all the different combinations of materials, and the unconfined compression strengths of the stabilized specimens are tested (den Haan, 1997,1998; Huttunen et al., 1996; Huttunen & Kujala, 1996; Kujala et al., 1996; Odajima et al., 1995).

The result is a correlation of a myriad of parameters with compressive strength. Such relations have been proposed by Odajima et al. (1995), Kukko and Ruohomaki (1995, in Rathmayer, 1996), Nagaraj et al. (1997); Babasaki et al. (1997), and other researchers. One of the problems with this approach is that the ground chemistry of each site is different and will result in different implications for construction. These changes are not examined thoroughly in typical laboratory mixture tests; therefore it is very difficult to extrapolate from one site to another when using deep mixing techniques.

This paper presents results from mixture tests conducted by the authors at the University of Wisconsin and Delft Geotechnics. Trends gleaned from these results show that traditional index properties are not suitable to base a reliable stabilization protocol, especially in consideration of soils with high organic matter contents. Rather, thorough knowledge of the chemical properties of the soil and the hydration properties of the binding agent are necessary.

BACKGROUND

Although mixing soils with stabilization agents has been used for thousands of years in surface applications, it has only been in the last 30 years that chemical stabilization of soil has been employed in deep in-situ mixing methods. In the mid

1970's Japan and Sweden concurrently developed methods using the in-situ forced stirring of cement or lime with soft, cohesive soil to create a stabilized soil column (Toth, 1993). In each of these mix-in-place methods, a mixing tool drills down through the soil to the desired depth of the column and a hardening agent is injected down the stem and mixed with the soil, resulting in the formation of a "soil-cement" column (Bruce, 1992).

It is important to note that even though these deep mixing methods may use cement-based grouts or reagent slurries, they are not considered soil grouting techniques. Grouting involves intrusion of the grout into cracks and fissures or replacement or displacement of soil with the grout, while deep mixing intimately blends the soil and the admixture to create a material with increased shear strength and bearing capacity and reduced compressibility. Due to the proprietary nature of certain equipment and techniques used in deep mixing operations, a general name for the technology has not been used consistently. Throughout this paper, "deep mixing" is a general term referring to both Japanese Deep Mixing Method and the Scandinavian lime-cement column method. The basic premise of deep mixing methods is that a new material is created in-situ by intimately mixing soil and a binding agent.

Methods of deep stabilization

The Japanese Deep Mixing Method was developed to improve the engineering properties of the soft littoral deposits for offshore construction of port and harbor structures. In this method, large diameter discontinuous flight augers are rotated to the desired depth of stabilization; a slurry of stabilizing agent is injected through the stem of the hollow auger during both penetration and withdrawal. Ordinary Portland cement and quicklime are the most common binding agents; however, the use of gypsum and slag cement is increasing. Stabilizer dosages range from 100 to 200 kilograms per cubic meter of unstabilized soil with resulting unconfined compressive strengths as high as 10,000 kPa.

Mixed in place lime columns and lime-cement columns have been used to increase strength and reduce compressibility of soft clays in Scandinavia for a variety of infrastructure projects. In the Swedish column method, a unique mixing tool resembling a pastry dough mixer drills down through the soil with a Kelly bar to the desired depth of the column. As the tool is slowly withdrawn (1.5-2.5 cm/rev), dry quicklime (CaO) and/or Portland cement is injected through the center of the tool by compressed air. Typical dosages are 50 to 100 kilograms per cubic meter of unstabilized soil. The soil is compacted as the tool is withdrawn due to the shape of the mixing tool (Tammirrinne, 1994). Design procedures call for soft to semi-hard columns of stabilized soil with unconfined compressive strengths of 100 to 150 kPa. The columns are expected to interact with the surrounding soil to create a flexible foundation. Design procedures for high strength columns are also being developed at this time (Carlsten and Ekström, 1997).

Strength gain in stabilized soils

Changes in the geotechnical properties of a stabilized soil are mainly due to the hydration products formed by reactions between binding agent and water, and reactions between binding agent, water, and soil particles. Studies have shown that the unconfined compressive strength of stabilized soft soil varies greatly, and is dependent upon a number of factors, such as soil type, type and amount of binding agent, and curing conditions.

When quicklime is added to a soft clay, its shear strength increases due to three reactions:
   (1) Isomorphous substitution of calcium in the clay particles decreases the interlayer spacing and causes coagulation of the clay particles, thus reducing the plasticity of the soil;
   (2) Quicklime reacts with excess water to form calcium hydroxide;
   (3) The increase in pH in the system promotes the dissolution of silica in the clay particles which then reacts with the calcium oxides to form calcium silica hydrate (C-S-H) cementing compounds. These compounds form bonds either between the binder particles and soil particles or between binder particles to form a stiff matrix.

These are the same reactions responsible for changes in the geotechnical properties of soft clays stabilized by deep mixing. Research has shown that organic compounds can alter the composition and structure of the C-S-H gel and the type and amount of other hydration products. Organic matter is known to hold ten or more times its dry weight in water, which may limit the amount of water available for hydration reactions to occur. Organic matter is also known to form complexes with aluminosilicates and metal ions which can interfere with hydration reactions of a cement. The manifestations of these mechanisms are not easy to isolate, and research dating back as far as the 1940's shows that organic matter content has a highly variable correlation with compressive strength of the stabilized product (Catton and Felt, 1943, Clare and Sherwood, 1956).

Japanese and Scandinavian studies examining the stabilization of organic soils reflect these difficulties in determining the effects of soil organic matter. A literature search conducted by the authors found that most studies are very difficult to compare, and even present conflicting results (Hampton, 1997). Research is being undertaken at the University of Wisconsin and at Delft Geotechnics to determine the stabilization potential of organic ground such as organic clay and peat. This research program is composed of a compressive strength investigation and a physicochemical investigation.

SOILS AND BINDING AGENTS

Soils used in this study include three Wisconsin peats of different botanical origins and degrees of decomposition. These peats were chosen to represent a range of materials exhibiting different stabilization potentials similar to the research by

Kujala et al. (1996). The sphagnum moss peat from northern Wisconsin has the lowest pH and lowest degree of decomposition. The reed sedge peats are significantly more decomposed than the sphagnum peat. As humification, or decomposition progresses in a peatland, the soil's pH, mineral content, bulk density, and cation exchange capacity increase. Sphagnum peat has significantly more hemicellulose than a decomposed reed sedge peat. The decomposed reed sedge peat has a considerably higher percentage of humic acids than the sphagnum peat. An organic clay and a peat from two sites in the Netherlands were also used in this research. The initial compressive strengths of all of these soils are under 20 kPa. Properties of these soils are shown in Table 1.

	Wisconsin			Netherlands	
	Sphagnum	Reed Sedge A	Reed Sedge B	Reed Sedge D	Organic clay
Origin	Northern WI	Southwest WI	N. Central WI	Abcoude	's-Gravendeel
Classification	Fibrist	Saprist	'Ultra Saprist"	Saprist	OC
water content	1500-2000%	200-300%	300-450%	600-700%	100-200%
pH	2.88	5.61	4.35	5.8	6.7
Loss on ignition	92%	59%	46%	83%	13%
bulk unit weight g/cm^3	1.01	1.15	1.10	1.01	1.31

Table 1 Soil Properties

The experimental programs used many different binders, including special cements made with blast furnace slag (hoogoven cement) from the Netherlands. These binders are under investigation through EuroSoilStab, a European Union project on deep mixing in organic soils. Table 2 shows the composition of several of these binders. The specific surface area of the dry particles is expressed as a Blaine value in cm^2/g.

	SiO_2 (%)	Al_2O_3 (%)	CaO (%)	MgO (%)	SO_3 (%)	Fe_2O_3 (%)	Blaine cm^2/g
Ordinary Portland Cement (OPC)	21.2	5.8	62.0	6.0	3.0	2.0	3200
High calcium quicklime (CaO)	1.0	0.5	97.0	1.0		0.5	
Blast furnace cement (FSC IIIB)	28.3	9.4	46.4	7.1	4.4	2.4	4375
Blast furnace cement (FSC IIIA)	27	11	49	7	3	2	5500
Binder F	21.8	10.9	47.3	6.1	10.8	1.6	5280

Table 2 Composition of Binding Agents

INFLUENCE OF SAMPLE PREPARATION ON COMPRESSIVE STRENGTH

Stabilized soil is not concrete, although most research treats it as such. Typically, cylinders are cast using procedures similar to concrete testing, but many different techniques have been used to cast and cure the cylinders. Standard procedures for mixing and preparation of test samples; storing and investigation of stabilized soils samples have been developed by the Swedish Geotechnical Institute for use in the European Union project, EuroSoilStab. At Delft Geotechnics, mixture tests were conducted on the organic clay to investigate the effects of three variables on unconfined compressive strength, and to determine the reproducibility of the unconfined compressive strength within a test batch using current specimen preparation techniques. The following conditions were investigated:

Reproducibility: Four replicates per set.

Sample size: 50 mm diameter specimens versus 66 mm specimens.

Sample preparation technique: hand tamping and curing the samples sealed versus hand tamping and curing the specimens submerged in water under a sustained load of 14 kPa.

Addition of binder in dry powder or slurry form: Wet (2 levels: water/cement = 1:1 and 1:1.5) vs. Dry Mixing

Organic clay from 's-Gravendeel, the Netherlands, at its natural water content was mixed for five minutes with 200 kg/(m^3 natural soil) of Binder F (80% blast furnace slag cement + 20% anhydrite) using a Hobart dough-mixer. The mixture was placed in the molds in three lifts to a total height of approximately 17 cm. Each lift was hand tamped 25 times with a 50 mm diameter tamper. The specimens were sealed and stored at a temperature of 18°C. The specimens of Set 5 were not sealed, but stored under water at 18°C and a load of 14 kPa was applied to the top of the specimen. After 28 days of curing, the samples were trimmed to a length of 15 cm and tested in unconfined compression. The water content of the sample trimmings was determined using a 105°C oven. The stabilized clay specimens were all considerably moist after 28 days of curing. The stress-strain curves of the specimens all show brittle post-peak softening consistent with previous work by den Haan (1997). Table 3 summarizes the results of this testing program.

Sample reproducibility

The average 28 day unconfined compressive strength determined by combining the results from all specimens was 1150 kPa ± 172 kPa with a 95% confidence interval. Within each set of specimens, the confidence interval was always less than 15% of the mean. Results from this set of experiments also agree with previous results of den Haan (1997).

Sample size

By comparing the results of Set 3 and Set 4, it appears that smaller sample diameters, while maintaining the same aspect ratio, result in significantly lower compressive strengths. The compressive strength of the 50 mm specimens was 50% to 60% of the strength of the 66 mm specimens cast in the same fashion. This is contrary to what was expected before the tests were completed.

Set	W/C	Specimen Diameter (mm)	Curing conditions	Number of Samples	Unconfined Compressive Strength (kPa)		
					Mean	Std. Dev.	95% Confidence
1	1:1.5	66	sealed	4	1094.77	143.9	162.8
2	1:1	66	sealed	4	842.8	54.6	75.7
3	Dry	66	sealed	4	1745.7	214.1	209.9
4	Dry	50	sealed	4	949.4	41.6	47.0
5	Dry	66	Load + immersed	4	1118.6	131.3	128.7
All				20	1150.0	372.2	171.9

Table 3 Influence of Sample Preparation on Unconfined Compressive Strength

Sample Preparation

Curing under water with a sustained load did not increase compressive strength of the samples. In comparing Set 3 and Set 5, curing under water actually resulted in a decrease in strength.

Wet vs. Dry Mixing

Mixing the binder as a slurry instead of as a dry powder made homogeneous mixing easier. However, the compaction of the wet samples was much less uniform than the dry mixed specimens. Two specimens of Set 2 had very large air voids; the lifts in which the material was placed were clearly visible. These specimens were not used to calculate the average compressive strength of the set. By comparing the means of Sets 1, 2, and 3, strength decreases with increasing amounts of mixing water.

This short experimental program was an indicator of the variability that can be expected within a single set of mixture tests. It also served as a reference study to set uniform standards for further mixture tests. Further experiments using these variables are underway at Delft Geotechnics.

STRENGTH GAIN OF PEAT/BINDER MIXTURES

The first phase of mixture tests conducted at the University of Wisconsin was used to screen seven different binding agents. Samples of the three Wisconsin peats were mixed at natural water content with varying dosages of binding agent. Simple penetrometer tests were conducted to measure the compressive strength of stabilized Wisconsin peats at several points in time during curing. A common representation of strength data is to plot compressive strength against a ratio of water content to stabilizer content. Since organic matter ties up a large amount of water, the water content was adjusted to reflect only the amount of water that can be extracted from the soil by 200 cm head of suction, and defined as 'free water'. Water held at higher suction head may or may not be available for hydration reactions, but this method provides a better reference point than the water content determined by drying in a 105°C oven. Figure 1 illustrates the variation of strength with the free water/stabilizer ratio for sphagnum peat stabilized with several different binders. The trend of increasing strength with decreasing water/stabilizer ratio is reasonable. There is not an obvious trend differentiating the three different types of peat in this manner, however. For all the Wisconsin peats, a 50:50 mixture of ordinary Portland cement and Na-bentonite achieved the highest compressive strength. Ordinary Portland cement alone achieved higher compressive strengths than a 50:50 mixture of OPC and quicklime. The highest 14-day compressive strengths were on the order of 500 kPa, and the lowest, 20kPa.

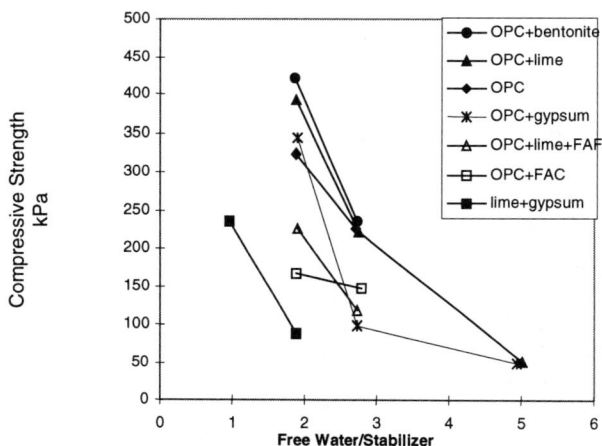

Figure 1 Compressive Strength of Sphagnum Peat stabilized with Various Binding Agents (14 days curing time)

Figure 2 presents a different method of comparison. Strength is plotted versus the ratio of the weight of the soil particles to the weight of stabilizing agent. In this case, the strength achieved in stabilized sphagnum peat is very sensitive to changes in the ratio of soil solids to stabilizer. From these results, it can be implied that characteristics of the solid soil particles dominate the stabilization potential of the organic soil. These could be mechanical or chemical characteristics.

The results of these experiments support the hypothesis that an important quality of the stabilizing agent is its ability to form strong hydration products. Since there is a very small fraction of clay minerals in the peats, adding lime will not induce

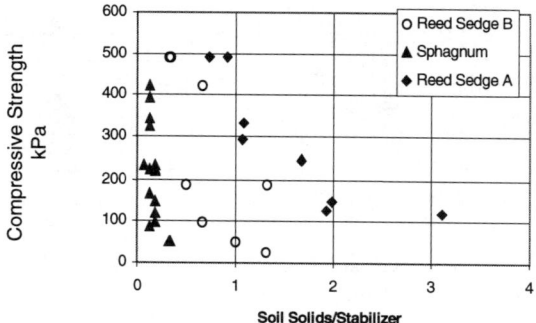

Figure 2 Compressive Strength versus Soil Solids Stabilizer – Wisconsin Peats

a pozzolanic reaction. The only hydration product formed in this case is calcium hydroxide. Adding bentonite provides two beneficial mechanisms. First, some excess water is bound by the bentonite; second, the bentonite serves as a filler material and source of silica to facilitate the development of C-S-H gel.

INFLUENCE OF WATER/CEMENT RATIO ON STRENGTH GAIN OF ORGANIC SOILS

Wisconsin Peats

The next batch of experiments had two objectives. First, to determine if the unconfined compression strength of a stabilized soil could be determined from the initial water content of the soil and the binder content, similar to Abram's law for concrete. Second, to determine if calcium chloride pretreatment or increased gypsum content increases compressive strength. One theory of organic interference in the formation of hydration products is that the organic matter binds calcium ions, and alters the system chemistry so the calcium silica hydrate gel can not form. Calcium chloride is used as an admixture to attempt to satisfy the calcium affinity of the organic matter.

Results from this set of experiments showed that the compressive strength of a stabilized peat can not be deduced from the contents of water, cement, soil solids and organic matter. A factorial experiment proved that there is too much interaction between the parameters to attribute the effects of a single parameter. Therefore, other properties of the soil must be examined such as the calcium affinity of the organic matter. Table 4 shows the results of part of a mixture test program in which several variables were examined: type of peat, water/cement/soil solids ratio (W/C/S), calcium pretreatment, increased gypsum content, and curing time.

The 95% confidence interval of each result reported is less than ±15% of the mean. When compared to strengths achieved from plain ordinary Portland cement stabilization, excess calcium had mixed results. While calcium clearly increased strength in Reed Sedge peat B at a w/c of 1.0 (compared to OPC alone), it had

	Reed Sedge A		Reed Sedge B		Sphagnum	
W/C/Sθ	1/1/1	1.4/1/1	1/1/1	1.4/1/1	1.4/1/1	2/1/1
OPC (200 days)	3265	1393	238	1398	2011	263
OPC (40 days)	1600	800	158	720	40	75
OPC+CaCl$_2$, 100:1 (200 days)	2577	2498	2159	1293	2083	328
OPC+Gypsum, 2:1 (200 days)	1914	643	368	389	1153	250

Binder (cure time)

Table 4 Unconfined Compressive Strengths of Stabilized Wisconsin Peats (kPa)

virtually no effect when the w/c increased to 1.4. The contrary is observed in the sapric reed sedge peat (A). At a w/c of 1.0, excess calcium decreased compressive virtually no effect when the w/c increased to 1.4. The contrary is observed in the sapric reed sedge peat (A). At a w/c of 1.0, excess calcium decreased compressive strength, while at a w/c of 1.4 compressive strength increased considerably. In all long-term specimens, adding gypsum significantly reduced compressive strength. This is an observation not seen from short term experiments, and may imply durability problems. For all of the 200 day specimens, reed sedge B achieved the lowest compressive strengths even though this soil has the highest mineral content.

In Figure 3, the final water content of the 200 day cured specimens of stabilized Wisconsin peat is plotted against compressive strength. If the mechanism of poor strength gain is simply whether water is being consumed by hydration reactions, compressive strength would increase as final water content decreases. There is enough scatter in the data to show that this is not the case, and the composition and structure of the hydration products must be considered.

Dutch Peat and Organic Clay

Mixture tests on the 's-Gravendeel organic clay by Delft Geotechnics and the Swedish Geotechnical Institute compare the effects of different binders on the compressive strength of the clay. All of these tests were performed with dry mixing techniques at the clay's natural water content and using similar sample preparation procedures. Table 5 summarizes these results. One can conclude that the blast furnace cement and anhydrite mixture (Binder F) increases the strength of the organic clay significantly more than lime-cement mixtures. The addition of a small amount of high aluminum cement appears to increase the compressive strength even more.

Similar mixture tests were conducted on the Abcoude peat. Again, the blast furnace cement and anhydrite mixture (Binder F) increases compressive strength significantly, to approximately 350 kPa.

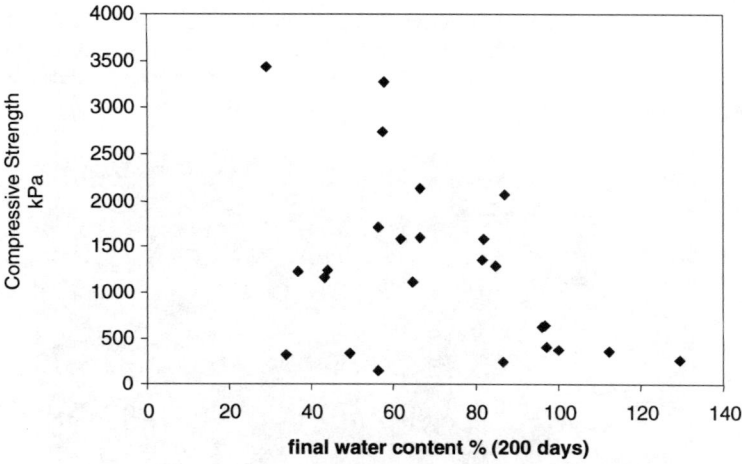

Figure 3 Compressive Strength versus Final Water Content of Stabilized Wisconsin Peats

Binder	28 day compressive strength (kPa)
Binder F, 200 kg/m^3	1460
50% OPC + 50% Lime, 100 kg/m^3	40
80% OPC + 20% Lime, 150 kg/m^3	145
80% FSC A+ 14.5% Anhydrite + 5.5% High Aluminum Cement, 200 kg/m^3	1827

Table 5 Unconfined Compression Strength of 's-Gravendeel Organic Clay

MICROSTRUCTURE OF STABILIZED PEATS AND ORGANIC CLAY

Fourier Transform Infrared Spectroscopy (FTIR) studies were conducted at Delft Geotechnics to track changes in chemical bonds as a cement paste hydrated, and to track changes in an organic soil or a soil amended with organic material treated with a binder. Particular changes that the investigators were looking for were indications of increased calcium hydroxide content, increased ettringite content, and indications of silicate polymerization. GeoCon (USEPA, 1990) conducted FTIR studies on stabilized organic contaminants, and attributed specific shifts in the peaks of the treated organic mainly to hydrogen binding of the organic to the stabilizer and reduction of the O-Si-O bond order. FTIR scans conducted in the Netherlands suggest that polymerization of silicates is severely hampered in soils with highly decomposed organic matter such as the Dutch peat from Abcoude. This is evidenced by the absence of peak shifts associated with the cement hydration from 925 cm^{-1} and 965cm^{-1} to approximately 1000 to 1100 cm^{-1}. Increased ettringite formation was found in stabilized organic soils as evidenced by the high FTIR peaks in the 1100 to 1160 cm^{-1} range. SEM and EDX studies of stabilized organic soil confirmed the abundance of spiky ettringite needles and absence of calcium silica hydrate amorphous gel. Ettringite needles can be seen in Figure 4.

Figure 2 SEM Picture of Reed Sedge Peat (A) Stabilized with Ordinary Portland Cement (60 day cure)

CONCLUSIONS

Despite the similarities between soil-cement, concrete, and deep mixing, relations such as Abram's Law that rely on the proportions of water, binder, and soil solids, will not accurately predict the geotechnical behavior of a stabilized soil. This is because the soil solids are essentially reactive aggregate, and the chemical interactions are numerous. The stabilization potential of an organic soil is dependent upon the ability of strong, stable hydration products to fill voids and take up excess water. The organic substances present in highly decomposed peats interfere with the development of calcium silica hydrate gel, and promote the growth of ettringite. Ettringite is not as strong as the calcium silica hydrate gel, and may not be stable in certain environments. Binding agents most suitable for organic soils should contain higher percentages of silica, such as with the addition of blast furnace slag. This reactive form of silica will aid in the development of the C-S-H gel phase. Further research is needed to determine the extent of silicate polymerization from the hydration of binding agents mixed with soils, and how the chemical characteristics of the system influence the structure and composition of the hydration products.

ACKNOWLEDGEMENT

Financial support for this research was provided by a National Science Foundation Graduate Research Fellowship, and the Federal Highway Administration. Additional support and technical advice from Dr. Evert J. denHaan of Delft Geotechnics, and Mr. Göran Holm of the Swedish Geotechnical Institute is greatly appreciated.

REFERENCES

Babasaki, R., Terashi, M., Suzuki, T., Maekawa, A., Kawamura, M., Fukazawa, E. (1996) "JGS TC Report: Factors influencing the strength of improved soil." Grouting and Deep Mixing. Proceedings of IS-Tokyo '96, The Second International Conference on Ground Improvement Geosystems. Eds. Yonekura, R., Terashi, M., Shibazaki, M. 913-918.

Bruce, D.A., (1992) "Current Technologies in Ground Treatment and In-Situ Reinforcement" Proceedings, Second Interagency Symposium on Stabilization of Soils and Other Materials. Metairie, Lousiana. 2-79 to 2-81.

Carlsten, P. and Ekström, J. (1997) Lime and Lime Cement Columns. Swedish Geotechnical Society SGF Report 4:95E.

Catton, M.D. and Felt, E.J. (1943) "Effect of Soil and Calcium Chloride Admixtures on Soil-Cement Mixtures." Portland Cement Association. 497-529.

Clare and Sherwood (1956) "Further Studies on the Effect of Organic Matter on Setting of Soil-Cement Mixtures." Journal of Applied Chemistry. Vol. 6, pt. 8, 317-324.

den Haan, E. (1997) Strength Gain of a Dutch Organic Clay and a Peat by Stabilization with Various Cement-type Binders. Report No. CO-366340. Civieltechnisch Centrum Uitvoering Research en Regelgeving. Netherlands.

den Haan, E. (1998) "Cement-type stabilizers for Dutch organic soils." Proceedings of the International Symposium on Problematic Soils. Japanese Geotechnical Society {in press}.

Hampton, M.L. (1997) "Deep Chemical Stabilization of Soft Organic Ground," Proposal for Dissertation Research. University of Wisconsin-Madison, Department of Civil and Environmental Engineering.

Huttunen, E. and Kujala, K. (1996) "On the Stabilization of Organic Soils." Grouting and Deep Mixing. Proceedings of IS-Tokyo '96, The Second International

Conference on Ground Improvement Geosystems. Eds. Yonekura, R., Terashi, M., Shibazaki, M. 411-414.

Huttunen, E., Kujala, K., Vesa, H. (1996) "Assessment of the Quality of Stabilized Peat and Clay." Grouting and Deep Mixing. Proceedings of IS-Tokyo '96, The Second International Conference on Ground Improvement Geosystems. Eds. Yonekura, R., Terashi, M., Shibazaki, M. 607-612.

Kujala, K., Mäkikyrö, M., Lehto, O. (1996) "Effect of Humus on the Binding Reaction in Stabilized Soils." Grouting and Deep Mixing. Proceedings of IS-Tokyo '96, The Second International Conference on Ground Improvement Geosystems. Eds. Yonekura, R., Terashi, M., Shibazaki, M. 415-420.

Nagaraj, T.S., Yaligar, P.P., Miura, N., Yamadera, A. (1996) "Predicting strength development by cement admixture and on water content." Grouting and Deep Mixing. Proceedings of IS-Tokyo '96, The Second International Conference on Ground Improvement Geosystems. Eds. Yonekura, R., Terashi, M., Shibazaki, M. 431-436.

Odajima, H., Noto, S., Nishikawa, J., Yamazaki, T. (1995) "Cement Stabilization of Peaty Ground with Consideration of Organic Matters." Proceedings of International Workshop on Engineering Characteristics and Behavior of Peat. Sa oro, Japan.

Okumura, T. (1996) "Deep Mixing Method of Japan." Grouting and Deep Mixing. Proceedings of IS-Tokyo '96, The Second International Conference on Ground Improvement Geosystems. Eds. Yonekura, R., Terashi, M., Shibazaki, M. 879-887.

Rathmayer, H. (1996) "Deep Mixing Methods for Soft Subsoil Improvement in the Nordic Countries." Grouting and Deep Mixing. Proceedings of IS-Tokyo '96, The Second International Conference on Ground Improvement Geosystems. Eds. Yonekura, R., Terashi, M., Shibazaki, M. 869-877.

Tammirinne, M. (1994) "Ground Improvement for Engineering Purposes: Soil Stabilization." US-Scandinavia Geotechnical Workshop, NSF, Trondheim, Norway.

Toth, P.S. (1993) "In Situ Soil Mixing." Ground Improvement. Ed. Moseley, M.P. Blackie Academic and Professional CRC Press, Inc. 193-204.

USEPA (1990) "International Waste Technologies/Geo-Con In-Situ Stabilization/Solidification: Applications Analysis Report," EPA/540/A5-89/004.

THE FREEZING OF SOIL MASSES
AS AN AID TO ENGINEERING CONSTRUCTION

John F. Donohoe[1], Derek Maishman, P. Eng.[2], Paul C. Schmall, P.E.[3]

Abstract

Typical applications of ground freezing involve the creation of a peripheral frozen structure to provide excavation support and ground water control for construction of a deep shaft or large excavation. Several projects undertaken have been of a different nature, providing stabilization to massive volumes of soil rather than the creation of relatively thin walls external to an excavation or work area. The technique is uniquely applicable to cases where a broad spectrum of subsurface conditions exist, the greatest assurance of water control must be provided, and minimal disturbance of the existing subsurface condition is desired. Three projects are referenced:
- Emergency freezing of a collapsing sewer line outside of Detroit, Michigan.
- Freezing to arrest subsurface erosion evidenced by a surface sinkhole and possible mine failure at a salt mine in Louisiana.
- The mass stabilization of soils to facilitate a tunnel jacking operation of unprecedented size on Boston's Central Artery.

Introduction

Traditionally, ground freezing has been employed to put in place, in advance of excavation, ground support and groundwater cut off systems which facilitate the construction process until such time as permanent systems are installed and functional. In soils, temporary support and groundwater control are achieved by freezing around (not inside) the proposed excavation. A wall of frozen soil, founded on an underlying aquiclude, allows the creation of an impermeable barrier to groundwater seepage. In addition, the frozen thickness and frozen strength of the wall can be made appropriate to resist the geostatic and hydrostatic pressures which will act externally on the wall during construction.

For civil engineering applications, freeze walls more than ten feet in thickness are rarely required. The formation of these walls, and their maintenance, are interrelated with the project construction schedule. Generally, freeze wall support systems are designed to allow the excavation contractor to complete most of his soft ground excavation in unfrozen soils. This paper is concerned with non-typical applications where masses, rather than containment barriers of frozen soil are required.

---
[1] John F. Donohoe, President and CEO, Moretrench American Corporation, P. O. Box 316, Rockaway, NJ, (973) 627-2100
[2] Derek Maishman, P. Eng., Consultant to Moretrench American Corporation, P. O. Box 316, Rockaway, NJ, (973) 627-2100
[3] Paul C. Schmall, P. E., Chief Engineer, Moretrench American Corporation, P. O. Box 316, Rockaway, NJ, (973) 627-2100

The earliest reported example, a successful pioneering effort, addressed a specific problem which threatened the Grand Coulee Dam construction in 1936. On the east flank, glacial silts overlying the granite became progressively less stable. Grant Gordon[1], describes how a 5 cu. yd. Shovel at the toe of the slope "could make no headway and was forced to retreat or be buried." A frozen mass, arched between two rock faces, was designed to restrain the sliding material "for such time as was required to remove the desired excavation and concrete a sufficient height of dam to get out of danger from further slides." 337 freeze pipes of average length 13m were installed and the resulting frozen mass "stopped a slide of 200,000 cubic yards of earth into the excavation on the east side of the Columbia River."

Since 1936, examples of mass freezings have remained infrequent, but three projects undertaken in the last 20 years fall into this category. Two of these, like the Grand Coulee project, came in response to emergencies and did not require excavations in frozen soils. The most recent example, now in preliminary stages of installation in Boston, is probably unique in that large volumes of soil will be frozen to increase the full face stability of tunnel excavations at shallow depth.

(1) Suburban Sterling Heights, (Detroit, Michigan). Early in 1980, the surface above a 15m deep, 4.6m OD, concrete lined, interceptor sewer was observed to be subsiding. Over a length of 30m, settlement was observed to be as much as 1.5m, and inspections through an adjacent manhole revealed severe cracking and displacement of sections of the roof and ovalization of the sewer lining. These conditions pointed to imminent collapse, with severe environmental repercussions.

Several simultaneous steps were instituted to prevent a catastrophe. Emergency pump stations were set up to pump the raw sewage directly to surface waters if necessary. Construction of a bypass was initiated, although it would take several months to complete. However, the immediate focus was halting further deterioration of the tunnel. Settlement pins indicated continued settlement in measurable amounts almost daily. Voids encountered during test drilling alongside the tunnel confirmed the extent of soil disturbance over the damaged sewer tunnel lining. Grouting of the voids appeared to accelerate the surface settlement, and the effort was suspended. Ground freezing was determined to be the most viable option, posing minimal additional stress to the already highly disturbed conditions.

To stabilize the soils above the tunnel and permit rehabilitation of the damaged structure, a massive frozen canopy; 30m long, 11m wide and 11m high, was formed in horse-shoe fashion in the soils surrounding the sewer (See Figure 1).

In light of the urgency required by this remedial freezing operation, liquid nitrogen was used to accelerate the formation of the frozen block, and afford the most critical 12m long tunnel "break" area some protection at the earliest possible date. At a temperature of -185°C, liquid nitrogen was a more effective cooling agent than a conventional chilled brine at temperatures of -30°C. Analyses indicated that a series of arches of frozen earth over the structure, founded on undisturbed soil on either side of the tunnel, would substantially remove overburden loading from the badly damaged structure (See Figure 1). A system of five individual frozen arches, each 1.0 m thick, and spaced 2.5 m on centers, was created by the process over the most distressed part of the tunnel. The interim arches in conjunction with the natural arching strength of the soil spanning the gaps between the frozen arches supported

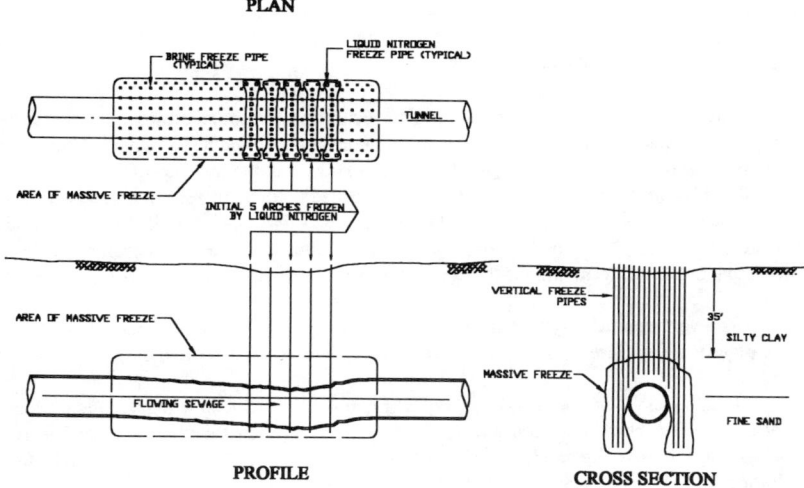

Figure 1. Detroit Emergency Sewer Remediation Project

the overburden pressure sufficiently to arrest further subsidence and buy time for the formation of a single arch canopy freeze over the complete area of concern. A relatively dense freeze pipe configuration, with pipes two feet apart was used to accelerate the formation of the individual frozen arches with the liquid nitrogen. Due to the fragility of the damaged liner, the pipes (installed vertically from the surface) were kept away from the damaged tunnel surfaces.

It was possible to monitor conditions between the nitrogen frozen arches using intermediate pipes not yet in service. After five days of nitrogen freezing, the individual arches were formed. Two days later, the settlement readings indicated movement of the tunnel was arrested. Nine days after the start of nitrogen flow, the temperature midway between the arches was below 0°C indicating the gap between the arches had closed.

Once the critical 12m section of tunnel was relieved by the quickly frozen liquid nitrogen arches, the frozen condition could be maintained for the repair work by a less expensive brine recirculation process. Several months would pass before the construction of the emergency sewer bypass would allow repair work within the tunnel to commence. During the liquid nitrogen application, a system of freeze pipes using the more conventional brine coolant was completed. A brine refrigeration plant to maintain the freeze was mobilized concurrently with the liquid nitrogen equipment. From its inception, the project was a unique combination of liquid nitrogen and refrigerated brine. As shown in Figure 1 the plan was to fill in between the liquid nitrogen arches with brine freeze pipes to provide more reliable support and also stabilize an adjacent, less critical 18m section of tunnel and maintain it with the brine equipment.

Brine pipes were installed to freeze the soil above and alongside the less critical 18m long tunnel section. Calcium chloride brine was recirculated at temperatures between -30°C and -37°C, extending the frozen block to its full 30m length in the span of one month. Brine freezing began three weeks after mobilization.

The freeze pipes and connecting piping were installed to be compatible with either process. The size of nitrogen freeze pipes and connecting piping is small (50mm), in comparison with that of brine systems (100mm). However, the spacing of freeze pipes is closer, for efficient use of the process. Pipes were spaced 0.6m apart for the nitrogen freezing of the arches with every second pipe 100mm in diameter, so that it could be subsequently converted to brine. The smaller diameter pipes were designed to accommodate the evaporating liquid nitrogen, and the larger diameter ones cold nitrogen gas.

To avoid unnecessary heat loading (and the creation of excessive frozen ground), the freeze pipes were insulated within the shallower depths, well above the area of concern.

Drilling of holes for the freeze pipes initially proved to be a significant problem. Work began with rotary soil boring rigs, using bentonite drilling fluid. The method proved slow, particularly with the careful procedures necessary to avoid damage to the tunnel. Loss of drilling fluid into the voids was troublesome, and there was concern over the verticality of the holes. A better solution proved to be a specially adapted-pile rig, with a continuous flight auger mounted in leads and powered by a hydraulic tophead drive. It operated on mats so that leveling was convenient to assure verticality, and to spread the potential surcharge load on the tunnel as much as possible.

The work advanced rapidly. Within 12 days after the decision to proceed was made, liquid nitrogen was flowing. Four days later, monitoring thermocouples showed the arches were approaching completion. After two more days, eighteen days after work began on site, movement of the tunnel was arrested.

After two months of freezing, the bypass was completed, and the tunnel bulkheaded and dewatered. By this time much of the concrete in the distressed area had collapsed, forming a pile of rubble in the invert. The frozen arch and its buttresses formed the only support over what had become a cavern, roughly 5.5m high, and 12.8m long. With the frozen arch in place, it was possible to reconstruct the existing tunnel, rather than to drive a new tunnel on a different alignment, resulting in a significant saving.

(2) Weeks Island, Louisiana. At the Department of Energy's Strategic Petroleum Reserve in Weeks Island, Louisiana, more than 2 billion gallons of crude oil has been stored in an abandoned salt mine cavern 175m below ground for more than 20 years. In 1994 a surface sink hole about 10m in diameter was discovered indicating erosion was occurring to the protective salt dome approximately 61m below ground surface due to leakage from overlying fresh water aquifer through mine-induced fractures in the salt. Similar fresh water leaks in other salt mines have resulted in quick erosion of the salt and subsequent mine failure. This developing condition raised concerns about the preservation of both the environment and the strategic crude oil reserve and led to the decommissioning of the site. Ground freezing was utilized to halt further deterioration during the estimated three year period required to pump out the oil and close the site.

The sinkhole and leakage path from the overlying alluvial aquifer to the top of the salt was mapped successfully with the use of angled drill holes. Additional investigations indicated that the ongoing erosion, displacing approximately three cubic meters of material per day, was due to a leakage flow of 12 liters per minute. Brine injection into the throat of the sinkhole mitigated further erosion, but a structural support to the overlying area was required for removal of the oil. Ground freezing was selected over chemical grouting to stop the flow of water into the cavern and to stabilize 61m of alluvial soils above the salt dome since it is well suited to deal with the highly disturbed ground conditions, variable geology, complex groundwater chemistry, and access limitations at the sink hole.

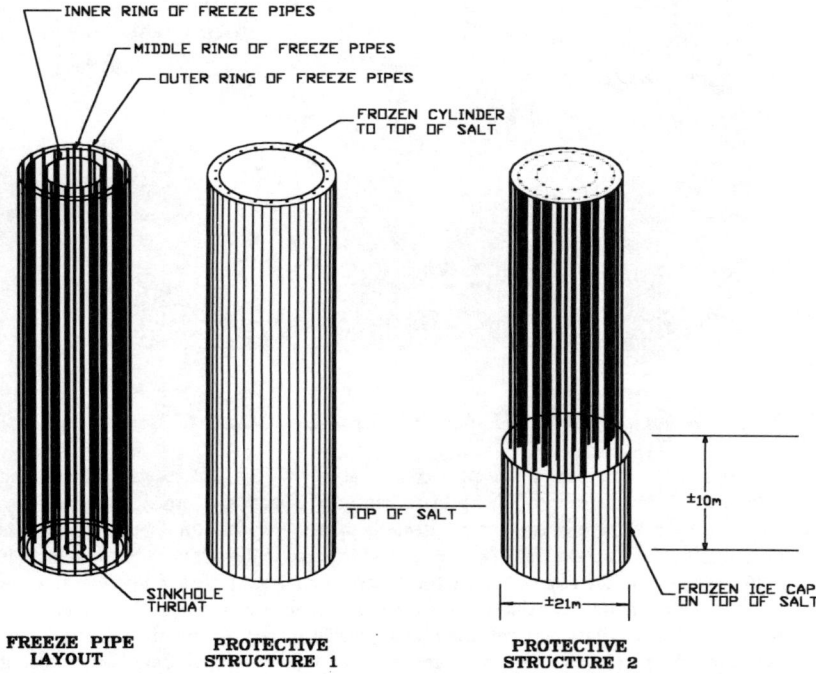

Figure 2. Weeks Island Project - Configuration of Structural Freezes

The ground freezing system was designed to form essentially two frozen protective structures. Figure 2 illustrates the configurations of the two structures. The first, a cylinder of frozen soil seated down into solid salt, was designed to prevent ground water inflow into the sink hole area and afford emergency structural support to the surrounding soils in the event of further significant ground subsidence. This first type of structural freeze wall formation has been used to sink mine shafts into deep salt deposits for the last 100 years. The secondary formation and ultimate objective was the creation of a more energy efficient massive frozen soil cap approximately 21m in diameter, 10m thick over the entire sink hole / salt contact to prevent further erosion for the duration of the oil removal.

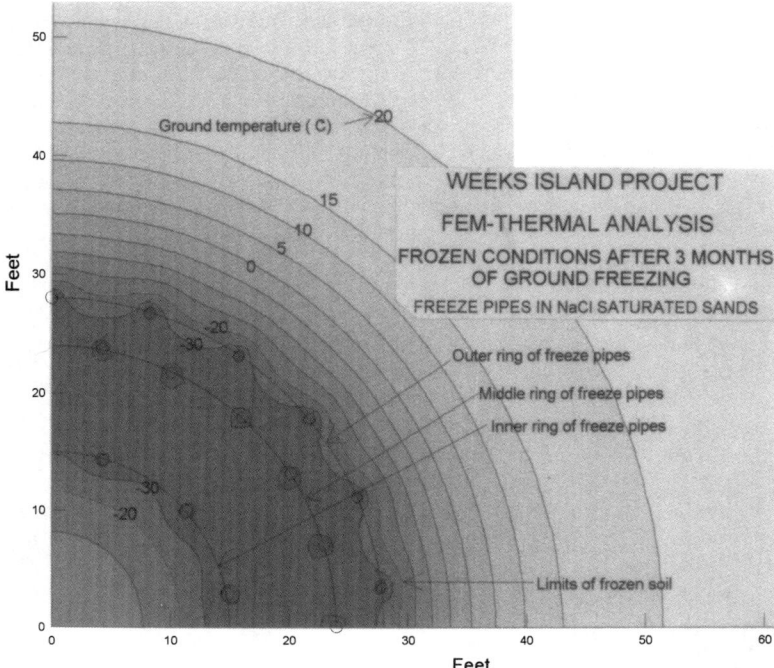

Figure 3. Weeks Island Project - Frozen Conditions After 3 Months of Ground Freezing

To arrive at the final freeze pipe configuration, evaluate freeze wall growth, and estimate thermal loads to meet the project schedule demands, thermal analyses were performed using a finite element ground freezing model. For structural purposes, it was determined that at least a 2.4m thick freeze wall was required in the salt water saturated sands above the salt prior to the removal of any oil. Thermal analyses indicated that two rows of freeze pipes were required to achieve the structural thickness of 2.4m in the allotted time period. Special considerations were made in the model to account for the lower freezing points and thermal properties of the salt-water-saturated soils. Isotherms through a section of salt-water-saturated soil after 3 months of freezing are shown on Figure 3. Figure 4 shows the modeled time growth of the freeze wall.

A total of 54 freeze pipes were installed in three concentric rings about the center of the sinkhole. An outer ring of 22 freeze pipes were installed on a 16.5m diameter circle, drilled and socketed 3m into the salt to give the freeze wall a positive seal with the top of the salt. Cores were taken of the salt, and borehole permeability tests performed to verify the competency of the material. A middle ring of 22 freeze pipes were installed to the top of salt on a 14.6m diameter circle. The middle and outside ring of pipes provided the formation of the cylindrical freeze with adequate thickness to temporarily support full-depth soil loading in the event of a complete collapse of the interior sinkhole backfill material. An inside ring of 10 additional freeze pipes were installed on a 12.2m diameter circle into the backfill of the sink hole itself. Five of the inside ring freeze pipes did not encounter salt, and were believed to be immediately inside the sinkhole cavity. The inner ring of freeze pipes was installed to propagate growth of the freeze inward and ultimately form the solid plug of frozen material inside the frozen cylinder over the sinkhole throat. Upon closure of the solid frozen plug over the sinkhole, the upper sections of the outside ring freeze pipes were insulated to change the focus of freezing efforts from the full depth cylindrical freeze to the lower frozen plug freeze.

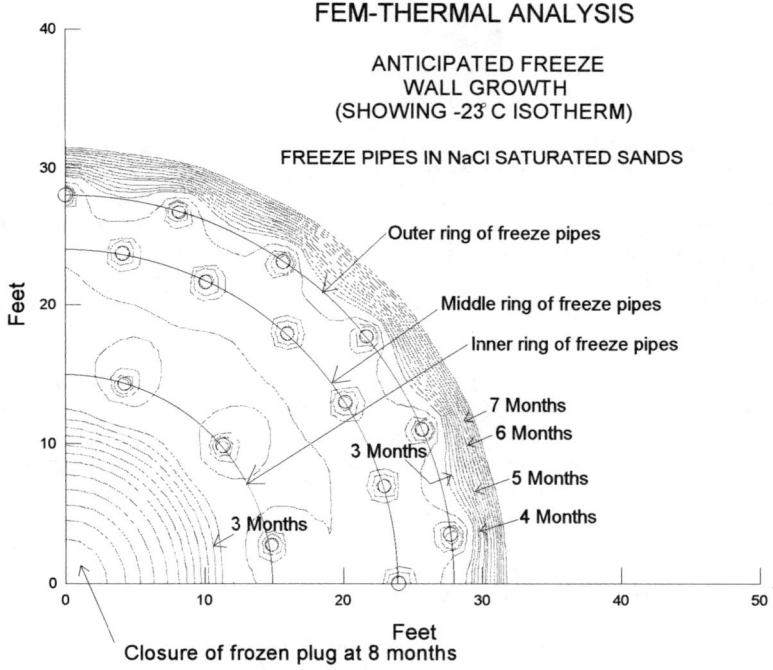

Figure 4. Weeks Island Project - Anticipated Freeze Wall Growth

Execution of the drilling phase of the work required several unique provisions. Due to the sensitivity of the salt formation to fresh water, all drilling fluids, grouts, etc., had to be saline solutions. The middle and inner rings of freeze pipes were installed immediately adjacent to and within the sinkhole rim. For these internal holes a special rail mounted, moveable bridge structure was designed to support the drilling equipment and to transfer its weight to stable soils outside the sinkhole area. This allowed freezing to commence on the outer ring while the inner ring wells were being drilled and prepared. At times, there were three drill rigs in operation simultaneously within the relatively small area. Verticality of the holes was carefully maintained so that at a 60m depth a maximum spacing between pipes of 2m to 2.5m would not be exceeded.

In addition to the freeze pipes, seven monitor holes were installed within the frozen soil limits to gather piezometric and temperature data to demonstrate the rate of growth and final integrity of the frozen structures. After each temperature monitoring casing was installed to the required depth, the pipe was filled with brine and temperature sensors were placed in the pipe at specific elevations inside and above the frozen limits. Based on the results of the freeze pipe alignment surveys, some of the monitors were located to sensitive areas of the freeze wall between divergent freeze pipes. Others provided typical values from which temperature gradient curves could be constructed and future trends extrapolated. In combination, these records characterize the rate of growth of the freeze and were used to predict milestones in its development. A central relief hole (piezometer) was installed with screens set to intercept upper, brackish water and saturated salt water. Closure of the cylindrical freeze wall was indicated by the independence of the internal piezometric levels compared with the external groundwater regime.

The actual installation was directed to complete the outer ring of freeze pipes first and start the formation of the full depth cylindrical freeze during the installation of the inner row. Freezing of the outer ring of pipes began in mid-July, 1995. Five large freeze plants were mobilized, employing ammonia as the refrigerant and calcium chloride brine as the coolant circulating through the freeze pipes. 1,800 refrigeration horsepower was required to chill the ground below the freezing point of the salt saturated groundwater immediately above the salt contact. A target coolant temperature of -39°C was designed for, and is now being maintained.

Monitoring indicated that the first objective of creating the full depth cylindrical freeze was reached in October, 1995. The cylindrical freeze wall configuration was an ice cylinder with an outside diameter of 21m between the ground surface and the top of salt. Pump testing procedures confirmed closure of the wall, and the first barrel of oil was pumped out of the mine workings beginning on November 8, 1995. The freeze pipes were modified in April 1996 to concentrate freezing at the lower depths, near the top of salt (-24m to -38m MSL). The resultant ice cap plug achieves the same ultimate purpose as the ice wall cylinder, but requires less energy to maintain. This cap will be maintained until the crude oil storage chambers have been emptied of crude oil and filled with brine, scheduled for late in 1998.

The modeling efforts led to not only the most efficient configuration of freeze pipes, but also accurately indicated the freezing effort and formation times required. The model time period for the formation of the full depth cylindrical freeze and ice cap plug were 3 and 9 months respectively. In the field, the formation of the full depth cylindrical freeze and "ice cap" required 3.5 and 10 months respectively, very close to the model values.

To date the oil has been removed from storage without incident, and replenishment of the mine with brine is underway.

(3) Boston Central Artery Contract Section C09A4. A tunnel jacking project of unprecedented magnitude in the United States is currently underway on Boston's Central Artery tunnel project as part of the work associated with bringing Interstate 90 / Interstate 93 underneath the Fort Point Channel. Three massive tunnels will be jacked a few feet underneath active rail lines of the busy North East Corridor through a myriad of underground obstructions (Ref figure 5). The largest of these tunnels is 107m long, 24.4m wide, and 10.7m high. The other two tunnels are only slightly smaller. A full face compartmentalized jacking shield (full width and height of the tunnel sections) has been designed to be driven into a vertical, self-supported face of fill, organic silt, and Boston Blue Clay (see figure 6). A stable mining face and groundwater cutoff for tunnel jacking is considered key to the success of the project.

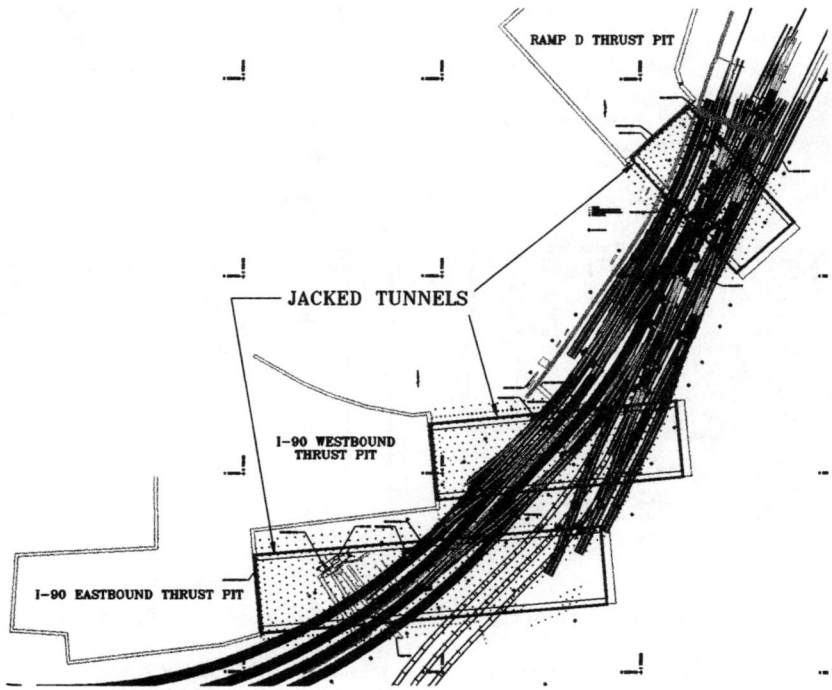

Figure 5. Boston Central Artery Contract C09A4 - Overall Site Plan

Designers initially envisioned the area directly ahead of the jacked face to be stabilized with a combination of dewatering, jet grouting, and soil nails. The general contractor, a joint venture of Slattery, Interbeton, J. F. White, and Perini, opted for an alternate of mass ground freezing of the full tunnel footprint to provide full face stability and water cut-off. Ground freezing was chosen over the various grouting alternatives for numerous reasons:
- Ground freezing will be applied to the area from the surface prior to excavation work, and will permit continuity of the jacking operation without the disruption of operations conducted from the face.
- Ground freezing will provide complete stability of the face, eliminating heavy support which would otherwise be required, and minimizing settlement effects on the tracks above.
- The technique is applicable to a wide range of geological conditions as well as large buried obstructions such as wood piles from earlier wharfs and building foundations. Additionally, freezing allows obstructions to be incorporated into the massive frozen ground matrix, and held in place so that they can be excavated with a minimum of disturbance

- The effectiveness of the freeze can be accurately evaluated with appropriate instrumentation such as borehole installed temperature sensors, eliminating the need for extensive probing ahead to evaluate subsurface conditions.
- The effects of the freeze can be confined to the area being frozen. There is no threat of sudden ground heave at the train rails or contamination of the ballast with cementicious material. There are no long term detrimental effects to the subsurface strata and environmental impact is negligible.
- Ground freezing eliminates the need for dewatering which would be required to supplement grouting.

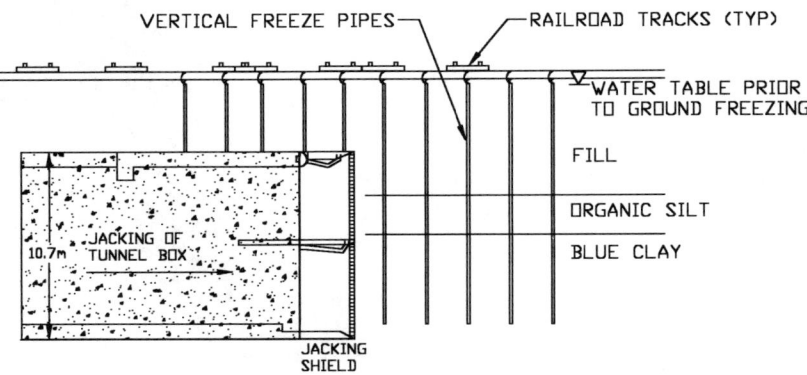

Figure 6. Boston Central Artery Contract C09A4 - Sectional View of Tunnel Jack

The ground freezing design is intended to provide both a stable mining face and groundwater cutoff during tunnel construction. The ground through which each box will be jacked will be frozen solid from three feet below the base of the tracks (near ground surface) to three feet above box subgrade and provide a stable, full height, free-standing vertical face, which allows mining without the need for face support. Face stability of the various strata types (both coarse granular material and clay) will be modified to that of a weak rock by freezing.

To accomplish such a task, chilled brine will be recirculated through as many as 700 vertical freeze pipes for three to four months in advance of jacking of each tunnel. At each location a volume of soil slightly greater than the finished tunnel volume will be frozen solid. The entire project will require the freezing of over 60,000 cubic meters of soil.

The frozen ground along side of the advancing box and above the box at the entry headwall will act as a groundwater cutoff and prevent water seepage along the sides of the box towards the tunnel face or entry portal. As each box structure advances, the freeze pipes installed through the face will be turned off. This will initiate a thaw above the box, but since the thawing process is slow, frozen ground will remain in the area of the cutting edge, over and along the shield and for some distance back along the box.

The most important parts of a design for ground freezing are the thermal analyses. These examine the transient heat flow from the ground to the freeze pipes in order to evaluate freeze pipe disposition, freeze formation period, and freeze plant capacity. The bulk of the excavation work will occur within the Boston Blue Clay, which in turn, needs the most refrigeration effort per unit volume to render it stable. Therefore, all design work and detailed thermal analyses conservatively were designed for this stratum. Numerous finite element models have been constructed to simulate various freeze pipe configurations under a variety of soil and temperature conditions. Figure 7 shows the proposed freezing pipe layout for Ramp D.

Littered with abandoned foundations and debris fill, the hydrological nature of the upper fill material is anticipated to be highly variable, with potential rapid groundwater movement in open work zones and voidaceous debris, which may result in subsequent freeze wall formation difficulties. Two measures will be taken to address this potential problem; 1) pipe spacing will be decreased along the sides of the tunnels to achieve an earlier peripheral freeze wall closure and water cut-off so the interior area of the jack may freeze without water movement in the fill, and 2) a grouting program will be implemented during freeze pipe installation to grout voids in the fill when encountered.

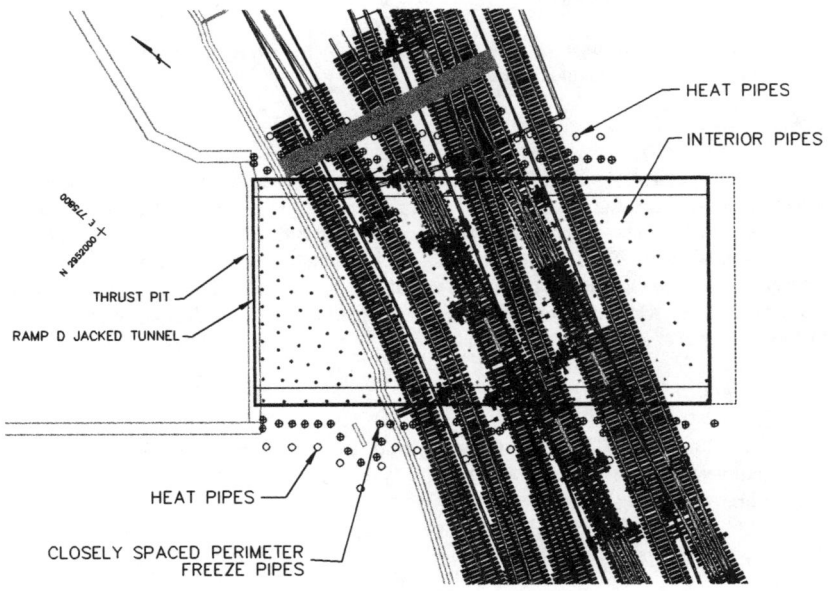

Figure 7. Freeze System Configuration - Tunnel Box for Ramp D

This project entails unique thermal objectives which involve adding heat to the systems before and during the jacking of each box. Artificial heating is proposed to limit possible adverse effects ground freezing may have on the installation procedures associated with advancing the pre-cast tunnel structures through frozen soils. There are two concerns; 1) frost adhesion at the box / ground interface and subsequently increased resistance to sliding movement of the structure, and 2) an increase in lateral pressure and "pinching" against the sides of the jacked box due to continued growth of the freeze wall during jacking.

Two heating systems will be implemented to mitigate those possible effects. At the vertical concrete box surface, a source of heat will be located within the concrete but as close to the outer surface of the concrete box. As the leading edge of the box enters the frozen mass of soil, sufficient heat will be available to melt the interface in a reasonable time period. A convenient and flexible method of applying, measuring, and controlling the amount of heat supplied to the soils via the box will be provided by setting electrical heating cables, within the external formwork when the structures are built. To limit continued growth of the freeze outside the limits of the jack area, and the possible increase in lateral pressure on the box, a perimeter series of leakproof steel pipes (similar to freeze pipes) will be installed where boundary heat is required, at the outside limit of the frozen ground cutoff on either side of the boxes. Heat transfer to the soil will be by heat cables immersed in a liquid antifreeze (weak brine) contained within the pipe. Vertical ground heating pipes shall not be heated until the leading edge of the jacked box has passed. Finite element modeling of both heat systems has been performed to estimate the power demands and spacing of heating elements.

Currently freeze pipes are being installed on site. Freezing is anticipated to begin in December of 1998 and continue through 2000.

Conclusions

The three mass ground freezing projects sited herein are characterized by difficult ground conditions and the need for a complete and certain groundwater cut-off which could not be provided by other techniques. Also, the freezes could be considered as non-invasive ground modifications. The Detroit sewer and Weeks Island projects sited herein relied on the controlled time growth of the ground freeze to stabilize fragile and highly disturbed areas with the installation of relatively remote and relatively small devices. In Boston, unprecedented massive volumes of soil will be frozen with essentially a minimum of intrusion.

Acknowledgments

- Slattery, Interbeton, J. F. White, Perini, A Joint Venture, Boston Central Artery Project
- Department of Energy and Dyn McDermott, Weeks Island Project
- Walbridge Aldinger, Detroit Sewer Project

References

1. Gordon, Grant, "Arch Dam of Ice Stops Slide", Engineering News Record, Feb 11, 1937.

Freeze-Thaw Effects on Boston Blue Clay

Christopher Swan and Christopher Greene[1], Associate Members, ASCE

Abstract

The Central Artery / Third Harbor Tunnel (CA/T) project in Boston, MA will transform the existing elevated viaduct system into a subsurface, multi-lane highway. Given its urban location, the project requires a number of complex construction techniques which minimize disturbance of existing transportation patterns and adjacent structures. Ground freezing is planned to temporarily stabilize soils beneath existing railroad tracks so that tunneling for new subsurface ramps can be performed. However, questions exist about the "thawed" behavior of subsurface soils, particularly the Boston Blue clay (BBC) formation, and how such behavior may influence ground freezing operations and future ramp performance.

This paper presents preliminary results of a laboratory study on undisturbed samples of natural BBC and BBC subjected to one freeze-thaw cycle. Freezing was performed to simulate the case of inward, radial freezing under no vertical confinement. Tests performed included Atterburg limits, consolidation, UU-triaxial tests, and direct shear tests. Test results indicate 1) both the plastic and liquid limits decreased for the frozen-thawed BBC; 2) the consolidation response of frozen-thawed BBC shows disturbance compared to unfrozen BBC; and 3) while both soils have similar direct shear strengths, there is a significant decrease (60%) in undrained shear strength with freezing.

These preliminary test results strongly indicate that the engineering properties of BBC are significantly changed with one cycle of freeze-thaw, and that this change in behavior in the BBC may effect geotechnical designs and post-construction behavior for the CA/T project.

---

[1]Assistant Professor and Graduate Student; respectively, Department of Civil and Environmental Engineering, Tufts University, 113 Anderson Hall, Medford, MA 02155, Ph. 617-627-2212, cswan@tufts.edu

Introduction

The Central Artery /Third Harbor Tunnel (CA/T) project will transform the major highway artery running through Boston, MA to a subsurface, multi-lane highway with eight miles of tunnel, viaduct and bridge construction scheduled for downtown Boston. To complete construction, the project requires a number of complex techniques to minimize disturbance to existing transportation facilities and adjacent structures. Towards this aim, ground freezing of subsurface organic soil and marine clay will be used to stabilize soils beneath existing railroad tracks so that new, subsurface ramps can be constructed without interfering with existing railway traffic. However, questions exist about the post-freezing or "thawed" behavior of the marine clay formation, commonly called Boston Blue clay (BBC), and how such changes in behavior may influence the proposed ground freezing operations and the future ramp's design and performance.

This paper presents results of laboratory tests on undisturbed samples of unfrozen BBC and BBC that experienced one cycle of freeze-thaw. Tests performed included index tests including liquid and plastic limits, grain size distribution, and specific gravity; one-dimensional consolidation; unconsolidated-undrained triaxial compression tests; and consolidated-drained direct shear tests. The following sections present the sampling and freezing processes, laboratory test procedures and results, a discussion of the results, and conclusions on how the results may influence future work in ground freezing of cohesive soils.

Background and Overview

The existing I-93 passes through downtown Boston, MA on an elevated viaduct. The Central Artery/Third Harbor Tunnel Project, CA/T, begun in 1989, consists of two major components, constructing a tunnel to connect Logan airport with I-90 (Massachusetts Turnpike) and placing a portion of the existing, elevated I-93 (Central Artery) below existing ground levels. The tunnel component has been completed; however, work to depress the artery continues. Another element adding to the size and complexity of the project is the long history of residential, commercial, and marine development in the area, along with accompanying above- and below-ground civil infrastructure systems. The complex design and construction of the CA/T project are further complicated by the presence of poor foundation soils, most notably the Boston Blue clay (BBC) formation which underlies a number of critical sections.

The CA/T project requires major urban excavations and unusual construction methods rarely used in North America, and perhaps never in these soil conditions, to complete the project. Contractors and engineers continually consider and re-evaluate alternative construction techniques which are technically feasible and cost-effective. One proposed technique is ground freezing for the temporary stabilization of soils prior to placement of a pipe-jacked tunnel (Angelo, 1997). The proposed ground freezing technique involves circulating a refrigerated brine solution through a series of vertical pipes placed into the ground. The subsurface soils will freeze in a radial direction from the pipes and form a monolithic mass of frozen, stabilized soil through which the tunneling (pipe-jacking) can occur.

Sampling and Freezing Procedures

BBC Sampling

Laboratory tests were performed on an undisturbed block sample of BBC recovered from on-going excavations associated with the CA/T project. The block sample, 33 cm long by 20 cm wide by 20 cm deep, was recovered from a depth of 21 m beneath existing grade, 6.5 m beneath the top of the BBC formation. The sample was intact with no noticeable fissure or other defects, but closer visual inspection showed thin, horizontal stratifications alternating between more silt and less silt content. An average water content of 39.4% was measured for the sample on the day of excavation. The block sample was vertically separated into two halves, one used for tests on unfrozen BBC, the other used for tests on frozen-thawed BBC.

Sample Freezing

The sample was frozen to simulate the possible field ground freezing scenario of a near-surface block of soil located inside a series of freezing pipes. That is, the sample was frozen radially from its outside in with no vertical confinement and no access to water during the freezing process. For freezing, the half-block sample was trimmed into a 15 cm diameter, 17.5 cm high, cylindrical steel mold. The ends of the mold were capped with foam insulation to reduce any freezing from the top and bottom of the sample and thus enhance the desired radial direction of freezing. The mold was then placed in a cold room facility and the ambient temperature allowed to decrease from approximately 20°C to a set temperature of -15°C. A thermocouple temperature probe was periodically placed in the sample at four locations to measure temperatures versus time during freezing (Figure 1). As shown in the Figure 1, the sample froze from outside in with the temperature in the middle of the sample reaching 0°C within 180 minutes. After 960 minutes, all points in the sample had reached -15°C, and the sample was left in the functioning cold room facility for additional 32 hours. The facility was then shut down and allowed to reach room temperature over a 48 hour period. The thawed sample was then wrapped in cellophane and placed in a controlled humidity room for approximately one week to ensure complete thawing before shear and consolidation test specimens were to be trimmed.

Comparison of pre- and post-freezing sample dimensions indicate less than 2% decrease in average sample height and no lateral shrinkage of the sample from the wall of the mold after one freeze-thaw cycle. Extrusion of the sample during test specimen trimming indicated no changes in BBC color or texture but subsequent trimming exposed a number of horizontal and vertical defects which lead to soil separation and "crumbling" during the trimming process (see Test Procedures and Results section below).

Test Procedures and Results

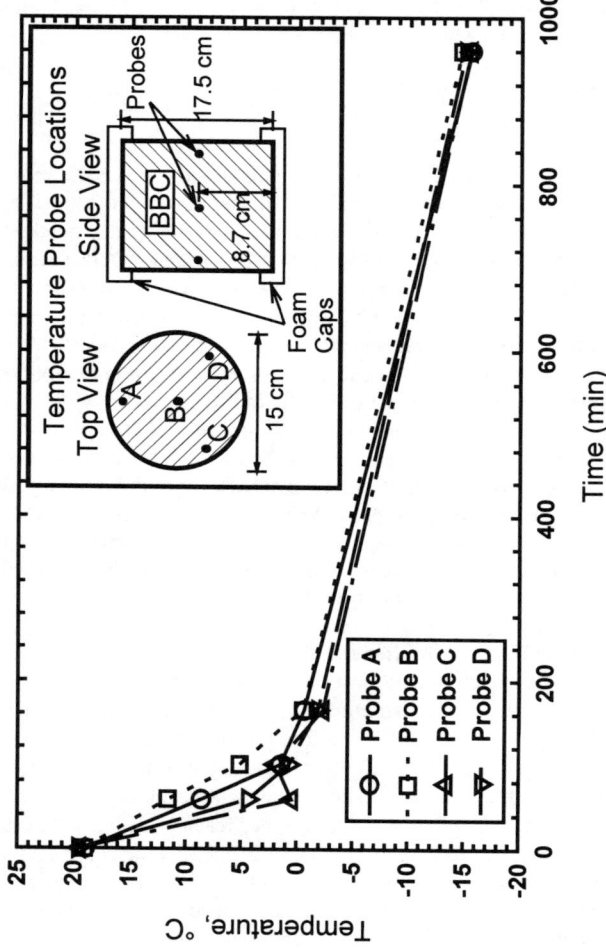

Figure 1 Temperature vs Time During the Freezing Phase of the Frozen-Thawed BBC Sample

A series of laboratory tests were performed on both the unfrozen BBC (tests designated with an UF) and the frozen-then-thawed BBC (tests designated with a FT). Laboratory tests are grouped into three categories: index tests, compression tests, and shear tests. Index tests included determination of natural water contents, plastic and liquid limits, grain size distributions by hydrometer analyses, and specific gravity. Compression tests performed were incremental, one-dimensional consolidation tests. Shear tests consisted of undrained-unconsolidated triaxial shear tests and consolidated-drained direct-shear. All tests were performed in general accordance with appropriate ASTM standards.

Index Tests

Table 1 summarizes the index test results for both unfrozen and frozen-thawed BBC soils. Three grain size analyses (ASTM D422) were performed, all on the unfrozen BBC, and gave similar results (see Figure 2). An average specific gravity of 2.715 was found based on four test results, two each on the unfrozen and frozen-thawed BBC samples. The average plastic and liquid limits (ASTM 4318) for the unfrozen BBC were 28% and 59%, respectively, leading to a plasticity index of 31%. Based on a natural water content of 39%, the liquidity index was 0.4. The average plastic and liquid limits for the frozen-thawed BBC were 25% and 56%, respectively, leading to a plasticity index of 31%, and, based on an average post-freezing water content of 42%, a slightly higher liquidity index of 0.5. Note that both the plastic and liquid limits decreased with freezing and thawing, yet have almost identical plasticity indices. It is also interesting to note that the post-freezing water content increased from the unfrozen BBC natural water content though the sample did not have access to water during freezing or thawing. This could be due to natural variability in BBC but seems to be a rather large difference. Another possible explanation is that during freezing, sample pore water will migrate from the center of the sample out towards the freeze front. Since specimens were trimmed from around the outside of the sample, higher water contents in test specimens are possible. Based on the Unified Soil Classification System (ASTM D2487), both the unfrozen and frozen-thawed BBC's are classified as CH clay (plotting just above the A-line).

Compression Tests (ASTM D2435)

A total of five incremental-loading, one-dimensional, consolidation tests were performed; two on unfrozen specimens and three on frozen-thawed specimens. A summary of consolidation test results is presented in Table 2. Unfrozen and frozen-thawed BBC consolidation specimens were prepared using the same procedures. The 6.3 cm diameter consolidation ring used in the consolidometer was measured and weighed before specimen trimming. The ring was then slowly pushed into a section of the block sample with excess soil trimmed away from the edge of the ring as it advanced. Once 3 mm or more of soil extended out the other end of the ring, the exposed faces of the specimen were then trimmed flush with the top and bottom edges of the ring. The trimmed specimen and ring were then weighed and placed in the consolidometer. Tests were performed by placing and maintaining incremental loads

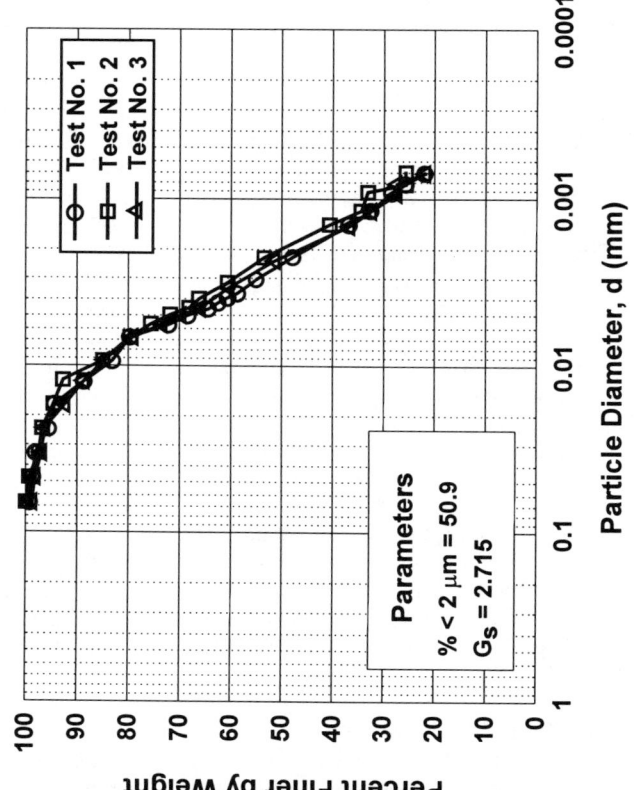

Figure 2 Grain Size Distribution of Unfrozen BBC

SOIL IMPROVEMENT FOR BIG DIGS 167

Table 1 Summary of Index Tests for BBC

Soil Condition	Natural Water Content, w Avg ± STD	Plastic Limit PL Avg ± STD	Liquid Limit LL Avg ± STD	Plasticity Index, PI	Liquidity Index, LI	% Particles < 2μm Avg ± STD	Activity, A	Specific Gravity, G_s Avg ± STD
Unfrozen (UF)	39.4 ± 2.1	28.3 ± 0.4	59.4 ± 3.3	31.1	0.36	50.9 ± 2.8	0.61	2.72 ± 0.01
Frozen Thawed (FT)	41.8 ± 2.5	25.4 ± 1.1	56.6 ± 3.6	31.3	0.53		0.61	2.72 ± 0.02

Table 2 Summary of Consolidation Parameters for BBC

Consolidation Parameter	Unfrozen BBC			Frozen-Thawed BBC				For All Tests	
	Test C1-UF	Test C2-UF	Average	Test C3-FT	Test C4-FT	Test C5-FT	Average	Average	Std. Dev.
e_o	1.064	1.176	1.120	1.143	1.171	1.144	1.153	1.136	0.041
C_c	0.314	0.332	0.323	0.332	0.307	0.364	0.334	0.329	0.020
CC	0.152	0.153	0.152	0.155	0.141	0.170	0.155	0.154	0.009
C_r	0.025	n/a	0.025	n/a	n/a	0.059	0.059	0.036	0.020
CR	0.012	n/a	0.012	n/a	n/a	0.027	0.027	0.017	0.009
C_α	0.002	n/a	0.002	0.002	0.002	0.002	0.002	0.002	0.000
$C_{\alpha\epsilon}$	0.001	n/a	0.001	0.001	0.001	0.001	0.001	0.001	0.000
σ_p' (kPa)	565.0	600.0	582.5	650.0	600.0	700.0	650.0	616.3	49.9

until the end-of-primary consolidation was reached. Vertical displacements were manually recorded via dial gauges and by a linear variable displacement transducer (LVDT) and recorded on computer data acquisition software. The square root of time method was used to evaluate end-of-primary conditions. Two tests, one each on unfrozen and frozen-thawed specimens, were subjected to an unload-reload cycle. The stress prior to any unloading sequence (either for an unload-reload cycle or for final unloading) was maintained for at least 24 hours, thus providing secondary consolidation data.

The trimming of consolidation specimens from the frozen-thawed block sample was more difficult than for the unfrozen BBC sample. The frozen-thawed specimen tended to "crumble" or be unstable during trimming, breaking randomly along horizontal and/or vertical planes. In some cases, obtaining a representative specimen was virtually impossible by the constant deterioration of the specimen during the trimming process. Therefore, more frozen-thawed soil was required to develop the same number of consolidation specimens developed for the unfrozen BBC.

Figure 3 compares the vertical strain (ε_v) versus effective stress (σ'_c) results from a test on unfrozen BBC (C1-UF) and frozen-thawed BBC (C4-FT). These results are indicative of the results measured in the other consolidation tests on unfrozen and frozen-thawed BBC. As shown in the figure, test C4-FT shows a higher compressibility (C_r) in the overconsolidated range, lower compressibility (C_c) in the normally-consolidated range and less distinct maximum past pressure (σ_{vm}) compared to test C1-UF. This suggests that the freeze-thaw cycle has caused sample disturbance. However, the swelling (rebound) portions of the responses are parallel indicating that the two soils have similar structures after being loaded to over 2400 kPa effective stress.

<u>Shear Tests</u>

Shear tests consisted of six unconsolidated-undrained triaxial compression (ASTM D2850) tests (three each on unfrozen and frozen-thawed BBC) and four consolidated-drained direct shear (ASTM D3080) tests (two each on unfrozen and frozen-thawed BBC). Unconsolidated-undrained compression (UUC) tests were conducted on 3.5 cm-diameter, cylindrical specimens trimmed from their respectively BBC material. Once trimmed, the specimen dimensions were measured, the specimen weighed, and then placed into the disassembled triaxial device. A single thin membrane was used to seal the specimen before the triaxial cell was re-assembled, filled with de-aired water and pressurized. All tests were conducted at a confining pressure of 152 kPa. Within minutes after pressurizing the cell, the specimen was sheared at a shearing rate of 0.013 cm/s to at least 20% axial strain. Changes in axial load and vertical displacement were measured using electronic transducers and recorded via computer data acquisition.

Direct shear (CK_oDS) specimens were prepared in a manner similar to the consolidation specimens since the diameter of the specimen to be placed in the shear box were the same. The shear box was prepared so that the specimen would be sheared in half (horizontally). After trimming of a specimen into a consolidation ring as described in the previous, the ring was placed on top of the prepared shear box opening,

SOIL IMPROVEMENT FOR BIG DIGS

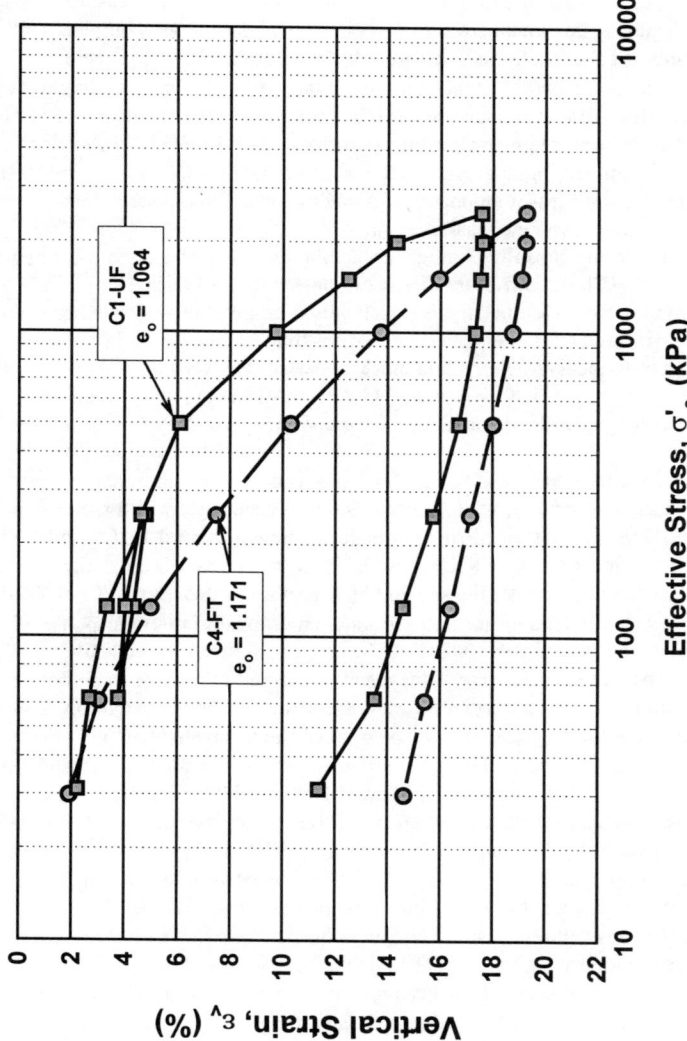

Figure 3 Comparision of Vertical Strain versus Effective Stress for Unfrozen BBC Test C1-UF and Frozen-Thawed BBC Test C4-FT

and the specimen slowly extruded directly from the consolidation ring into the shear box. The specimen was then vertically loaded, via a hanger system, to a vertical effective stress of 163 kPa. Based on consolidation tests, as well as measured vertical displacements during this loading phase, primary consolidation was completed in less than 10 minutes and shear deformation begun. To ensure drained shear, specimens were then sheared at a nominal horizontal displacement rate of 5×10^{-5} cm/s, leading to times of failure of at least 1080 minutes. Shear load, horizontal displacement and vertical displacement were measured with electronic transducers and recorded via computer data acquisition software during the test. After approximately 20 hours of horizontal deformation, the shearing was stopped, the horizontal and vertical loads removed and the specimen removed, weighed and oven dried.

As noted in the consolidation test preparation section, specimen trimming for frozen-thawed BBC tests was more difficult than for the unfrozen BBC. This was especially true for the UUC specimens which would crumble upon trimming without special care. In fact, in some cases, the frozen-thawed block specimen, with the freezing-induced vertical and horizontal cracks/weak planes, was not stable enough to even initiate trimming for UUC specimens. Given this difficulty, two of the three UUC specimens had length-to-diameter ratios of 1.4 and 1.5, far below the recommended value of 2.

Table 3 presents the results of the UUC tests and Table 4 presents the result of the CK_oDS tests. In general, UUC shear results indicate a significant decrease, an average of 60%, in the undrained shear strength of frozen-thawed BBC compared to unfrozen BBC. In contrast, CK_oDS results indicate similar drained direct shear strengths, regardless of whether the soil had been previously frozen.

Figure 4 shows the deviator stress ($\sigma_1-\sigma_3$) versus axial strain (ε) response of two UUC tests, one on unfrozen BBC (UUC1-UF) and one on frozen-thawed BBC (UUC4-FT). These responses are indicative of the results measured in other tests on both BBC conditions. Comparison of the two test results shows that 1) the undrained strength (s_u) and strength to confining pressure ratio (s_u/σ_{cell}) of the unfrozen BBC is higher than those of the frozen-thawed BBC; 2) the strain to failure (ε_f) for the unfrozen BBC specimen is one-fifth that of the frozen-thawed BBC; and 3) the overall stress-strain response of test UUC1-UF is one of early build up to a peak strength followed by significant post-peak strain-softening while for test UUC4-FT, the stress-strain response is an early build up to an initial yield point followed by post-yield strain-hardening to a peak strength at a higher strain level then post-peak strain-softening.

Figure 5 shows the shear stress (τ) versus horizontal displacement (δ) responses for CK_oDS tests on unfrozen BBC (DS2-UF) and frozen-thawed BBC (DS3-FT). The figure shows that the drained shear behavior of both soils are nearly identical with similar drained friction angles (ϕ') and peak shear strengths (τ_{max}) which occur at large displacements.

Discussion and Conclusions

The results of the index, compression, and shear tests clearly show that the

Table 3 Summary of Unconsolidated-Undrained Tests on BBC

Parameter	Unfrozen BBC				Frozen-Thawed BBC			
	UUC1-UF	UUC2-UF	UUC3-UF	Average	Standard Deviation	UUC4-FT	UUC5-FT	UUC6-FT	Average	Standard Deviation
w (%)	37.5	39.9	35.6	37.6	2.2	43.2	44.1	43.5	43.6	0.4
e_o	1.02	1.00	0.94	1.0	0.0	1.13	1.12	1.15	1.1	0.0
ϕ' (°)	24.8	25.8	24.7	25.1	0.6	12.6	15.5	10.2	12.8	2.6
s_u (kPa)	54.8	58.6	54.4	55.9	2.3	21.2	27.5	16.4	21.7	5.6
ε_f (%)	2.3	3.0	3.9	3.1	0.8	12.8	11.0	10.5	11.4	1.2
s_u/σ_{cell}	0.36	0.39	0.36	0.37	0.02	0.14	0.18	0.11	0.14	0.04

Table 4 Summary of Direct Shear Tests on BBC

Parameter	Unfrozen BBC			Frozen-Thawed BBC			
	DS1-UF	DS2-UF	Average	Standard Deviation	DS3-FT	DS4-FT	Average	Standard Deviation
w (%)	44.1	47.3			44.0	44.4	44.9	1.6
e_o	1.26	1.30			1.27	1.16	1.25	0.06
S_r (%)	95.2	98.6			93.8	>100	95.9	2.5
ϕ' (°)	22.4	21.0			21.7	22.6	21.9	0.7
τ_{max} (kPa)	73.5	66.9			70.2	73.5	71.0	3.1
τ_{max}/σ'_v	0.45	0.41			0.43	0.45	0.44	0.02

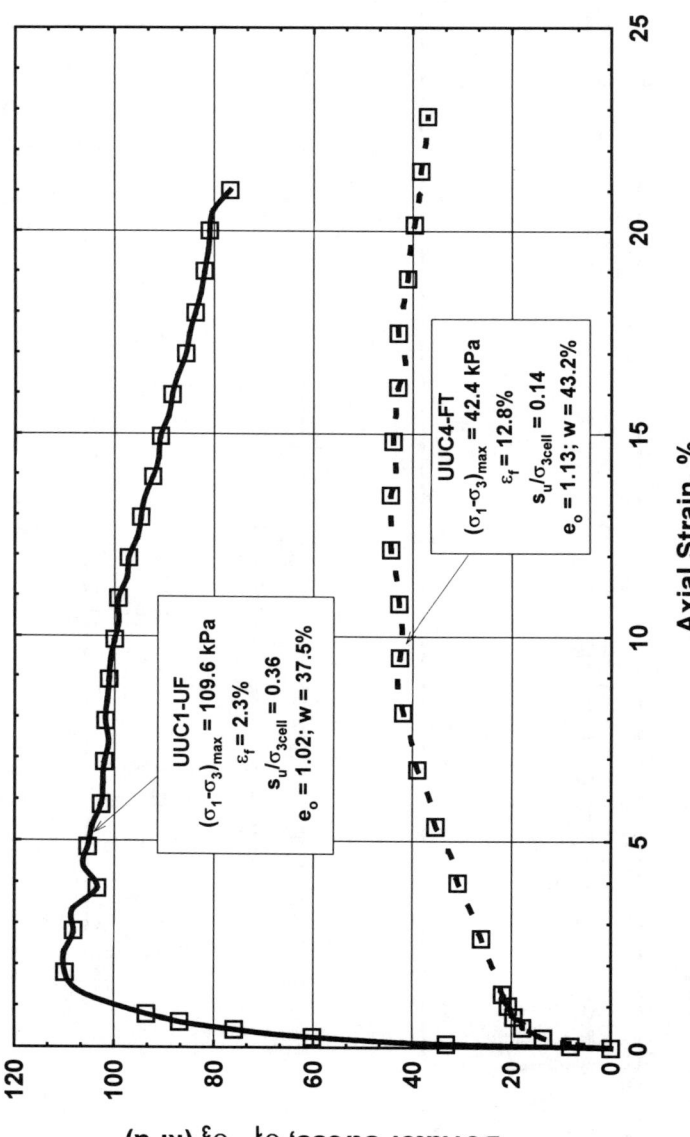

Figure 4 Deviator Stress vs Axial Strain for Unconsolidated-Undrained Triaxial Compression Tests UUC1-UF (Unfrozen BBC) and UUC4-FT (Frozen-Thawed BBC)

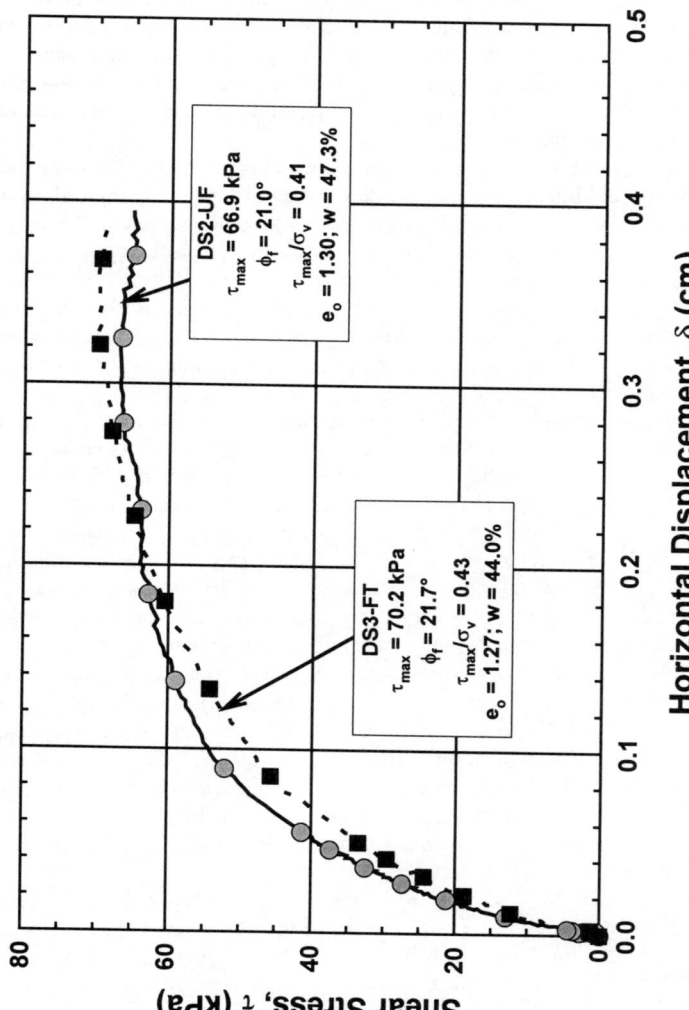

Figure 5 Comparision of Consolidated-Drained Direct Shear Tests on Unfrozen BBC Test DS2-UF and Frozen-Thawed BBC Test DS3-FT

engineering properties of BBC are changed after one cycle of freeze-thaw under conditions of radial freezing from the outside towards the center, no vertical confinement, and no access to water during freezing. These changes in engineering properties can be attributed to the macroscopic structural changes noted in the soil mass as well as microscopic changes in the soil. Microscopic changes are associated with the physical and chemical characteristics of the clay and silt particles, their fabric (orientation and distribution), and inter-particle forces. Macroscopic changes are generally visually observable changes in the soil mass such as the development of defects (e.g., the fractures noted during specimen trimming, etc.) or a change in color, texture or grain size distribution.

It may be possible to infer microscopic changes from the results of the index tests. The plastic and liquid limit tests are indirect indicators of clay mineralogy and clay particle interactions. Test results show relatively small changes in plastic limit (11% decrease) and liquid limit (5% decrease) indicating that freezing does cause a slight change in BBC mineral properties. It could be hypothesized that the electrostatic double layer of the clay particles is reduced by the freezing process; thus leading to decreases in the plastic and liquid limits yet no change in plasticity index. However, given the lack of true microscopic observations, this observation is speculative but deserves further study.

Macroscopic changes were observed directly during the trimming of test specimens and in the results of consolidation and shear strength tests. As noted earlier, the trimming of tests specimens from the frozen-thawed sample was significantly more difficult than trimming the unfrozen BBC sample as separation or crumbling of the soil along horizontal and/or vertical planes of weakness was observed. The development of these cracks or defects appear to be caused by the freezing process since this crumbling behavior was not observed during trimming of unfrozen BBC test specimens. The authors believe that the freezing of water trapped by the radial freezing process lead to the development of both horizontal and vertical micro-cracks in the soil mass.

The disturbance caused by these cracks is evident in both the consolidation and undrained shear strength tests. As shown in Figure 3, the shapes of the ε_v-logσ'_c loading response of test C4-FT (frozen-thawed BBC) indicates the soil's disturbance (Casagrande, 1932). The significant reduction, up to 60%, of the undrained shear strength is also indicative of this disturbance. The micro-cracks may act as pre-determined failure planes to which water can migrate during undrained shear and initiate specimen failure, and/or, the micro-cracks may lead to an overall weakening of the structural integrity of the entire soil mass.

Work described by Wood (1976) indicates a similar reduction in undrained strength (approximately 45%) occurs after one freeze-thaw cycle on saturated Leda clay. After four cycles, the undrained strength reduced 66%. Using a scanning electron microscopy, Wood determined that the loss in strength was due to micro-fissures expanding during pore water freezing. Broms and Yao (1964) studied the effects of freezing rate, surcharge loading and drainage conditions on compacted Dunkirt silty clay and found an approximately 50 to 70% reduction in ultimate strength after one freeze-thaw cycle for specimens frozen at a freezing rate of 7.6 cm/day, under a surcharge of

11.5 kPa, and closed to drainage.

Another interesting aspect of the current work is the change in σ-ε response for the two soil conditions. The unfrozen response (Test UUC1-UF) exhibited a brittle-type failure of an cohesive material with the peak strength occurring at small strains followed by strain-softening; however, the frozen-thawed response (Test UUC4-FT) is similar to that for undrained *cohesionless* soil with a initial yield at small strains followed by strain-hardening to a peak strength (e.g., see Swan 1996).

The similarity in drained shear behavior of unfrozen and frozen-thawed BBC, as shown in Figure 5, indicates the importance of drainage in controlling the detrimental shear behavior induced by the micro-cracks. With drainage, the migration of pore water to the micro-crack during shear is of no consequence in the shearing of specimens; thus the strength measured represents soil particle-to-soil particle strengths which are the same for both BBC conditions.

Summary

The test results indicate that the engineering behavior of BBC subjected to one cycle of freeze-thaw is significantly altered from that of unfrozen BBC. The noted changes in consolidation and undrained shear behavior due to the re-structuring of the soil mass by the freezing and thawing process are potentially very significant, especially with respect to how ground freezing is to be done and post-construction soil movements and strengths. Soil mass disturbance caused by the process may lead to substantial post-construction ground movements since the soil will now follow the more disturbed consolidation response. More significantly is the reduction in undrained strength of the soil mass which could potentially lead to underdesign or un-conservative analysis of adjacent foundations or other structures.

However, it must be realized that the freezing process used in this study represents only one ground freezing scenario and actual freezing conditions may vary. Future research efforts will examine the effects of vertical pressure on the freezing process as well as examine the effects of temperature, rate, and direction of freezing on BBC behavior.

Acknowledgements

The authors wish to thank Jennifer Stone, an undergraduate student at Tufts University, for performing some of the index tests. The authors also thank Bechtel/Parsons Brinckerhoff (B/PB), Perini-Kerwitt-Cashman (joint venture) and the Massachusetts Highway Department for granting permission and allowing access to the site for soil sampling. The Technical Service department of B/PB provided assistance in performing specific gravity tests.

References

Angelo, W. J. (1997) "Frozen Soil May Provide Solid Venue for Huge Jacked Tunnel" ENR, v. 238, April 7, 1997, p. 12.

ASTM (1995), *Standard Test Method for Soils and Rock,* ASTM Designation: D 4318, Annual Book of ASTM Standards, Vol. 4.08, Philadelphia, 560p.

Broms, B.B. and L.Y.C. Yao (1964) "Shear Strength of a Soil After Freezing and Thawing", Journal of Soil Mechanics and Foundation Division, ASCE, Vol. 90, No. SM4, pp. 1-25.

Casagrande, A. (1932) "The Structure of Clay and Its Importance in Foundation Engineering", Contributions to Soil Mechanics, BSCE 1925-1940, pp. 70-112.

Swan, C. (1996), "Behavior of a Sand in Frozen and Unfrozen States", Proceedings of the Eighth International Conference on Cold Regions Engineering, ASCE Technical Committee on Cold Regions Engineering, ed. Robert Carlson, pp. 483-493.

Wood, J.L. (1976), "Influence of Repetitious Freeze-Thaw on Shear Strength of Leda (Messena) Clay", Ph.D. Dissertation, Clarkson College of Technology, May, 1976, 152p.

A CASE STUDY OF TIMBER PILE IN-SITU SOIL REINFORCEMENT

Ching L. Kuo[1], Associate Member, ASCE; Wing Heung[2], Member, ASCE
Francisco J. Tejidor[3], Member, ASCE; and John Roberts[4], Member, ASCE

ABSTRACT: The Turnpike District of the Florida Department of Transportation is constructing the Polk Parkway, which is a multi-lane toll facility expressway looping around the southern extent of Lakeland, Florida. Section 3A of the Polk Parkway traverses a heavily mined area. Owing to the uncertain soil conditions, a surcharge program was designed to eliminate potential excessive total and differential settlements before constructing Mechanically Stabilized Earth (MSE) walls. During the construction of the surcharge embankment, a localized slope failure occurred in an area where a deposit of soft phosphatic waste clay was not detected during the original field exploration program. A stabilization berm was placed for temporary remediation while soil improvement alternatives were evaluated. A timber pile reinforcement alternative was selected to overcome the slope stability and settlement problems. After removing the surcharge embankment, approximately 1,100 40-foot long timber piles and high strength geotextiles were installed and followed by the MSE wall construction. Results of the instrumentation program indicated that minor settlements and lateral movements occurred after the timber pile reinforcement system was installed.

INTRODUCTION

The Turnpike District of the Florida Department of Transportation is constructing the Polk Parkway, a multi-lane limited access toll facility around Lakeland, Florida. The total length of the Parkway is about 24.5 miles, beginning at the Interchange of I-4 and Clark Road on the western side of Lakeland, and terminating at the interchange of I-4 and

---

[1] Chief Engineer, PSI, Tampa, FL

[2] District Geotechnical Engineer, Florida's Turnpike/Parsons Brinckerhoff, Pompano Beach, FL

[3] District Geotechnical Engineer, Florida's Turnpike, Miami, FL

[4] CEMC Program Director, Parsons Brinckerhoff Construction Services, Pompano Beach, FL

Mount Olive Road, northeast of Lakeland as shown in Figure 1. The entire project was divided into seven (7) sections in the design phase. Section 3 is about three (3) miles in length traversing in a west to east direction. The alignment of Polk Parkway crosses many areas where phosphate mining was active in 1940's to 1960's. The excavated areas were disposed with mining spoils or waste clays which are by-products of phosphate mining. Subsequently, the spoils were leveled and the clay settling areas were covered with overburden soils.

Historical aerial photographs revealed that large areas along the roadway from Station No. 375+00 to 439+00 were subjected to mining activities which were subsequently reclaimed in the 1980's. Due to the heterogeneous soil conditions encountered in the reclaimed area, a surcharge program was designed to eliminate potential excessive differential settlement. Instrumentations including settlement plates, vertical inclinometers and piezometers were installed to monitor the surcharge performance. The design required Mechanical Stabilized Earth (MSE) walls to be constructed after removal of surcharge fill. During the surcharge embankment construction, a mud wave was observed bulging from the bottom of an adjacent detention pond (SMF-3) between Stations 378+00 and 380+00. Tension cracks were also observed on the surcharge embankment.

This paper summarizes the findings of additional soil explorations, the selection, design, and construction of an in-situ soil improvement system, and the performance monitoring of the MSE wall during and after construction.

SOIL EXPLORATION AND SUBSURFACE SOIL CONDITIONS

During the design phase, soil borings were generally spaced at 100-foot intervals. Five (5) soil borings (Borings AB3-16, AB3-18, and TB3-38 to TB3-40) were performed in the vicinity of the failed slope (PSI, 1994) and did not reveal any phosphatic waste clays. Boring locations are shown in Figure 1. Subsoils in this mined area consisted of a surficial sand cover of 4 to 8 feet thick overlying mixed sands, clayey sands and clay spoils 30 to 40 feet in depth.

After the failure of the surcharge embankment, field exploration programs including a total of 13 additional soil borings (Borings BH-1 through BH-6 and Boring WB-1 through WB-7) were performed to explore the subsoil conditions in the problem area. The majority of the borings encountered a very soft waste clay layer ranging from Elevation +135 to +110 (NGVD) between Stations 377+50 to 380+00, as shown in Figure 2. The Standard Penetration Test (SPT) N-values ranged from weight of rod to 10 blows per foot. The laboratory index test results indicated that the natural moisture content of the waste clay ranged from 115% to 135%, the liquid limit from 130% to 192% and the from 98% to 150%. The undrained shear strength of the waste clay was estimated to be 300 to 500 psf.

Based on the additional soil borings information, it was concluded that the surcharge embankment was constructed along the edge of a settling pond filled with phosphatic waste clays, which was not detected during the initial field exploration program in the design phase. Figure 3 presents a typical roadway cross section in the failure area. Slope stability analysis using computer program PCSTABL6 indicated a safety factor of 0.7 to 1.0 when the initial failure occurred.

INSTRUMENTATION MONITORING PROGRAM

To monitor the stability of the failure area, three (3) vertical inclinometers, VI-A, VI-B and VI-C were added to the existing vertical inclinometer VI-2 in that area. Also, an additional settlement plate SP-4A and a vibrating wire piezometer PT-A were installed on the embankment at Station 379+00, as shown on Figure 4.

INITIAL STABILIZATION EFFORTS

In order to stabilize the embankment, an earth berm approximately 60 feet in width and 7 feet in height was constructed against the slope and atop the mud wave in the pond. Initial monitoring of the vertical inclinometers after the construction of stabilization berm indicated that no further movement occurred. The slope stability analysis indicated that the safety factor was on the order of 1.0 to 1.1 after the earth berm was constructed. However, as the construction of the surcharge embankment continued, tension cracks on the embankment enlarged and the monitoring of inclinometers indicated that the waste clays had excessive lateral movements. Embankment construction was stopped when the lateral movement exceeded 0.5 inches per day and no earthwork activity was allowed until the vertical inclinometers measured 0.1 inches per day or less. The intention was to allow the dissipation of excessive pore pressure in the waste clays and to gain in strength to support the additional embankment fill. These resting periods lasted up to five days. The vertical inclinometer casings failed after cumulative displacement reached 22 to 28 inches and they were replaced shortly after failure. When the top of surcharge embankment reached EL.+162, some inclinometers had failed two to three times. It was concluded that the embankment construction using the resting period approach was not effective due to the very slow dissipation rate of excessive pore pressures, as measured by piezometer PT-A. Because of excessive lateral movements and continued enlargement of tension cracks on the embankment, different soil improvement alternatives were evaluated.

ENGINEERING EVALUATION OF IN-SITU SOIL IMPROVEMENT

A total of five (5) in-situ soil improvement alternatives were evaluated as follows.

1. <u>Wick drains</u> Remove the surcharge embankment to EL.+143 (NGVD) and install wick drains with a sand blanket and strip drains. A layer of high strength geosynthetic reinforcement would be installed atop the sand blanket and followed by surcharge embankment construction. This

alternative would accelerate the consolidation of underlying waste clays and sufficiently increase the embankment stability. However, the major disadvantages were that the construction schedule would be delayed and the cost was estimated to be on the order of $1,250,000. The high cost is attributed to the large quantity of earthworks related to the re-installation and subsequent removal of surcharge fill.

2. <u>Soil Embankment and Realignment</u> The proposed MSE wall would be replaced by soil embankment and the South Frontage Road would be realigned to the location of the stabilizing earth berm. Wick drains would be installed at EL.+162 on the highway and at EL.+137 under the realigned South Frontage Road. A layer of high strength geotextile would be required under the South Frontage Road as reinforcement against potential stability problem. However, post-construction settlement of this alternative could be excessive unless a surcharge program is added to this scheme. The estimated cost was $1,000,000.

3. <u>Pile Reinforcement System</u> Remove surcharges to EL. +143 (NGVD) and install timber piles at 5 to 7-foot spacing. Two (2) layers of high strength geosynthetic reinforcement would be required above the timber piles. This alternative permits the construction of the MSE wall immediately after timber pile installation and would accelerate the construction schedule. No additional cost would be associated with the handling of embankment fill since the removal of existing surcharge fill was part of the original scope of work. Estimated cost was $1,000,000.

4. <u>Stone Column</u> Install stone columns at EL. +162 (NGVD) at a grid spacing of 7 feet. The concept was to increase the stability of the waste clay and to accelerate the rate of consolidation through radial drainage using stone columns. However, this alternative was not selected due to the lack of experience using stone columns in Florida and the uncertainties of its long term performance. Estimated cost was $700,000.

5. <u>Excavation</u> Remove the surcharge waste clays (slimes) to EL. +110 (NGVD) and backfill with sand. After that, the MSE wall could be constructed immediately. This alternative would involve a shoring support system. A large quantity of waste clay excavation and disposal are expected. The cost of this alternative was the highest among all the alternatives and was estimated to be $1,400,000.

In considering the long term performance of the proposed MSE wall, construction cost, schedule, constructability and reliability, the timber pile reinforcement alternative was selected. The final configuration of the reinforcement system consists of 5-foot square grid pattern under the MSE wall vicinity, supporting approximately 20 to 26 feet of fill height. The pile spacing is increased to 7 feet under the South Frontage Road, corresponding to a lower supported fill height of 5 to 10 feet. It is estimated that each pile will be subjected to a design load of 30 tons. Calculations showed that an ultimate capacity of 45 tons can be developed using the 40-foot timber pile at the site. The factor of safety was about 1.5, which was considered adequate for the purpose of the reinforcement system.

The function of the two layers of high strength geotextile is to transfer the weight of the MSE wall to the timber piles, preventing stability and settlement problems (Huat, et al. 1994). In our opinion, if these reinforcement layers were not installed, the most probable failure would involve the punching of soils around the pile top as the ground settles relative to the piles. In fact, the largest concern of the reinforcement system at the design stage was the potential tensile failure of the geotextile through punching failure. Using the method described by Broms (1987), the maximum tensile force anticipated in the geotextile is approximately 12.5 kips per foot at each of the two layers, as shown in Figure 5. The maximum vertical deflection of the geotextile at the center of the grid pattern was estimated to be 1 inch relative to the locations above the pile supports. Using the method presented in Naval Facilities Engineering Command (NAVFAC) Design Manual (1986), the timber piles were expected to settle approximately 0.5 inch under the 30-ton design load. Therefore, the maximum anticipated settlement of the system was 1.5 inches.

CONSTRUCTION OF PILE SUPPORTED REINFORCEMENT SYSTEM

The surcharge embankment was excavated to Elevation +143 (NGVD), except in locations where storm drainage pipes and structures were to be installed. In these locations, the excavation reached one foot below the bottom of the pipe and structures, in order to prevent conflicts between the geotextile during the subsequent pipe installation. After removal of embankment, approximately 1,100 treated timber piles were installed using Kobe K-13 and ICE 30S single acting diesel hammers. These piles were treated with 0.8 pcf Chromated Copper Arsenate (CCA) and were 40 feet in length. The minimum tip diameter was 7 inches. The piles were installed at 5-foot spacing in a square pattern in the vicinity of MSE wall and at 7-foot spacing at a distance greater than 20 feet from the wall, as shown on Figure 5. The piles were cutoff at Elevation +143 feet (NGVD) and followed by the placement of 6 inches of sand. The first layer of high strength geotextile was installed above the compacted sand. The long term allowable design strength of the geotextile was 1550 lb/in (18.6 kip per foot). The cross-machine direction of these geotextiles was sewed together in the field to enhance installation. Generally, the geotextile was overlain by 12 inches of sand, and topped with the second layer of geotextile in an orthogonal direction to the first layer. After the construction of the timber pile supported reinforcement system, the MSE wall was constructed according to the original design.

INSTRUMENTATION MONITORING

In order to evaluate the performance of the pile supported reinforcement system, one vertical inclinometer and four vibrating wire settlement cells were installed. The inclinometer VI-D was installed immediately in front of the MSE wall at Station 378+98 and measured a maximum of 0.7 inch lateral movement during construction of the MSE wall and embankment. This measurement confirmed that the pile supported reinforcement system stabilized the lateral movement within the waste clays.

Two settlement cells (SC-1 and SC-2) were installed below the MSE wall and the remaining two settlement cells (SC-3 and SC-4) were installed below the South Frontage Road in front of the wall, as shown in Figure 4. These settlement cells were located in the center of square pile clusters. Settlement cells SC-1 and SC-3 were damaged by the construction activities approximately two months after they were installed. The remaining cells, SC-2 and SC-4 indicated that 3 to 4 inches of total settlement occurred in about six months. However, it was concluded that the settlement cells had a system problem because the cell readings fluctuated considerably and in unison, as shown on Figure 6. Despite efforts to evaluate all accountable factors, including temperature of the fluid in the exposed tubings, atmospheric pressure, and possible inclusion of air bubbles in settlement cell systems, the cause of problems could not be identified. Because of the observed fluctuations, the monitoring program was supplemented by elevation surveys of the MSE wall panels. The elevation surveys revealed a settlement of approximate 0.6 inches on the wall panels. Since the settlement that occurred during the first two weeks of wall construction was not accounted for by the elevation surveys, the total settlement of the MSE wall was estimated to be approximately one inch which was within a reasonable range comparing with the prediction of 1.5 inches.

CONCLUSION

Construction problems related to the unexpected encounter of a waste clay settling pond in Section 3A of Polk Parkway were resolved by the design and construction of a timber pile supported reinforcement system. The system consisted of 1,100 treated timber piles and two layers of high strength geotextiles. A monitoring program indicated that lateral movement of the waste clays was stabilized and the maximum settlement was estimated to be approximately one inch.

ACKNOWLEDGMENT

The authors express their appreciation to Henri V. Jean of PSI and Richard A. Hawkins of Consulting Foundation Engineers for their significant contributions to the success of this project. Thanks also are given to Jim Moulton Jr. of the Florida's Turnpike, Rhett Leary of Parsons Brinckerhoff Construction Service and Joe Chao of Kisinger Campo and Associates for their excellent construction supervision and management.

SOIL IMPROVEMENT FOR BIG DIGS 183

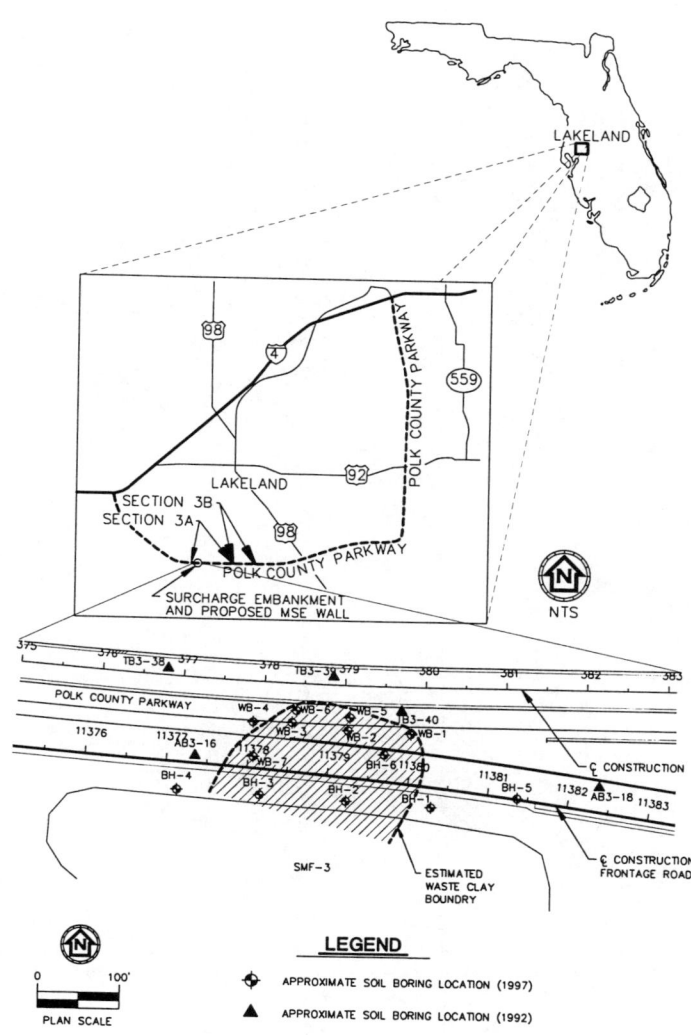

FIGURE 1   SITE VICINITY MAP AND BORING LOCATION PLAN

184   SOIL IMPROVEMENT FOR BIG DIGS

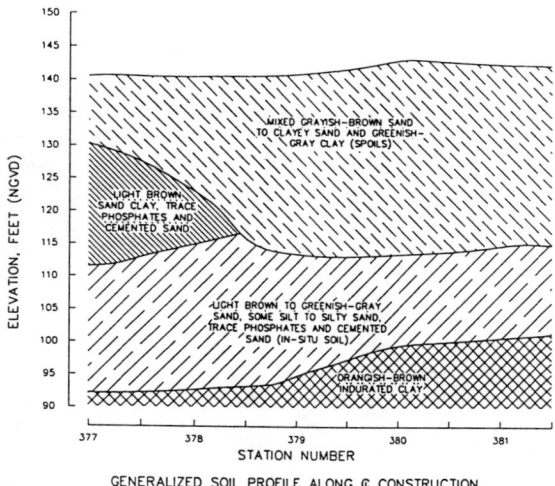

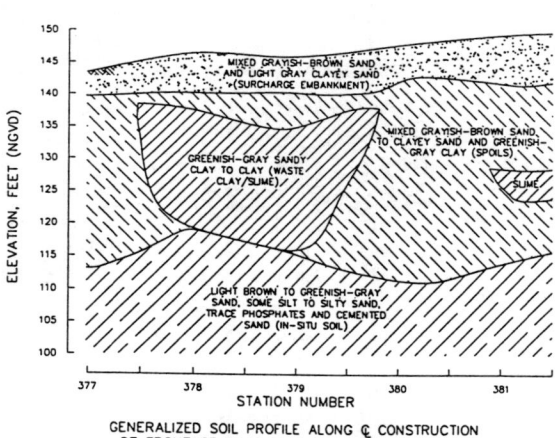

FIGURE 2   GENERAL SOIL CONDITIONS AT FAILURE AREA

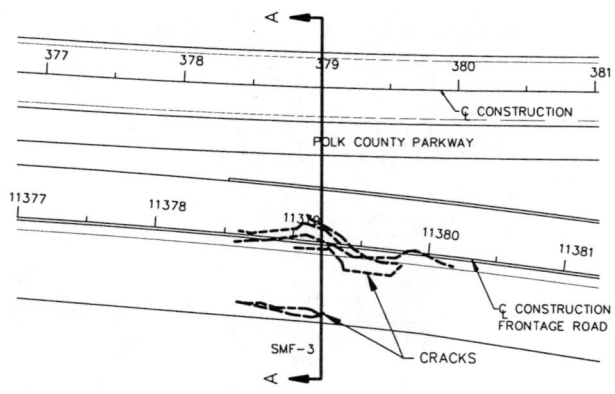
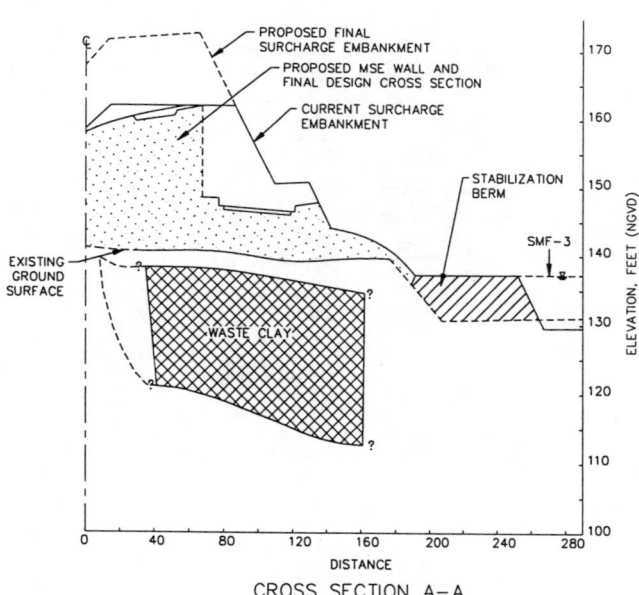

FIGURE 3 TYPICAL SECTION OF ROADWAY EMBANKMENT AT FAILURE AREA

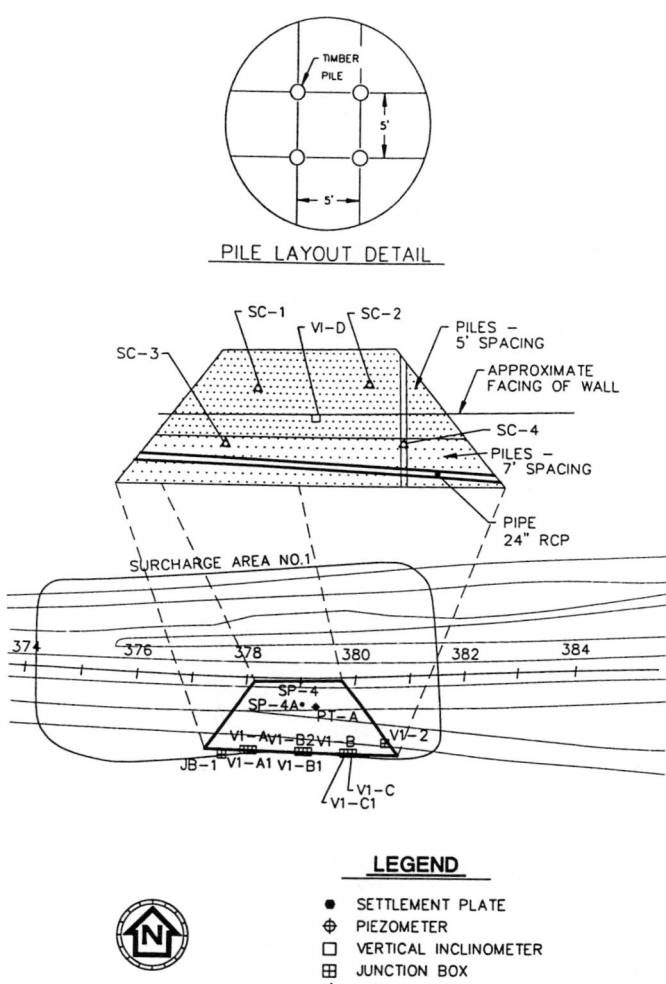

FIGURE 4 INSTRUMENTATIONS LAYOUT AND PILE LAYOUT PLAN

SOIL IMPROVEMENT FOR BIG DIGS 187

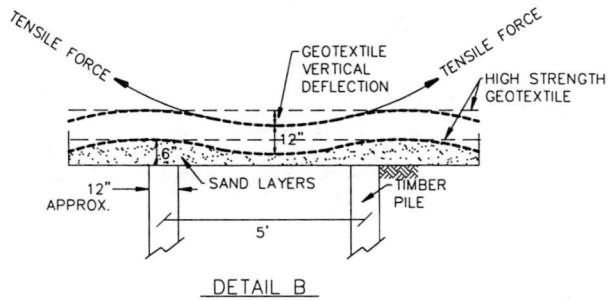

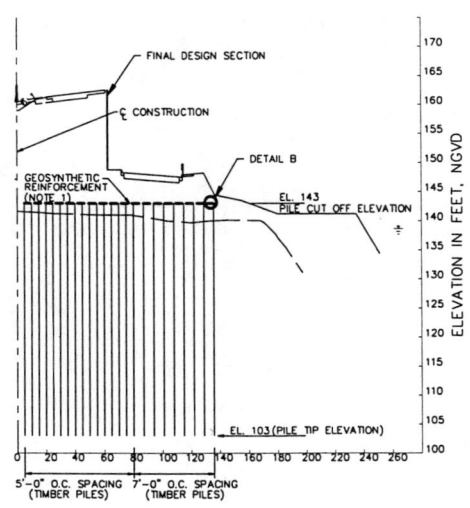

FIGURE 5   TIMBER PILE INSTALLATION LAYOUT

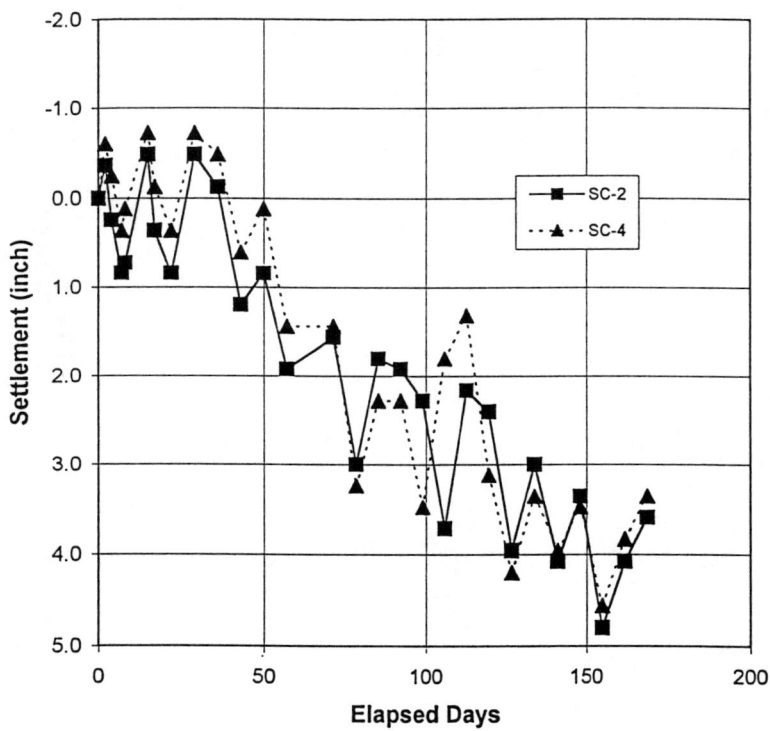

FIGURE 6 MONITORING RESULTS OF SETTLEMENT CELLS

REFERENCE

1. Broms, B.B. (1987). "Stabilization of Very Soft Clay using Geofabric." Geotextile and Geomembranes, Vol. 5, No.1.

2. Huat, B.B.K., Craig, W.H. and Ali, F. (1994) "The Mechanics of Piled Embankment." Proceedings, International Conference on Design and Construction of Deep Foundation, Orlando, Florida.

3. Naval Facilities Engineering Command (1986) "Foundations and Earth Structures." Design Manual 7.2, Department of the Navy.

4. Professional Service Industries, Inc. (1994). Roadway Soil Survey - Polk Parkway Section 3. Submitted to Reynolds, Smith & Hills, Inc./Florida Department of Transportation Turnpike District.

Augered Minipiles: A Cost-Effective Foundation for Light Structures

Bashar S. Qubain[1] and Jianchao Li,[2] Members ASCE

Abstract

A simple yet innovative pile foundation is proposed for lightly loaded structures. The piles are constructed using a standard soil boring rig, thus eliminating the high mobilization costs usually associated with conventional piles. Design curves are developed in advance between the standard penetration test N-values and the targeted capacity of the minipiles for later use in the field to reliably determine the termination depths of the piles. Comparisons with traditional piles, caissons, as well as other minipiles show that substantial savings can be achieved in using the proposed piles. Detailed construction steps are included with emphasis on dealing with various soil and groundwater conditions. Advantages and limitations are clearly indicated. This foundation was successfully utilized to support five canopy columns of a warehouse in Norristown, Pennsylvania, where a nearby sewer line prevented the use of spread footings.

Introduction

Deep foundations are generally used for heavy structures to transmit the structural loads below weak subsurface soils. Consequently, the emphasis has been traditionally placed on developing high capacity piles or caissons. Under certain circumstances, however, deep foundations have to be used for light structures. For example, existing utility lines may be overstressed if the structural loads are not transmitted below them. Also, miscellaneous highway structures such as noise walls and signs have light vertical loads in combination with relatively high lateral loads

---

[1] Chief Geotechnical Engineer, Valley Forge Laboratories, Inc., 6 Berkeley Road, Devon, PA 19333. E-mail: bqubain@valleyforgelabs.com. Tel.: (610) 688-8517.
[2] Project Geotechnical Engineer, Valley Forge Laboratories, Inc., 6 Berkeley Road, Devon, PA 19333. E-mail: jcli@valleyforgelabs.com. Tel.: (610) 688-8517.

and moments, which may make shallow footings uneconomical. Furthermore, when floor slabs are supported on piles, they usually transfer light loads to the piles.

In general, any of the available deep foundation systems can be used for light structures. Nevertheless, the following questions arise: (1) How cost effective are the available systems? (2) Can a simpler, more effective method be easily developed? The analysis and response to these questions is the subject of this paper.

Conventional Deep Foundations

Driven, auger cast, and bored piles as well as minipiles, caissons, and proprietary piles are among the widely used deep foundation systems (see Teng 1962; Navy 1986). Driven piles may include timber, steel pipe- or H-sections, precast concrete, and concrete-filled steel shells. These are installed by a driving hammer, and their design capacity is typically greater than 200 kN. During installation, the design capacity can be verified using empirical dynamic pile formulas or analytically by the wave equation (Polous and Davis 1980).

Auger cast piles are installed by drilling a hole with a continuous flight hollow-stem auger (minimum 300 mm diameter), which has a plug at the tip of the auger. Once the desired depth is reached, high-pressure grout is pumped through the hollow stem, forcing the plug out and filling the hole as the auger is withdrawn. (See Deep Foundation Institute 1993.) Pile capacities, which range from 350 to 900 kN, are developed mainly due to ability of the pressurized grout to fill and expand the hole.

Minipiles, also called micropiles, are grouted cast-in-place piles. They are constructed by driving a small diameter casing (100 to 300 mm) or by rotary/percussive drilling after which a temporary or permanent casing is inserted. Various grouting techniques are then employed to fill the hole, resulting in a wide range of pile capacities. For example, there are pressure grouted, compaction grouted, and jet grouted minipiles (see Welsh 1987; Flick et al. 1992). Minipiles can be installed under restricted access and low overhead clearance.

Bored piles, also referred to as small caissons, typically have diameters that vary between 450 to 600 mm. They are installed by drilling a hole to the required depth and filling it with concrete. Depending on the soil conditions, drilling and concrete placement may be achieved by the dry method, casing method or slurry-displacement method (Johnson and Gardella 1995). Design capacity varies depending on the subsurface conditions along the pile as well as below its base. A cleanout bucket is used to remove loose materials at the bottom of the hole. This is very important because these piles are too small to be entered and manually cleaned.

Caissons or drilled shafts are generally greater than 600 mm in diameter. They are constructed by augering the overburden and rock coring below (see Woodward et al. 1972; Greer and Gardner 1986). The loose soils at the bottom are removed either manually or by a cleanout bucket and the conditions of the bearing surface are checked through down-hole inspection, video camera or probing from the surface. Large diameter caissons can provide capacities greater than 18,000 kN. A caisson drill rig normally requires a large work area and high overhead clearance.

There are many proprietary types of piles, which incorporate a combination of various installation techniques. These include compacted concrete piles (also called pressure-injected footings or Franki piles, see Demcsak 1975); Monotube piles; and helical piles. For Franki piles, low slump concrete is placed in lifts and compacted by repeated hammer blows to form an enlarged base and shaft. This process densifies the surrounding soils and develops a high pile capacity (Nordlund 1982; Neely 1990). The proprietary aspect of the other two types of piles is primarily due to the materials used; their installation is standard. Monotube piles utilize specially fabricated corrugated steel shells that are regularly driven and then filled with concrete. Helical piles consist of specially-made augers, which are drilled and left in place to form the pile section.

The foundation systems discussed above provide more capacity than is typically needed for light structures. Special equipment and/or specialty contractors are required which make the installation cost high. In general, the mobilization costs for these systems are relatively high.

Augered Minipiles

A new pile foundation referred to as "augered minipiles" is proposed. The specific installation steps are schematically illustrated in Figure 1. Certain steps are similar to those of the bored piles. However, the proposed piles are simply installed using a standard soil test boring rig rather than a specialized drilling machine. The hole is drilled to the required depth using hollow stem augers, typically 200 mm in outside diameter.

After pulling out the augers, a 100-mm diameter split-spoon sampler attached to drill rods is lowered to the bottom of hole. Through repeated hammering on the spoon, any loose auger cuttings at the bottom are retrieved. After cleaning the bottom of the hole, 150 to 300 mm of crushed stone are placed in several lifts and compacted with a solid rammer attached to drill rods. These two steps are easy and simple yet crucial for this technique. By doing so, a clean and compacted bottom is assured. Concrete is then placed by inserting the pump hose or pipe into the bottom of the hole. Unlike minipiles, high-slump concrete instead of cement grout is placed under atmospheric pressure. The pipe is gradually pulled out while concrete is placed

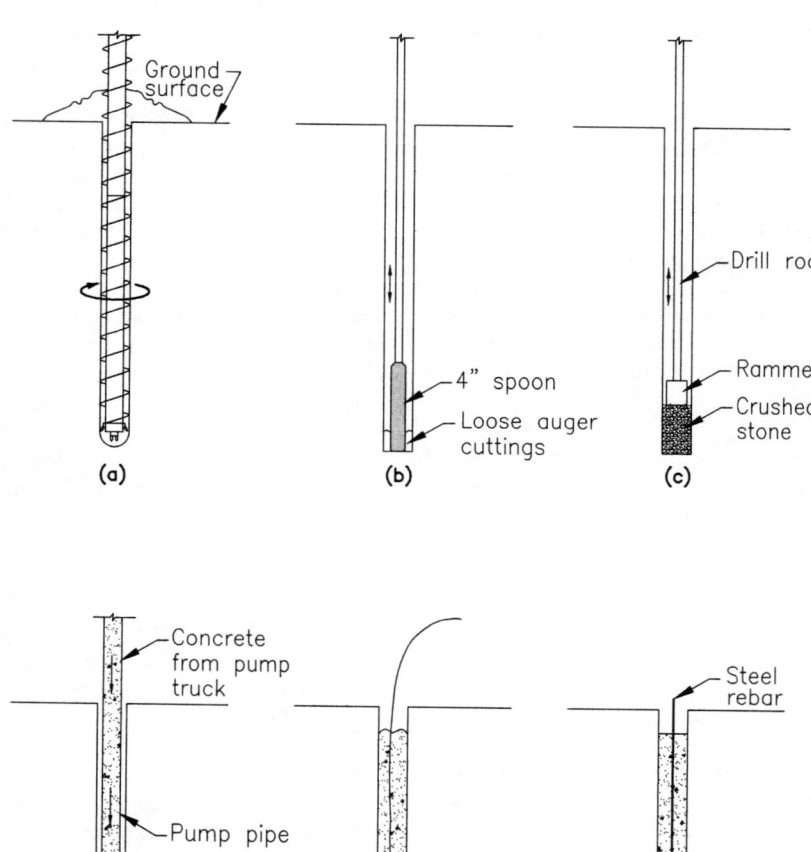

Figure 1. Augered minipile installation procedure. (a) Augering; (b) Retrieving auger cuttings using 4" spoon; (c) Placing crushed stone and compacting; (d) Pumping concrete; (e) Vibrating; (f) Inserting reinforcement.

until the hole is filled to the specified cutoff level. Finally, one or several steel bars are inserted into the fresh concrete to achieve the required structural strength of the pile.

In severe situations where the hole does not remain open even for a very short period of time (hours) due to loose or granular soils, a casing or drilling mud or slurry may be used. The advantage of using a casing is that it easily accommodates the cleaning and compaction steps. The disadvantage: it must be left in place. The use of slurry, on the other hand, does not permit cleaning the bottom of the hole, although it allows compaction of crushed stone. Once concrete is placed it is vibrated to ensure that all voids are completely filled. It is important to note that moderate groundwater flow does not require casing or slurry. The water can simply be pumped out of the hole and the cleaning and compaction steps can proceed unaffected. Significant water flow that erodes the sides of the hole would require the use of a casing or slurry.

An important aspect of augered minipiles is its utility to perform the standard penetration test (SPT) as an integral part of the installation process. With little increase in cost and effort, SPT sampling while drilling the hole of an augered minipile provides necessary subsurface information and establishes with certainty the termination depth of the pile as will be discussed later in the paper.

Depending on the pile length and soil conditions, a vertical capacity ranging from 50 to 200 kN may be achieved. This is very well suited for the light loading situations. Evidently, the cost of mobilizing a soil boring drill rig is substantially lower than that of any conventional or specialized deep foundations.

Table 1 presents a comparison of various deep foundation systems with respect to approximate mobilization cost, equipment, typical capacity range, verification/quality control, and installation speed. For light structures, the advantages of the proposed piles over other systems is clearly illustrated in the table.

Analysis and Design Approach

In terms of analysis and design, an augered minipile is simply considered as a bored or drilled pile (Skempton 1959; De Beer 1964; Vesic 1970; Touma and Reese 1974; Meyerhof 1976; Tomlinson 1977; Poulos and Davis 1980). Empirical correlation between SPT N-values and pile capacities are widely used. According to Meyerhof (1976), the ultimate unit skin friction f_s in tsf (1 tsf $\cong$ 100 kN/m^2) of a bored or drilled pile in cohesionless soils is roughly given as:

$$f_s = \frac{\overline{N}}{100} \qquad (1)$$

Table 1. Comparison of Various Deep Foundation Systems

Foundation Type (1)	Typical Capacity (kN) (2)	Mobilization Cost (3)	Equipment & Materials (4)	Verification & Quality Control (5)	Installation Speed (6)
Driven piles	200 - 1,000	$10,000	Pile driving rig	Dynamic formula/ Wave equation	Fast
Auger cast piles	350 - 1,000	$25,000	Specialized rig & augers	Grout pressure & volume control	Medium
Minipiles	20 - 1,500	$15,000	Specialized rig & grout pump	Grout pressure & volume control	Medium
Bored piles	350 - 1000	$10,000	Specialized rig & grout	Cleaning/ probing	Slow
Caissons	1,500 - 18,000	$30,000	Specialized rig & regular concrete	Down-hole Inspection	Slow
Franki Piles	1,000 - 2,500	$30,000	Specialized rig & concrete hammer	Hammer blows & volume control	Slow
Monotube Piles	500 - 1,000	$10,000	Specially fabricated piles	Dynamic formula/ Wave equation	Fast
Helical piles	50 - 200	$5,000	Specialized rig & fabricated helicals	Installation torque	Fast
Augered minipiles	50 - 200	$500	Soil boring rig & regular concrete	SPT, cleaning & compacting	Fast

where $\overline{N}$ is the average standard penetration resistance within the embedded length of pile, in blows per foot (blows per 30 cm). The recommended upper limit of f_s is 0.5 tsf (or 50 kN/m²).

The ultimate point bearing q_p (Meyerhof 1976) in tsf (100 kN/m²) may be taken approximately as:

$$q_p = \frac{0.12 \, L \, N_b}{B} \qquad (2)$$

where N_b is the average standard penetration resistance near the pile tip, L is the pile embedment in the bearing soil, and B is the width or diameter of the pile. The limiting value of q_p is 1.2 N_b in tsf (100 kN/m²) for sand and N_b for silt. The above equations have been successfully used to estimate pile capacity in soils other than cohesionless materials, such as medium to stiff clay (Bromham and Styles 1971).

For a pile with point and shaft areas A_p and A_s, the ultimate bearing capacity Q_u is expressed by the sum of point resistance Q_p and skin resistance Q_s. Thus,

$$Q_u = Q_s + Q_p = f_s A_s + q_p A_p \tag{3}$$

For a targeted design capacity Q_0, a correlation between SPT N-value and the required pile length L can be simply derived by making

$$Q_0 = \frac{Q_u}{FS} \tag{4}$$

where a typical value of 3 is specified for the factor of safety FS and Q_u is as given in Eq. (3).

When the subsurface conditions are relatively homogeneous, a simplified relationship (design curve) can be developed between N-values and pile length L for a specified pile capacity. This is illustrated in Figure 2 for piles with a diameter of 200 mm, embedded in uniform sandy soils, and having a design capacity of 75 kN. This design curve accounts for shaft resistance and end bearing. In situations where the piles are driven through weak soils and bear on firm strata, only the tip contribution is considered in developing design curves. Other site-specific conditions can be addressed as well.

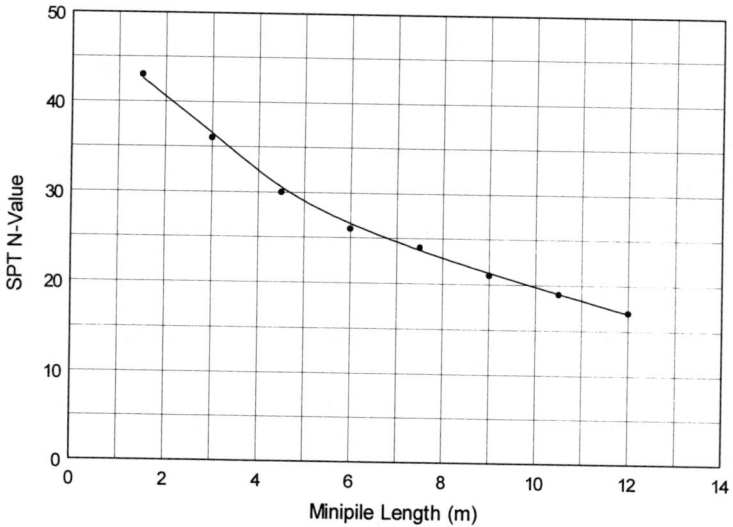

Figure 2. Design Curve for 200 mm Diameter Augered Minipiles with a Capacity of 75kN in Uniform Granular Soil.

Design curves like the one shown in Figure 2, which are developed in advance and utilized during pile installation, are fundamental to augered minipiles due to the ability of easily performing the standard penetration test as an integral part of the proposed installation technique. Therefore, not only the termination depths are established with certainty, but the pile capacities as well are confirmed in the field.

Practical Application

The proposed pile foundation was successfully utilized for a warehouse canopy in Norristown, Pennsylvania. The design drawings called for spread footings to support five canopy columns, whose loads varied between 75 and 150 kN. However, the presence of a nearby sewer line necessitated that the foundation bearing level be lowered 3 m to match the sewer invert elevation. Lowering the footings would have required deep excavation and costly shoring to protect the sewer line, not to mention the associated construction delay.

The subsurface conditions were investigated by performing 3 test borings using a truck-mounted drill rig. Hollow stem augers were used to advance the borings and the holes were observed to remain open and stable. A general subsurface profile is presented in Figure 3; it primarily consists of silty sand with gravel overlying very dense completely weathered sandstone.

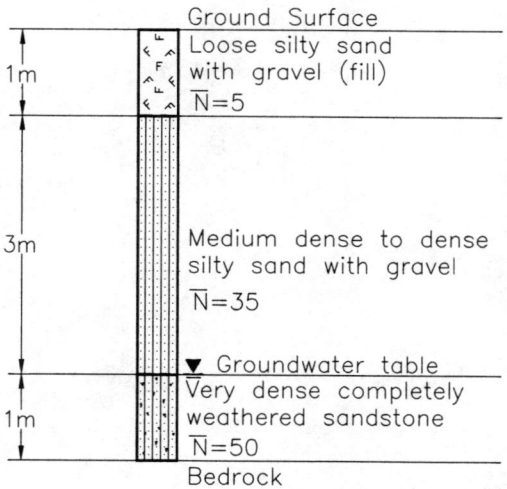

Figure 3. General Subsurface Profile.

Due to its effectiveness and low mobilization cost, augered minipiles were selected to support the warehouse canopy. The high mobilization fee for any of the other deep foundations would have been cost prohibitive. Augered minipiles drilled through the medium dense to dense silty sand with gravel provided a design capacity of 75 kN. The upper 1 m of loose fill was ignored in determining the pile capacity. Installation of the minipiles was performed using the same test boring rig (Figure 4). In one day, 12 augered minipiles were completed in two- and three-pile groups. The minipiles were terminated in the field based on the design curve in Figure 2 together with SPT verification during pile augering. As an added assurance the piles were taken at least 0.5 m into the very dense completely weathered sandstone. Pile lengths varied from 4 to 5 m. Figure 5 shows the concrete placement in one of the piles. The canopy structure has been functioning well since the construction was completed about one year ago.

Figure 4. Pile Drilling with a Standard Test Boring Rig.

Figure 5. Concrete Placement in an Augered Minipile

Conclusions

The proposed augered minipiles were successfully utilized to support a lightly loaded structure, where a nearby sewer line prevented the use of spread footings. Substantial savings were realized due to pile installation by a standard soil boring rig, which has very low mobilization cost in comparison with traditional piles, caissons, as well as other minipiles.

This method has a sound basis for reliably terminating the piles in the field and estimating their capacity by its inherent capability of performing the standard penetration test as an integral part of the installation process. Site-specific design curves, developed based on empirical correlation between SPT N-values and the targeted capacity, are utilized during pile installation to achieve this objective.

Further developments and refinements could be achieved through pile load testing in various subsurface conditions. This would be valuable not only to confirm the design equations and optimize the installation but also to extend this foundation type to a wide range of soil types.

In general, this foundation system works best in situations where the augered holes remain open without the use of a casing or slurry. This is not a major limitation, however, since most soils other than pure sands would be able to, at least, temporarily maintain an open hole. In the event that an augered hole does not remain open, a casing or slurry can still be used, albeit with some disadvantages for either one. Moderate groundwater flow does not require casing or slurry. Another obvious limitation to this technique is the relatively low pile carrying capacities which can be achieved. Therefore, this is indeed a method for light structures.

References

Bowles, J. R. (1988). *Foundation Analysis and Design, 4th Ed.* McGraw-Hill, New York, N.Y.

Bromham, S. B. and Styles, J. R. (1971). "An Analysis of Pile Loading Tests in Stiff Clay." *Proceedings, 1st Aust.-N.Z. Conference in Geomechanics*, Vol. 1, Melbourne, Australia, pp. 246-253.

De Beer, E. E. (1964). "Some Considerations Concerning the Point Bearing Capacity of Bored Piles." *Proceedings, Symposium on Bearing Capacity of Piles*, Roorkee, India, pp. 178-204.

Deep Foundation Institute. (1993). *Auger Cast-in-Place Pile Model Specification.* Englewood Cliffs, N.J.

Demcsak, M. R. (1975). "Franki Pressure Injected Footings and Their Use in the Middle Atlantic Region." *Proceedings, 1974 - 1976 Lecture Series: Geotechnics, A Bicentennial State of the Art 1976*, Geotechnical group, ASCE, Philadelphia, P.A.

Flick, L. D., Osborn, N. B. R., Graham, A. E., Marasa, M. J., and Tobey, F. T. (1992). "Minipile Milestone." *Civil Engineering*, September, ASCE, New York, N.Y.

Greer, D. M. and Gardner, W. S. (1986). *Construction of Drilled Pier Foundations.* Wiley & Sons, New York, N.Y.

Johnson, L. F. and Gardella, L. N. (1995). *Field Inspection Handbook, Chapter 8 - Engineering Control of Pile Installation.* D.S. Brock, S.M. Levy, and L.L. Sutcliffe, eds. McGraw-Hill, New York, N.Y.

Neely, W. J. (1990). "Bearing Capacity of Expanded-Base Pile in Sand." *Journal of Geotechnical Engineering*, ASCE, Vol. 116, No. 1, pp. 73-87.

Nordlund, R. L. (1982). "Dynamic Formula for Pressure Injected Footings." *Journal of the Geotechnical Engineering Division*, ASCE, Vol. 108, No. GT3, pp. 419-437.

Meyerhol, G. G. (1976). "Bearing Capacity and Settlement of Pile Foundation." *Journal of the Geotechnical Engineering Division*, ASCE, Vol. 100, No. GT3, pp. 196-228.

Poulos, H. G. and Davis, E. H. (1980). *Pile Foundation Analysis and Design.* John Wiley & Sons, New York, N.Y.

Skempton, A. W. (1959). "Cast In-situ Bored Piles in London Clay." *Geotechnique*, London, England, Vol. 9, pp. 153-173.

Teng, W. C. (1962). *Foundation design.* Prentice-Hall, Englewood Cliffs, N.J.

Tomlinson, M. J. (1977). *Pile Design and Construction Practice.* Viewpoint Publications, London, England.

U.S. Navy (1986). *Design Manual 7.2 - Foundation and Earth Structures.* Naval Facilities Engineering Command, Department of Navy, Washington, D.C.

Vesic, A. S. (1970). "Load Transfer in Pile-Soil Systems." *Proceedings, Conference on Design Installation of Pile Foundations*, Lehigh University, Bethlehem, P.A., pp.47-73.

Welsh, J. P. (1987). "Soil Improvement - A Ten Year Update." *Committee Report, Placement and Improvement of Soils*, ASCE Geotechnical Special Publication, No. 12, New York, N.Y.

Woodward, R. J., Gardner, W. S. and Greer, D. M. (1972). *Drilled Pier Foundations.* McGraw-Hill, New York, N.Y.

An Underpinning Solution to a Persistent Foundation Settlement Problem

Eric J. Seksinsky[1] and Bashar S. Qubain[2], Members ASCE

Abstract

A persistent foundation settlement problem affecting a fifty-year old house in Wynnewood, Pennsylvania was successfully resolved by intermittent underpinning. This system was selected due to cost and two unique conditions: (1) the bottom of the wall was deteriorated which would have required the added cost of constructing grade beams had pile underpinning been selected; and (2) conventional underpinning would provide access to point and repair the lower part of the wall. In designing the underpinning system, the size, sequence of construction, and spacing of the piers were selected around existing major vertical cracks to minimize structural distress during remediation. Monitoring gauges were installed at critical locations to evaluate the foundation's response during and after the underpinning work. Steel plates or boards were also secured below the bottom of the wall to prevent dislodged masonry units from falling into the underpinning pits. In spite of today's high labor rates and the latest innovations in foundation repair, conventional, labor-intensive underpinning was still the most appropriate and cost-effective remedial scheme to resolve this unique foundation settlement problem.

INTRODUCTION

A fifty-year old stone masonry house in Wynnewood, Pennsylvania experienced decades of progressive settlement along a back corner of its foundation wall. The latest owner had occupied the house for about six years when he realized the

---

[1] Senior Project Engineer, Valley Forge Laboratories, Inc., 6 Berkeley Road, Devon, PA 19333. E-mail: eseksinsky@valleyforgelabs.com. Tel.: (610) 688-8517.
[2] Chief Geotechnical Engineer, Valley Forge Laboratories, Inc., 6 Berkeley Road, Devon, PA 19333. E-mail: bqubain@valleyforgelabs.com. Tel.: (610) 688-8517.

seriousness of the problem. Like the previous owner, he had pointed the vertical cracks in the exterior face of the wall, but they simply opened again. Finally, with evidence mounting that the rate of settlement was not diminishing but actually advancing, the owner realized that the foundation had to be stabilized. Initial opinions indicated that poorly compacted fill was the culprit, since it was visually obvious that fill had been placed to build up the backyard. Yet this could not explain the progressive nature of the settlement. This paper describes the subsurface exploration work undertaken to identify the cause of the settlement and presents the considered remedial foundation alternatives as well as the details of the adopted system.

EXTENT OF SETTLEMENT DAMAGE

A field survey of the rear wing of the house revealed that about 14 m of a perimeter stone masonry foundation wall enclosing the east and south sides of a two-car garage were damaged and in need of repair. Figure 1 shows the location of the garage with respect to the main part of the house. (Note that the floor of the garage is at the basement level of the main part of the house.) The greatest movements were observed at the southeast corner of the garage, which underwent 80 mm of vertical subsidence as well as 100 mm of maximum outward tilt along two major cracks extending from top to bottom of the wall (see Figure 1). These cracks and other smaller ones in the exterior face of the wall had been pointed several times in the past, but continued to re-open along the same fracture. At the time of the survey, the major cracks were 16 mm wide on the exterior side. Old newspapers stuffed into open cracks inside of the garage dated back to the 1970's, which may have been the start of the problem.

Other settlement-induced damage to the structure includes distorted window and door frames and a family room floor that sagged toward the corner due to an estimated maximum angular distortion of nearly 1/100. The seriousness and progressive nature of these differential movements at the corner were punctuated by the owner's eerie stories of occasional snapping noises coming from the floor in the family room.

The general layout of the back yard behind the garage at the start of the field survey is also depicted in Figure 1. Of interest are a mild ground slope (10 H: 1 V downward toward the south), flagstone patio with exterior concrete stairs to the first floor, and a drainage swale along the eastern edge of the yard. The swale flows toward the south and appears to follow original grade. On average, the patio level is 1.2 m higher than the top of slab in the garage. It is also important to note that significant settlement cracks and subsidence were observed in the flagstone patio and concrete stair unit. (The patio and stairs were demolished shortly after the field survey.)

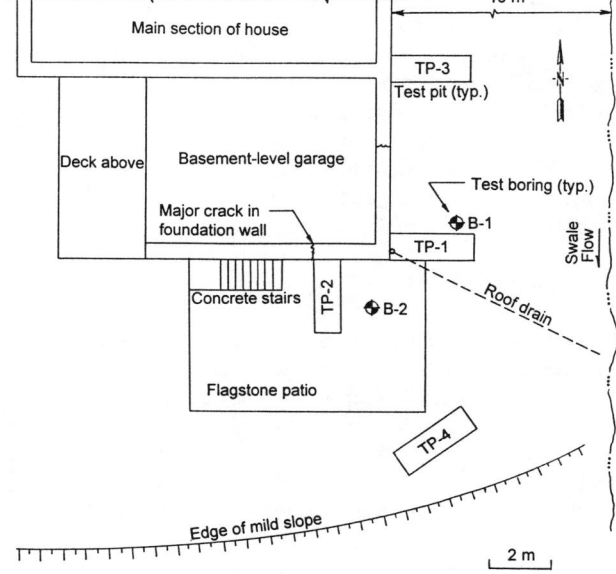

Figure 1. Site Plan.

SUBSURFACE EXPLORATIONS

After the field survey identified the extent of the settlement damage, the subsurface soil, rock, and foundation conditions were investigated by four test pits and two test borings. (Investigated locations are marked in Figure 1.) The test pits were excavated using a backhoe to a depth of 4 m to sample soils, document moisture levels, expose the bottom of the foundation wall, and inspect conditions in load-bearing strata. Three pits were opened next to the garage (TP-1 through TP-3), while another (TP-4) was opened outside of the subsidence area to gain information on favorable conditions. The test pits were terminated in dense soils.

The test borings taken near the southeast corner supplemented the test pit information and furnished cores of the underlying bedrock for study and testing. Marked as B-1 and B-2 on Figure 1, the borings were extended to a depth of 9 m, including 3 m of rock coring. While advancing through the overburden, the standard penetration test (SPT) blow counts were recorded to delineate weak zones.

Foundation and Subsurface Conditions

The test pits revealed that the stone masonry foundation wall of the garage was 0.5 m thick, had no footing, and was founded on very weak clayey sand with gravel fill at an average depth of 1.5 m below exterior grade (0.6 m below top of garage slab). In addition, the mortar of the lowest several courses of stonework in the wall was in a deteriorated condition. In fact, stones were easily dislodged by hand when examining exposed wall sections in the test pits. Also, perched water in the backfill against the wall seeped into the test pits during the study.

The thickness of fill revealed by the investigation (2.2 m) agreed with initial visual estimates of the amount of soil placed along the back of the house to create a level area for the patio. Test pit observations and SPT blow counts in the fill showed saturated conditions and very poor compaction, especially within the 0.7 m of material under the foundation. Concentrations of topsoil and decomposed leaves, as well as cobbles and a few boulders were also observed in the fill. On the other hand, the pit outside of the settled area (TP-4 in Figure 1) showed comparatively moist, medium dense, and better compacted fill than the area next to the garage.

Like the fill, the upper stratum of natural soil (0.8 m of clayey gravel) had also been weakened by excessive moisture. However, at a depth of 3 m below exterior grade the natural soils made a transition into better material; first a medium dense silty sand and finally at 4.6 m, saprolite or completely weathered rock consisting of dense, dry silty sand with gravel. Sampler refusal (50 blows per 75 mm) was experienced in the saprolite.

Top of rock was encountered 6.7 m below exterior grade in the borings. The cores were identified as hornblende-bearing mafic gneiss with the aid of the Lansdowne preliminary geologic quadrangle map by the Pennsylvania Geological Survey (Berg and Dodge 1981). Physically, the rock mass was characterized as dark gray, medium hard, moderately weathered, thinly banded, broken, and very poor quality. Average fracture spacing was 75 mm. There were also quartz veins and numerous zones of soft, highly weathered material. Figure 2 presents an idealized subsurface profile for the site, with depths referenced from exterior grade. The bottom of the foundation wall is also noted in the figure.

Cause of Continued Settlement

The foundation settlement was attributed to poorly compacted fill and excessive subsurface moisture in both the fill and upper natural soils. The source of the moisture was identified in TP-1 as a leaking underground roof drain. This line (see Figure 1) originated at the southwest corner of the garage at a depth of 0.6 m and extended eastward toward the swale, where it apparently discharged when originally installed. At the time of the subsurface exploration its end was completely blocked, permitting rainwater to back up in the pipe and saturate the ground at the corner of the garage. This situation seems to have existed for many years. With an ever-increasing water content, the clayey fill and natural soil underneath the foundation were progressively weakened and subjected to continual subsidence.

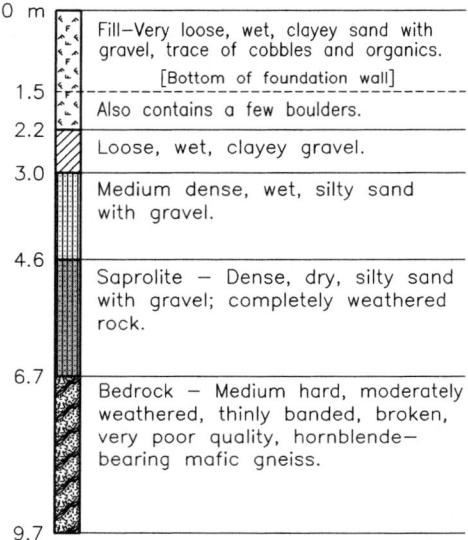

Figure 2. Idealized Subsurface Profile.

REMEDIAL FOUNDATION ALTERNATIVES

Of the various remedial foundation alternatives evaluated, soil improvement by compaction grouting was ruled out early on because the weak zone under the wall was too shallow to accommodate the high pressures needed for adequate densification (Welsh 1987). Jet grouting was also ruled out due to high mobilization costs (Furth 1997). Thus, only minipiles and conventional underpinning remained for further consideration.

Minipile Underpinning

Minipile options included 3 types: helical piles, driven piles, and drilled minipiles. Due to the cobble and boulder content in the soil, helical and driven minipiles were considered not to be feasible. This left drilled minipiles as the only feasible minipile system. In considering the use of this support system, it was recognized that the deteriorated condition of the lower part of the stone masonry foundation wall would complicate the design. While a concrete footing can handle discrete supports to transfer the wall load (estimated as 150 kN/m), the wall of this structure could not.

To deal with this situation, it was decided that the typical discrete supports at minipile locations be replaced with a continuous reinforced concrete grade beam on each side of the wall as shown in Figure 3. The reason for having a grade beam on each side of the wall is obviously to avoid eccentric loading on a wall that has already cracked and tilted. The cross section of the grade beams would be 0.4 m wide by 0.3 m deep. The bottom of the grade beams was positioned 0.6 m above the bottom of the wall to ensure that dowels would be installed in sound masonry. The connection to the foundation wall would be accomplished by installing dowels every 0.3 m.

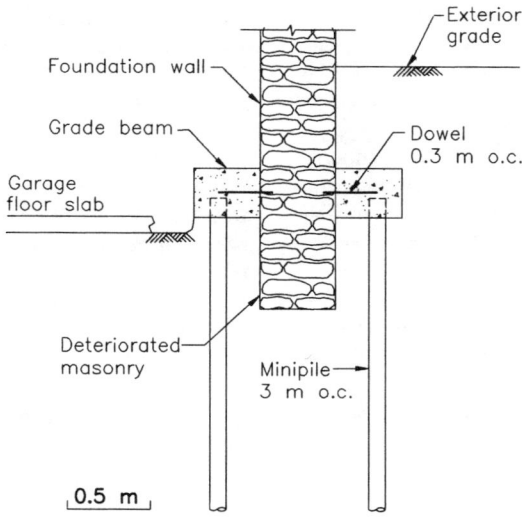

Figure 3. Minipile Detail.

The drilled minipile support option featured ten concrete-filled steel casings (100 mm in diameter) spaced every 3 m, or five on each side of the wall. The casings would be installed 0.3 m away from the wall to a depth of 5.2 m using rotary/percussion drilling, thereby providing an embedment of 1.5 m into the saprolite bearing stratum. The design allowable pile capacity was limited to 80 kN to keep the maximum span between piles as 3 m. At this low capacity the concrete would not be placed in the casings under pressure, and conventional geotechnical design methods apply (see Poulos and Davis 1980). Before pouring the grade beams, the dowels in the wall would have to be grouted and tested for pullout resistance. The drilled minipile system had one drawback—the continuous grade beam along the inside of the wall would encroach on the usable space of the garage.

Conventional Underpinning

Of the conventional underpinning options, continuous underpinning support on the medium dense silty sand at a depth of 2 m below the wall was first considered. Such a system was too conservative, however, considering that dense material was only 1 m deeper. Accordingly, the piers were lowered and an intermittent system of support was developed to take advantage of the dense saprolite. The refined underpinning design featured 3-m long piers with intervening lintels.

Figure 4 depicts a plan view of the intermittent underpinning scheme. The layout of the lintels and piers was established based on a safe unsupported wall length of 1.5 m for excavation purposes, and the need to have piers underneath the corner and the ends of the wall. To minimize structural distress to the house during remediation, the piers also had to avoid the two major cracks along where the corner was separating. Typically, the piers were designed to be 1.2 m wide, while the lintels spanned 1.5 m. Based on this pier and lintel arrangement, the bearing pressure at the bottom of the foundation wall was 350 kPa. The lintels were 0.5 m wide by 0.5 m deep, and were reinforced with three No. 5 steel bars. A major advantage of the underpinning option was that it afforded a means to point and repair the lower part of the wall. A drawback of this scheme when evaluated in terms of minipiles was a comparatively longer time for construction due to all of the hand labor.

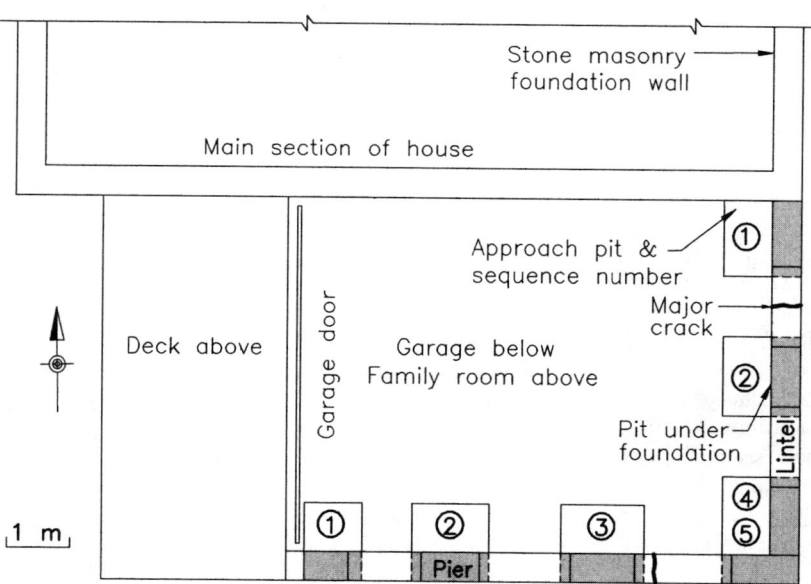

Figure 4. Intermittent Underpinning Plan.

Selected Alternative

Aside from the aforementioned advantages and disadvantages, both minipiles and intermittent underpinning were equally viable remedial designs from a technical point of view. Therefore, selection of the construction alternative ended up being based primarily on cost. Estimates for the intermittent underpinning system were about 30% less expensive than drilled minipiles. Consequently, it was selected for construction.

CONSTRUCTION DETAILS

Before the repair work on the foundation wall was initiated, monitoring gauges were installed inside the garage across the two major cracks in the east and south walls. This provided a means of recording the response of the structure during and after the underpinning. The approach pit method (White 1975) was used on the job. Figure 4 illustrates the as-built arrangement of the approach and underpinning pits, while Figures 5 and 6 show elevations of the south and east sides, respectively. All of the underpinning work was conducted inside the garage rather than outside, even though this required the edge of the garage floor to be sawcut and removed during construction. This had two advantages: (1) about 1.1 m of soil excavation and shoring

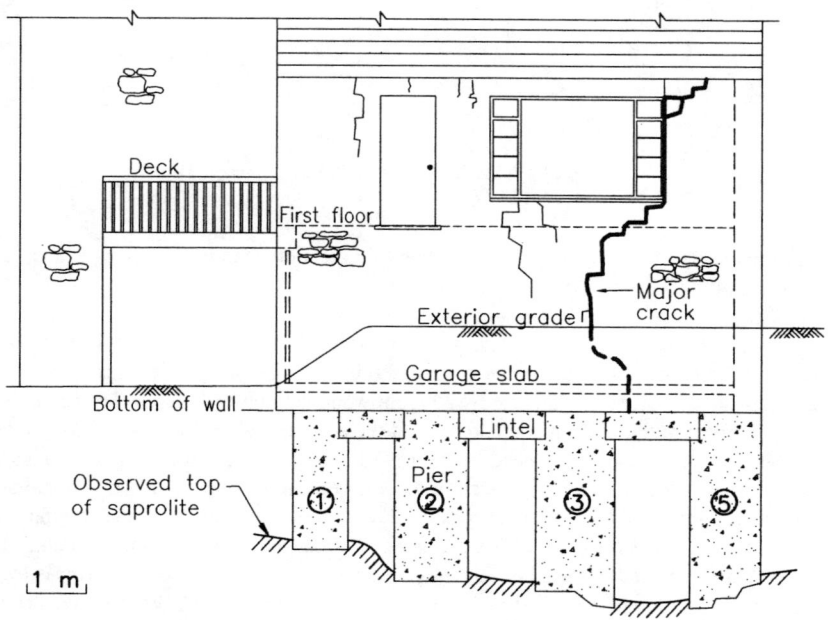

Figure 5. South Elevation of Underpinning.

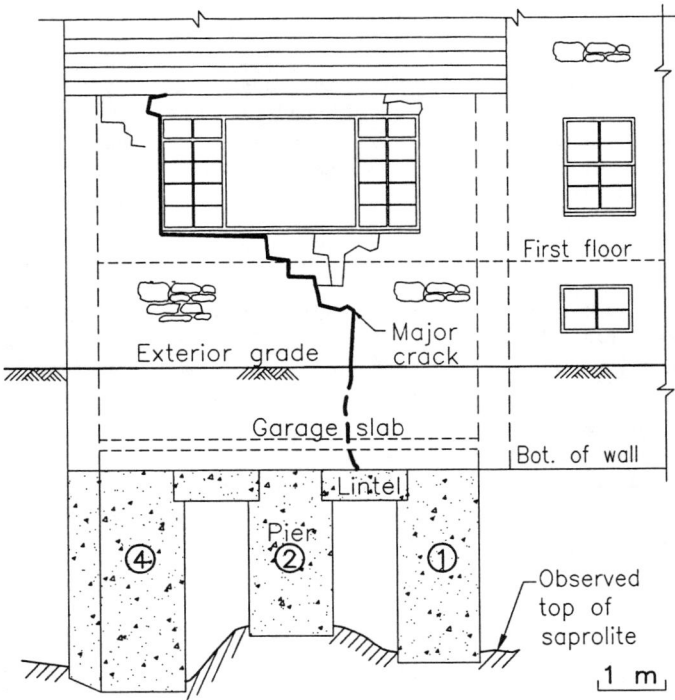

Figure 6. East Elevation of Underpinning.

were saved due to the grade difference between the floor and exterior grade; and (2) progress of the work was not hampered by rain.

All of the excavation work was done using hand tools. Figure 7 depicts a detailed view of the underpinning and shoring. The shoring of the pits was accomplished by installing 50 mm by 200 mm wood lagging with depth. To secure the inside corners of the shoring box and prevent the ends of the lagging from kicking into the pits, steel angles were fastened to the lagging using screws. Installing these angles was actually easier, more effective, and less time-consuming than nailing traditional 50 mm by 100 mm wood cleats to the corners. Figure 8 is a photograph of the shoring for Pier 1 at the northeast corner of the garage. As can be seen, the underpinning pit underneath the wall was the same thickness as the wall itself—about 0.5 m. This is less than the 0.9 m dimension called for on the construction drawing. Although it created a tight work space, this change by the contractor was approved because it minimized excavation and formwork. The completed pits had the appearance of a mine shaft. The

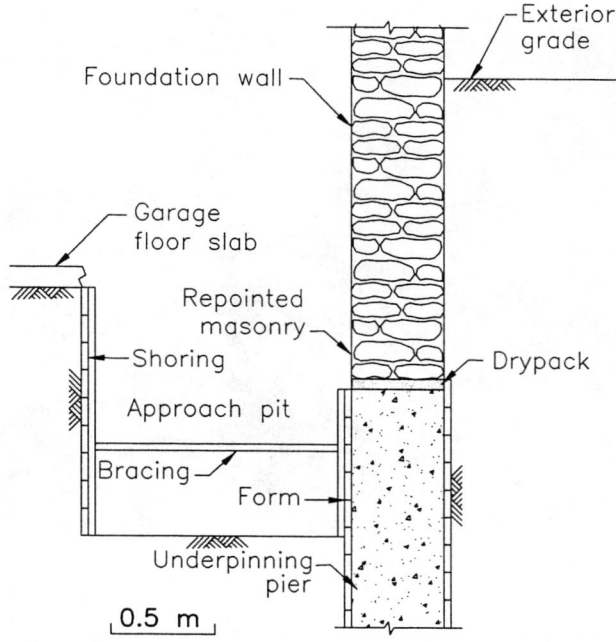

Figure 7. Underpinning Detail.

shoring boxes also served as the forms for the underpinning piers, so it was essential to keep the sides plumb as the pits were deepened. All pits were measured plumb to within 75 mm.

Another job safety measure was made necessary by the deteriorated condition of the mortar at the bottom of the masonry wall. To prevent dislodged stones from falling into the pits and injuring workers while the work was being done, the contractor temporarily secured boards or a 6 mm thick steel plate underneath each unsupported section of wall. This also facilitated the pointing and repair of the lower courses of stone.

As shown in Figures 5 and 6, the underpinning piers were taken at least 0.15 m into the saprolite. Due to undulations in the rock weathering profile the final constructed length of the piers varied from 2.35 m to 3.35 m, as compared to the design length of 3 m based on the idealized subsurface profile of Figure 2. The concrete mix

Figure 8. Shoring for Pier 1 at the Northeast Corner of the Garage.

used for the piers and lintels attained a 72-hour compressive strength of 27.5 MPa before the drypack (1 part cement: 1 part sand) was installed.

The 75-mm layer of drypack between the bottom of the wall and top of a completed pier is shown in Figure 7. Throughout the underpinning work there was very little distress to the structure, as evidenced by negligible movement (less than 2 mm) in the crack gauges. Also, upon completion of the underpinning, continued monitoring of the crack gauges showed zero movement. With the successful completion of the foundation underpinning work, the cracks in the wall could finally be pointed neatly without them opening again.

CONCLUSIONS

This case history demonstrates contrary to common practice that the conventional approach pit underpinning can be effectively applied to repair damaged stone masonry foundation walls. It further shows that under certain conditions, this old and labor-intensive method can even be a cost-effective and preferred alternative over the latest innovations.

In a broader sense however, the lessons behind this paper are the importance of detailed characterization at every site and the fact that every project, no matter how small, has its own unique geotechnical challenges. On the Wynnewood job, structural complications led to a costly dowel and grade beam feature for the minipile underpinning, while detailed information on the subsurface soils economized the conventional underpinning and made it work. Therefore, only diligent attention to the subsurface conditions and their relation to the structure can generate the optimal geotechnical design.

REFERENCES

Berg, T. M., and Dodge, C. M. (1981). *Atlas of Preliminary Quadrangle Geologic Maps of Pennsylvania*. Forth Series, Pennsylvania Geological Survey, Harrisburg, PA

Furth, A. (1997). Personal Communication. Hayward-Baker Company, Odenton, MD.

Poulos, H. G. and Davis, E. H. (1980). *Pile Foundation Analysis and Design*, John Wiley & Sons, New York, N.Y.

Welsh, J. P. (1987). "Soil Improvement - A Ten Year Update," *Committee Report, Placement and Improvement of Soils*, ASCE Geotechnical Special Publication, No. 12, New York, N.Y.

White, E. W. (1975). "Underpinning." *Foundation Engineering Handbook, Chapter 22*. H. F. Winterkorn and H.-Y. Fang eds. Van Nostrand Reinhold, New York, N.Y.

SOIL NAIL WALLS IN RESIDUAL SOILS

Mike Khalil[1],
Mark Rhodes[2],
Jim Daly[3], Member, ASCE,
Jeanine Ferris[4]

ABSTRACT

Construction of the Sandy Springs Subway Station of the Metropolitan Atlanta Rapid Transit Authority's (MARTA) North Line required temporary excavation support. The Station excavation was to be approximately 245 meters (800 feet) in length and varied from 15 to 46 meters (50 to 150 feet) in width. Excavation depths of up to 20 meters (70 feet) were required. Traditional construction practices in Atlanta would dictate that a soldier pile and lagging system be employed. Right-of-way easement limitations prohibited the use of the lengthy tiebacks, as required for a soldier pile and lagging system. An internally braced excavation was not considered feasible because of the width and depth requirements of the excavation. MARTA's design team elected to utilize a soil nail wall for temporary excavation support of the approximately 8300 square meter (89,000 square foot) excavation face. To date, this wall represents the largest and deepest known soil nail wall constructed in the Southeastern United States and represents a successful application of the soil nail technique in the Piedmont Physiographic Province.

INTRODUCTION

In evaluating the feasibility of a soil nail wall, consideration was given to the suitability of the local soils to soil nailing. The ability of the excavated soil face to stand vertically and of the uncased borehole to remain open during nail installation greatly affects the economy of the system. The selection of the design nail pullout resistance is typically

1- Associate & Senior Project Manager, Golder Associates Inc., 3730 Chamblee Tucker Road, Atlanta, GA 30341, 770-496-1893, Mike_Khalil@golder.com.
2- Construction Technical Services Manager, Parsons Brinckerhoff Tudor – Turner Associates, 2424 Piedmont Road, NE, Suite 300, Atlanta, GA, 404-848-5498, MarkRhodes@mindspring.com.
3- Project Manager, Golder Associates Inc., 3730 Chamblee Tucker Road, Atlanta, GA 30341, 770-496-1893, Jim_Daly@golder.com.
4- Geotechnical Specialist, Golder Associates Inc., 3730 Chamblee Tucker Road, Atlanta, GA 30341 770-496-1893, Jeanine_Ferris@golder.com.

based on past experience with similar installation procedures and similar soil types. Since there is a paucity of published pullout data from field tests available for walls installed in the residual soils and partially weathered rock (PWR) of the Atlanta area, the available pullout resistance was estimated based on published correlations and site specific soil test data. A summary of data generated from pullout testing of the soil nails, a summary of wall performance, a discussion of the suitability of typical Piedmont soils for soil nailing, and a discussion of lessons learned will be presented.

PROJECT DESCRIPTION

In the late 1980s MARTA decided to extend their heavy rail line into the heavily developed areas of north Fulton and northwest DeKalb Counties. The initial 13 kilometer (eight-mile) extension into the area, which included two Stations, was completed in time for Atlanta's hosting of the 100th Olympiad in 1996. The next two Stations, Sandy Springs and North Springs Stations, will extend the line an additional 3.2 kilometers (two miles) northward. These two Stations are currently under construction, to be completed for Revenue service by December 2000.

Metropolitan Atlanta is located in Georgia's Piedmont physiographic province. The typical ground topography, which

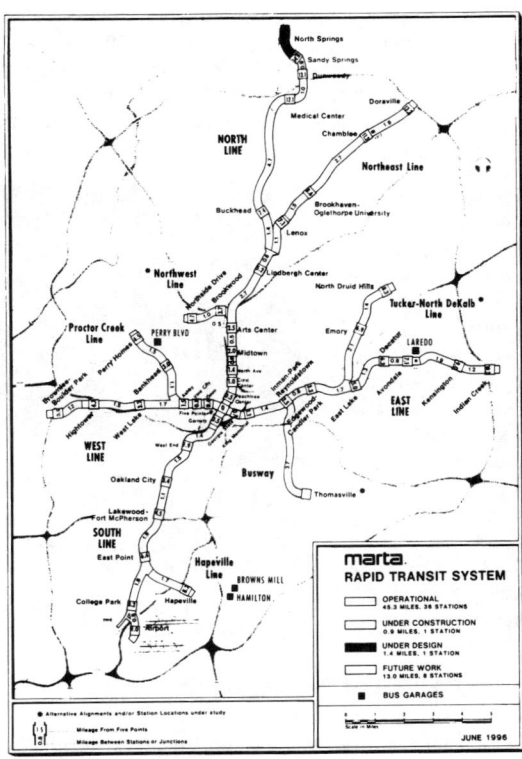

Figure 1 – MARTA System-Wide Map

includes areas of gently rolling hills with moderate elevation changes in relatively short horizontal distances, was a factor in the design of the Sandy Springs Station. The Station is located in a busy commercial zone, approximately one mile north of MARTA's new three level aerial Dunwoody Station near Perimeter Mall and adjacent to the intersection of Abernathy Road and Mount Vernon Highway, one of the busiest traffic areas in Atlanta. Due to the busy intersection and the topography change north of Dunwoody

Station, the Sandy Springs Station was constructed completely underground. Most of the Station will actually be under the traffic intersection.

Construction of the station required excavation depths of up to 20 meters (70 feet) below the existing ground surface. The Station construction schedule required that the excavation remain open for almost two years. Easement and property line restrictions and working space limitations necessitated vertical cuts and hence a substantial excavation support system. Several excavation support alternatives including tie-backs, structural bracing, and soil nailing were considered and reviewed by MARTA; Parsons Brinckerhoff Tudor - Turner Associates, the Authority's General Engineering Consultant and Station Designer; and Golder Associates, the Authority's Geotechnical Consultant. Bracing of the excavation, which measured up to 20 meters (70 feet) deep and 46 meters (150 feet) wide, was not practical because the density of bracing would interfere with construction of the Station The tieback and soil nail alternatives provide a completely open excavation for construction. The tieback lengths were found to be significantly greater than the soil nail lengths (approximately 23 meters (75-foot) maximum compared to 17 meters (55 foot) maximum). This additional length for tiebacks increased costs associated with acquisition of real estate for temporary subsurface easements. For this project, the soil nail option was found to be significantly less expensive, per square meter of installed excavation support, than tiebacks. For these reasons, the soil nail option was chosen.

Figure 2 – Completed soil nail wall in station with internally-braced excavation to north.

GEOLOGY OF THE PIEDMONT REGION

The Piedmont and Blue Ridge Physiographic Provinces form a band of crystalline rocks that extend from New Jersey southwest into Alabama (Sowers 1983). The Atlanta Area is located in the Piedmont Physiographic Province, which contains the oldest rock formations in the Southeastern United States. Rocks within the project area have been identified (McConnell and Costello, 1980, and Higgins and McConnell, 1978), as belonging to the Sandy Springs Geologic Group of the Northern Piedmont Geologic Province. This group is comprised of several formations containing gneisses, schists, amphibolites and quartzite. Weathering of the rock has typically resulted in a soil zone and a partially weathered rock (PWR) zone over the rock.

PWR is a regionally used term for residual material with Standard Penetration Test values of 100 or more, but which can be penetrated by soil drilling procedures. PWR typically retains the banding and texture of the parent rock, but generally breaks down during sampling to a very dense micaceous silty sand or a very hard micaceous sandy silt. PWR is commonly described as soft rock when exposed in excavations. In the Atlanta area, SPT values in PWR vary erratically, depending on the degree of weathering and composition of the parent rock. The erratic weathering of the parent rock leads to large variations in top of PWR elevation within short horizontal distances.

The upper and lower boundaries of the PWR zone may be poorly defined. In many cases the zone may be overlain by very dense residual soils similar in many respects to the PWR. Lenses of moderately hard or hard rock often exist within the PWR zone. These lenses may cause refusal to soil drilling methods prior to encountering relatively sound parent rock. In some cases, significant amounts of PWR were found to underlie such lenses.

SITE GEOLOGY

An extensive geologic and geotechnical assessment of the site was completed, prior to design of the wall, in two phases (Golder Associates, 1992 & 1996). Additionally, Golder Associates performed geologic mapping of exposed wall cuts as the excavation progressed. The purpose of the mapping was to assess whether subsurface conditions were consistent with interpretations made during the design phase. In general, the site stratigraphy, from top to bottom, consists of randomly located fill soils of varying quality, residual soils, PWR, and rock. Groundwater was encountered up to 11 meters (35 feet) above proposed excavation levels along portions of the Station area. Materials to be dewatered included residual soil (SM-ML), PWR, and rock. The volume of water inflow into the excavated area was estimated to be minimal.

The zone of fill materials was discontinuous and was likely placed as backfill over utility pipes, or to bring low areas up to the desired grades of existing parking lots, roads and landscaping. Residual soils were encountered below the natural ground surface and existing fills. These residual soils are the product of in-place weathering of the parent crystalline rock. The residual soils encountered varied from loose to very dense micaceous silty sands (SM) to stiff to very hard micaceous sandy silts (ML). The SPT values though somewhat erratic, generally tended to increase with depth and typically ranged from 20 to 50. PWR was encountered within a transition zone between the relatively sound parent rock and overlying residual soils. Within portions of the wall segment, significant thicknesses of either PWR or locally deep-weathered residual soils were found to underlie lenses of relatively fresh rock.

Mapping of the geologic exposure as the excavation progressed was used to augment the geologic and geotechnical data collected in the site investigation. This proved to be a critical part of the construction sequence. Orientation of foliation, characteristics of joint sets in lithologies observed in the excavation were mapped to assess the potential for unstable blocks to daylight in the excavation and potentially dislodge. In general, the mapping indicated no evidence of previous movements, such as slickensides, on the joints sets. Almost all of the joint sets were observed to persist less than 3 meters (10 feet), thus, the loads imposed on the completed wall by potentially unstable blocks were deemed inconsequential from a global stability point of view.

As with the transition from residual soil to PWR, the erratic weathering of the parent rock leads to large variations in the top of rock elevation. Since the soil nail wall was to terminate approximately at the top of rock, it was necessary to closely monitor and assess the rock conditions as work progressed along the wall. It was noted on this site that the typical depth of weathering varied in the different rock types at the site. Most notably, the deepest weathering is in the biotite schist layers and the least weathering is in the metagrawacke layers.

SOIL NAIL WALL SYSTEMS

Soil nail wall systems are a cost effective alternative to tieback and internally braced excavations for support of temporary and permanent excavations. Soil nailing is an in-situ soil reinforcement technique for retaining excavations and stabilizing slopes. Construction is completed in a "top down" fashion as the excavation proceeds. Non-pretensioned reinforcing members ("nails") are placed in the ground to improve the shear strength of the soil by limiting decompression and dilation, and thereby restraining its displacement. The reinforcing is developed through the nail-ground interaction as the ground deforms during and after construction. The nails work primarily in tension, and while some moment-resisting capacity within the nails also contributes to the improved stability of the cut, this is not usually relied upon in the design. A structural facing which is designed to act in concert with the nails is placed over the excavation face to maintain local stability between nails and to protect the ground from erosion. Typical elements of a soil nail system consist of the following:

Nails: Steel bars, placed in drilled holes, and grouted over their entire length. The grout serves as corrosion protection for the steel bars. For permanent applications or where conditions warrant, the steel bars are epoxy coated, thus a double corrosion protection system is employed. Typical hole diameters are four to eight inches. The result is a composite member that is capable of carrying tensile and shear stresses.

Facing: Typically, reinforced shotcrete facing is applied to the wall after the nails are installed. The shotcrete is placed in the appropriate thickness and with either wire mesh or larger tied reinforcing bars according to the load requirements. For permanent walls, cast in place or precast concrete facing may be used.

Drainage: Weepholes, typically consisting of 5 centimeter (two-inch) diameter PVC pipe extending through the shotcrete facing, or strips of drainage fabric extending vertically behind the facing and daylighting through the shotcrete at the bottom of the face, are used for drainage. These weepholes could be utilized in the permanent dewatering system if designed accordingly.

Sequencing: For excavation retaining applications, the nails are installed in a lift-by-lift sequence as the excavation progresses. A lift of soil is excavated, nails are drilled and grouted, reinforcing and drainage provisions are installed, and the face is shotcreted.

Right of easement for off-site anchors and access for construction equipment are some of the considerations that should be taken into account when doing a feasibility study for using soil nailing techniques. Dewatering must be accomplished prior to excavation and nailing. Effective dewatering prior to excavation is critical to the success of soil nailing to maintain the stability of the wall excavation face before support is in place, to obtain good contact between the soil/shotcrete interface and to keep soil nail boreholes open during drilling, nail installation and grouting operations.

SANDY SPRINGS STATION WALL DESIGN ASSUMPTIONS

The design of the soil nail wall for temporary excavation support was completed with full consideration of the easement, access, and utility constraints described previously in the Project Description. The height of the vertical cut to be supported varied from 11 to 20 meters (35 to 70 feet) and the length totaled 245 meters (800 feet).

The Design Engineer's intent was to install a soil nail system where soil, both fill and residuum, and PWR were encountered in the excavation. Rock anchors were designated for use in rock; thus a field determination of "top-of-rock" was required. Based on past experience with similar soils, it was assumed that the eight-inch diameter boreholes for nail installation and the unsupported vertical excavations of up to 2 meters (seven feet) would remain largely stable for the duration required for construction, approximately two days for boreholes and one week for excavations. Based on the geotechnical properties of the soils as determined in the geotechnical investigation, the following engineering properties were used for the analysis:

TABLE 1: ENGINEERING PROPERTIES FOR INTERNAL ANALYSIS

Soil Type	Typical Standard Penetration Resistance N-Values (Bpf)[6]	Internal Friction Angle (Deg.)	Internal Cohesion kPa (Psf)	Unit Weight kg/m^3 (Pcf)	Design Ultimate Bond Stress kPa (Psi)	Ultimate Adhesion Value Recommended By FHWA kPa (Psi)
Residual Silts and Clays (ML, SM)	20 - 50	28	19.2 (400)	1922 (120)	71.0 (10.3)	20.7-31.0[1] (3.0-4.5) 48.3-75.8[2] (7.0-11.0)
Partially Weathered Rock	> 100	40	7.2 (150)	2082 (130)	142.0 (20.6)	200-300[3] (29.0-43.5) 100-150[4] (14.5-21.5) 100-175[5] (14.5-25.5)

Notes:
1 Non-plastic silt
2 Medium dense sand and silty sand/sandy silt
3 Weathered Sandstone
4 Weathered Shale
5 Weathered Schist
6 Bpf – blows per foot, 1 foot = 0.3048 meters

The design was completed using a static ultimate analysis. The following safety factors, which were consistent with the recommendations of the FHWA soil nail manual (Byrne, et. al., 1996) for temporary excavation support were employed.

TABLE 2: SAFETY FACTORS FOR INTERNAL ANALYSIS

Analysis	Soil Strength	Adhesion	Nail Head	Tendon Capacity
Static	1.35	2.0	1.5	1.82

The site investigation indicated that the groundwater table was at levels up to 11 meters (35 feet) above the base of the wall. Therefore, an active dewatering system, to be operated during construction, was included as part of the design. A passive drainage system consisting of 40 centimeters (16-inch) strips of geosynthetic drainage composite running the full height of the wall centered vertically between nail rows was also included in the design to collect water from behind the wall. The passive geosynthetic drainage layer feeds into an undertrack drainage system and is overlain by an impermeable barrier designed to prevent seepage of groundwater into the completed Station.

CONSTRUCTION RELATED ISSUES

Inspection and Monitoring Program

Quality Assurance/Quality Control monitoring is essential to a successful completion of a soil nail wall in order to verify that it is being built in accordance with the plans and specifications. Past experiences indicate that the overall performance of a soil nail wall is highly dependent on the quality of the installation practices employed by the Contractor. Most reported failures are related to poor construction practices or changed conditions, as opposed to improper design. Thus careful field monitoring is essential.

Inspection and monitoring during construction of the soil nail wall were completed in general compliance with FHWA guidelines (Porterfield et. al., 1994). The Contractor's excavation sequencing, soil nail installation procedures, temporary shotcrete facing construction procedures, and rock anchor installation procedures were monitored and documented. In addition to construction monitoring, documentation of soil nail quality assurance pullout testing and the documentation of the geologic conditions exposed in the excavation were completed. The installation of observation wells and ground deflection monitoring (inclinometer) equipment was monitored and the data evaluated. Grout and shotcrete sampling and testing were completed on representative samples according to the project specification.

The different degrees of weathering of the parent rock in Piedmont geology typically lead to variations in top of PWR and rock over short horizontal distances. Therefore, careful monitoring of subsurface conditions was performed per FHWA recommendations (Porterfield et. al., 1994). Geologic mapping was conducted as excavation progressed.

The objective of the mapping was to document and evaluate subsurface conditions that might impact:

- the short term stability of unsupported vertical cuts in soil, PWR and bedrock;
- the effectiveness of the groundwater extraction system;
- the depth to which the soil nail wall should be constructed;
- the overall performance of the soil-nail wall; and
- the spacing and location of rock bolts required to locally stabilize bedrock faces.

The performance of the soil nail wall was also monitored at the top of the excavation by optical survey. Points were spaced at approximately 15 meter (50 feet) centers along the wall alignments.

Soil Nail/Utility Interference

As with most urban construction sites, there was a myriad of utility lines in the area of the Sandy Springs Station. These included several electrical mains, which feed two adjacent 20 plus story buildings as well as United Parcel Services' headquarters. There are also several natural gas lines, cable TV lines and fiber optic communication lines in the Station area. MARTA had to spend a considerable amount of money rerouting, supporting, and protecting these lines while building the Station. Some lines could not be rerouted away from the excavation and had to be supported to span the 40 meter (130 feet) wide excavation.

The utility situation made it very difficult to place the upper two to three rows of soil nails at normal design lengths and declinations. Initially, a clear zone of three feet above and below each utility was used as a constraint for design. However, during installation, two electrical power feed ductbanks were damaged by the drilling of the soil nails. These incidents proved to be costly and were potentially dangerous. These incidents necessitated more stringent monitoring of the soil nail drilling and some changes to the design. A heightened inspection effort was utilized to verify the location and inclination of each drilled hole in the upper two to three rows of soil nail. This was necessary in order to avoid damaging any additional utilities.

Cut Face Instability

One of the principal requirements for economical installation of a soil nail wall is that the unsupported cut face stand vertically long enough for nail installation and shotcreting of the face to be completed. Soils and PWR typical of the Piedmont often have varying weathering patterns and can retain relict structure of the parent rock. Stiffer residual soils and PWR on this site tended to behave more like fractured rock than soil; that is to say that local face stability tended to be controlled by sliding of wedges or blocks into the excavation along unfavorably oriented relief joints or foliation. During construction, there were several instances where 0.75 to 1.5 cubic meter (1 to 2 cubic yard) blocks of soil or PWR broke away from the face, resulting in an unsafe condition. The face instability resulted in additional expenses to the Contractor who was required to use shotcrete to bring the face back to original profile.

Figure 6 – Unsupported vertical excavation face prepared for shotcreting

In areas where geologic mapping or the performance of the previous lift indicated potential instability of the cut face, drilling was performed through a temporary soil berm in front of the final soil face. Where mapping and adjacent cut face performance indicated a great likelihood of face instability, (which typically occurred along relief jointing or foliation) the height of the excavation was reduced from 3 meters (six feet) to approximately 1 meter (three feet) per lift. The reduction of excavation height resulted in the requirement of two stages of excavation and shotcreting for one lift of nails, and an increase in costs to the Contractor.

Figure 7 – Wedge failure of vertical excavation face.

Soil Nail Testing

Soil nail pullout testing was performed to verify that the nails could carry the design loads without experiencing pullout or excessive creep under sustained loadings. The testing also verifies that the nail capacity provides the factor of safety assumed for design. In addition, testing was used to verify the adequacy of the Contractor's drilling, installation, and grouting operations prior to and during construction of production soil nails.

Verification testing was performed on 15 sacrificial nails, both at the start of the project and as the work progressed. Verification test nails were loaded at the least to the assumed ultimate nail load or 200% of the design load to confirm the adhesion value used in the design of the wall.

Proof testing was performed on approximately 5 percent of the production nails installed. Proof test nails were loaded to 130% of design load to verify that the production nails can safely withstand the design loads. Of the 151 proof tests performed on the project, four failed by pullout during testing. Two of the failed tests were in an area of highly micaceous residual soils. Additional nails were added in these areas to address the low adhesion values. The remaining two failing tests were performed in an area of existing, uncontrolled fill. The affected segment of the wall was redesigned using the lower adhesion values.

The design assumption of an ultimate bond stress of 71 kPa (10.3 psi) for the residual soils and fill and 142 kPa (20.6 psi) for PWR was selected based on the FHWA recommended values. The testing performed on the project confirmed the ultimate bond stress for the residual soils, with the exception of the highly micaceous soils, to be at least 71 kPa (10.3 psi). The ultimate bond stress for PWR was greater than 142 kPa (20.6 psi).

WALL PERFORMANCE

Lateral deflection of the soil nail wall was monitored by 9 inclinometers installed behind the wall (3 of which were damaged or destroyed during soil nail drilling) and by optical survey (points every 50 feet along the length of the wall). Additionally, the ground surface behind the wall was visually inspected on a routine basis for evidence of cracks or other signs of disturbance. Visual inspections also included checks for cracks, deterioration, or groundwater leakage in the shotcrete facing.

Published data (Byrne, et. al., 1996) indicates that soil nail walls installed in other geologic conditions have typically experienced lateral movements at top of the wall in the range of 0.1%H or less for weathered rocks and very competent soils (e.g. glacial tills), to 0.2%H for granular soils, and up to 0.4%H for fine-grained clay type soils, where H is wall height. The measured movements of the Sandy Springs Station soil nail wall range from 0.1% to 0.4% of wall height. Thus the majority of the movements are within the published expected ranges. However, a section, of the right wall, approximately 46 meters (150 feet) in length, experienced movements which exceeded the expected values (about

0.6% of wall height). The larger than anticipated movements appear to reflect the presence of highly micaceous materials within the weathered rock and a greater degree of weathering within this zone. These deflections do not appear indicative of a global stability problem or yielding of the wall. Observations of wall performance did not indicate evidence of distress in the shotcrete facing; however, tension cracks of up to 2 inches in width were noted to have developed behind the wall sections which experienced the greatest deflections. The tension cracks were sealed by the Contractor to prevent infiltration of surface water.

In the zone of maximum movement, rates of movements during the final stages of excavation were in the range of 0.2 to 0.5 inches per week. The movements essentially ceased upon completion of construction of the wall.

CONCLUSIONS:

Soil nailing is well suited for excavation support in Piedmont soils. Recommended ultimate bond stresses (Byrne, et. al., 1996) for low pressure grouted nails, which are based on experience with similar soils in other areas, appear appropriate for use in Piedmont soils. Geologic mapping during excavation proved to be especially useful. Because of the variability in weathering, borehole data alone was insufficient in establishing the location of top of sound rock (thus termination of the wall). The location and orientation of relict jointing and foliation in soil and PWR may affect the stability of excavation faces. The wall section constructed in zones of highly micaceous soils deflected more than expected.

Figure 8 – Completed soil nail wall.

Use of longer nails or tighter spacings may be necessary where zones of highly micaceous soils are present. Due to the erratic weathering profiles in residual soils, it is recommended that soil nail walls be designed for soil over the full excavation height. If sound rock is verified in the field, the design should be adjusted or the wall terminated as determined by the design engineer. Lastly, cooperation between the owner, designer, construction manager, contractors, and filed inspection personnel is essential in soil nailing. Changed conditions in the field may require design modifications be completed on a fast-track schedule.

ACKNOWLEDGMENTS

The general engineering consultant for MARTA was Parsons Brinckerhoff Tudor/Turner Associates. The geotechnical consultant for MARTA was Golder Associates Inc. The general contractor for the project was Archer-Western Contractors, Ltd. The soil nailing contractor for the project was Malcolm Drilling Co., Inc. The design engineer for the soil nail wall was Ground Enhancement Corporation.

The authors would like to express their appreciation to Chris Wolschlag for conducting an internal review, and Betty Triebert for compiling the paper and graphics.

REFERENCES

Byrne, R. J., Cotton, D.M., Porterfield, J.A, Wolschlag, C., and Ueblacker, G., 1996, "Manual for Design & Construction Monitoring of Soil Nail Walls", Federal Highway Administration, Washington, D.C., FWHA-SA-96-069.

Golder Associates Inc., 1992, "Report on Geotechnical Exploration Final Design Phase F410, Sandy Springs Station", Metropolitan Atlanta Rapid Transit System.

Golder Associates Inc., 1996, "Compilation of Letter Reports Associated with the F410/F420 Project", Sandy Springs Station, Metropolitan Atlanta Rapid Transit System.

Higgins,, M.W., and McConnell, K.I., 1978, "The Sandy Springs Group and Related Rocks in the Georgia Piedmont-Nomenclature and Stratigraphy": Georgia Geological Survey Bulletin.

McConnell, K.I., and Costello, J.O., 1980, "Guide to Geology Along a Traverse Through the Blue Ridge and Piedmont Provinces of North Georgia": Excursions in Southeastern Geology, Volume 1, the American Geologic Institute.

Porterfield, J.A., Cotton, D.M. and Byrne, R. J., 1994, Soil Nailing Field Inspectors Manual, Federal Highway Administration, Washington, D.C., FWHA-SA-93-068.

Sowers, G.F. and Richardson, T.L., , 1983 "Residual Soils of the Piedmont and Blue Ridge", Transportation Research Record 919, P.10-16.

MONITORING THE PERFORMANCE OF A SOIL NAILED WALL

BY NASSEF SOLIMAN[1], F. ASCE and KWANG RO[2]

Abstract

This paper presents data from measurement of an instrumented soil nailed wall (SNW) constructed in granular soil in Elmwood Park, New Jersey. The first to be constructed in New Jersey, it was designed as a permanent structure to retain a 15 foot (4.57 m) high excavation supporting a 20 foot (6 m) high embankment roadway. The wall was required for the construction of a 22 foot (6.7 m) wide ramp, to support the excavation under an existing abutment within the granular backfill. The main concern was the lateral movement of the abutment due to removal of the soil in front of it, which was estimated to be 1.75 inches (44 mm). Accordingly, the SNW and the abutment were instrumented to monitor movement during and after construction.

The SNW was constructed using the top down method and was completed with a series of 5 x 80-foot (1.5 x 25 m) excavated lifts. When each lift was excavated, drainage elements were placed at 7-foot (2.1 m) spacing and held in place by a wire mesh covering the excavated area. A 3- to 4-inch (75 to 100 mm) shotcrete layer was applied over the wire mesh and left to cure for 24 hours. Holes were drilled at an angle of 15 degrees from the horizontal. The soil nails (tendons) were inserted in the holes and grouted to the surface. A plate and nut were then placed on each 1-inch (25 mm) tendon and the nut was tightened. Upon completion of the nail installations, they were extended to be embedded in the final 12-inch (300 mm) concrete facing.

An instrumentation program consisting of optical survey points, tiltmeters, and strain gages was implemented and four types of measurements were made. Six optical survey points provided the horizontal and vertical movement of the abutment. Eight digital tiltmeters provided data regarding the tilting of the abutment and the soil nail wall, and 36 strain gages provided stress/strain data along the tendons. Data collected during construction and for one year after construction indicated that there was practically no movement of the abutment or the wall. Measured loads carried by the soil nails varied from 5 to 8 kips (22 to 35.5 kN) for the upper row of nails and 1 to 3.3 kips (4.5 to 14.7 kN) for the lower row of nails.

---

[1] Manager, Geotechnical Department ~ Parsons Brinckerhoff, Princeton, New Jersey (Phone: 609 734 7051)
[2] Geotechnical Engineer ~ Parsons Brinckerhoff, Princeton, New Jersey (Phone: 609 734 7059)

Introduction

A new ramp connecting the southbound Garden State Parkway to eastbound Interstate Route 80 in Elmwood Park, New Jersey, was proposed to improve traffic flow from the Parkway to Route 80 East and reduce congestion on local roads. The proposed ramp was to be located between the abutment and the first pier of the Route 80 Bridge over the Parkway. This location required the excavation of the fill in front of the abutment, which is supported on steel H piles driven to bedrock, to provide enough width and grade for the proposed ramp.

A retaining structure was required to retain the soil underneath the abutment during construction and during the life of the ramp. The originally designed wall consisted of a soldier pile and lagging system to provide temporary support during excavation and a reinforced concrete wall as the permanent structure. However, due to the shallow overhead clearance available for construction, the construction plans allowed the contractor to bid on an alternative design utilizing the Soil Nail technique. The contractor was to design the Soil Nail Wall in accordance with performance specifications. The winning bidder opted to use the Soil Nail Method because the contractor felt that it was more economical and practical for the site condition.

The designed Soil Nail Wall was constructed using the top-down excavation method, done in stages, each 5 feet (1.5 m) in height. The total number of tendons was 107, installed in two rows. The spacing of the tendons was designed to be 3'-8" (1.12 m) horizontally and 5 feet (1.5 m) vertically, with each tendon to be located at the center between two existing steel H piles.

A structural analysis indicated that the bridge abutment will experience a horizontal movement of 1.75 inches (44 mm) after excavation of the soil in front of it and without the construction of the soil nailed wall. As a part of the construction of the soil nail wall, an instrumentation program was developed to monitor the existing abutment movement and the performance of the Soil Nail Wall. The program consisted of optical survey, tiltmeters, and strain gages.

Site Condition

Subsurface conditions in the area behind, in front of, and beneath the existing abutment consisted of embankment fill material. The fill consisted mainly of medium dense fine to coarse sand, trace silt, and trace gravel, the amount of which increased with depth. The slope in front of the abutment was covered with a concrete apron. The construction was planned to be done while maintaining the heavy traffic on Route I-80 above the wall. Access for the construction of the wall did not present a problem.

Design of the Soil Nail Wall

The project was bid as a design/build contract in accordance with a performance specification. The design was reviewed by Parsons Brinckerhoff, Inc. There are different methods for the design of Soil Nail Walls, including: the Davis method (Ref. 5 & 6); the

French method (Ref. 8 & 9); and the German method (Ref. 3 & 4). However, the design was undertaken and checked using the method developed by CALTRANS which uses a 2 or 3 part wedge analysis for determining the minimum factor of safety in a one or two layer soil system with inter slice forces included. CALTRANS developed a computer program, SNAIL, which was used to determine the tendon length.

The boundary condition for the SNW was not ideal since the abutment was supported on steel H piles. Several questions were raised as to how much of the load would be carried by the tendon and whether there would be any arching effect between the piles. Because of the long term effect of soil relaxation and the importance of the wall to the stability of the abutment, the Soil Nail Wall was designed to carry both the soil behind the abutment and the surcharge due to vehicular traffic, Figure 1.

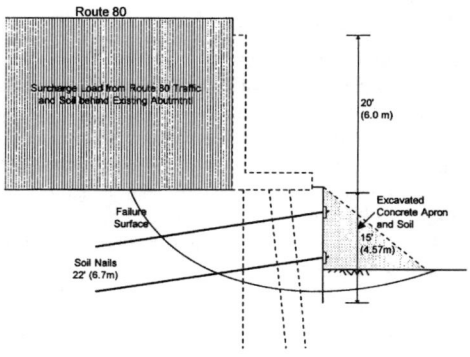

Figure 1 - Typical Cross Section through the soil nailed wall

The analysis indicated that two rows of nails would be required; each nail would carry a maximum load of 8 kips (35.5 kN). The total length of the nails was estimated to be 22 feet (6.7 m), with an assumed bonded length of 8 feet (2.4m) to carry the estimated load of 8 kips (35.5 kN). This is based on an allowable capacity of 1k/foot (4.5 kN/0.3 m). During construction, the entire length of the nail was typically grouted.

Sequence of Construction of the Soil Nail Wall

The soil nail retaining wall construction involved the following steps:

Step 1. Excavation of a small area 5 feet in height and a maximum of 80 feet in width (1.5 x 25 m).

Step 2. Placement of geocomposite drain: Wall drainage consisted of prefabricated vertical geocomposite drainage strips installed from the top to the bottom as the excavation proceeded downward. The drainage strips were approximately 24 inches wide (60 mm), and were centered between the vertical nail columns with a maximum 7-foot (2.1 m) center to center spacing. The strips were connected to the foundation drain at the wall base.

Step 3. Installation of a shotcrete layer: A layer of 3- to 4-inch (75 to 100 mm) thick shotcrete was used as a temporary support for the excavation which protected the soil from sloughing and facilitated the installation of the nails. The shotcrete facing was placed to cover the soil immediately after completing the excavation. The cement grout was applied under pressure and it was reinforced with a single layer of welded wire mesh. The shotcrete was left to cure for 24 hours prior to installation of the tendons, Photo 1.

Photo 1 - Placement of Shotcrete

Step 4. Installation of the Soil Nails: A hole was drilled for nail installation, using a 4.5 inch (114 mm) OD steel casing to the predetermined depth. The steel casing was required to avoid collapsing of the hole due to the nature of the sandy material, Photo 2.

Step 5. Cleaning the casing from soil cuttings using compressed air pumped from the bottom of the hole.

Photo 2 - Installation of the Lower Row of Nails

Photo 3 - Preparation of Tendons with Strain Gages

Step 6. Installing and grouting tendons: A galvanized deformed steel reinforcing bar (Grade 60) was inserted into the hole. In order to keep the bar in the center of the hole, each bar was equipped with 3 centralizers (5 for tendons with strain gages), made of expanded PVC material that is not detrimental to the tendon steel. The centralizers were placed at equal spacing to provide a minimum specified grout cover, Photo 3. The grout was pumped through a tremie grout pipe inserted to the bottom of the hole. The grout mix consisted of Portland cement with a water/cement ratio of 55 gal./94lb. bag (208L/42.65kg), and strength of 4000 psi (27.58MP). The steel casing was withdrawn and the hole was re-grouted as needed. Bearing plates and nuts were installed, and strain gages were installed on certain tendons selected for instrumentation, Photo 4.

Step 7. Repeating the process of excavation and installing nails to final grade, Photo 5.

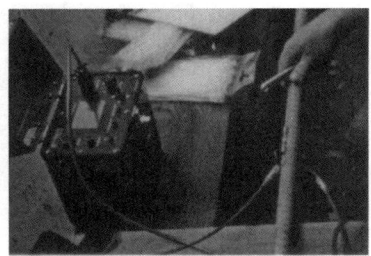

Photo 4 - Calibration of Strain Gages

Photo 5 - Soil Nail Wall prior to Construction of Concrete Wall

Placing final vertical reinforced concrete facing (permanent wall). A structural wall facing was installed for soil confinement, protection of the retained soil against weathering and erosion, and resistance to lateral earth pressures. Upon completion of the nail installation, the tendons were extended to be embedded in the final wall facing. The tendons were provided with bearing plates attached to their heads to transfer the pressures to the nail. The bearing plates were attached to the nails with a hex nut typically turned with a wrench to a tight position. The tendon head/bearing plate was structurally connected to the permanent wall. The wall consisted of 12 inch (300 mm) thick Class B reinforced concrete placed against the shotcrete.

Field Load Testing

Four non-production, "sacrificial" tendons were installed for load testing to determine the ultimate soil/grout pull-out capacity and the tendon capacity at which excessive creep occurred.

The load tests were performed by loading the tendon until pullout failure occurred in accordance with the recommendations in FHWA's guidelines (Ref. 10) for sequence of loading. The soil nails were constructed with a temporary unbonded length that was backfilled with grout after testing was completed, except tendon test No. 1, which was grouted for the entire length.

Creep tests were performed as

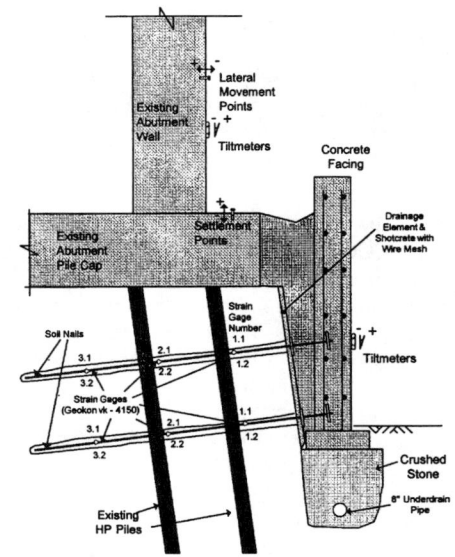

Figure 2 - Schematic Diagram of the Instrumentation Setup and Wall Arrangement

part of the ultimate load test. Creep testing was conducted at a constant load of 200 percent of the design capacity, with movement recorded at specified time intervals. Test acceptance criteria were:
1) Total movement at the maximum test load (200 percent of design load) must exceed 80 percent of the theoretical elastic elongation of the unbonded length.
2) Creep movement between the 0.5 and 5-minute readings, at the specified test load must be less than 0.08 inch.
3) Pullout failure must occur at more than the 200 percent of the design load.

The acceptable movement was defined as follows:
measured movement in inches AL should be larger than

$$(>)\ 0.80 \times P \times UL \times 12/(A \times E)$$

where:

$P =$ Maximum applied test load (kips)
$UL =$ Length from the back of the reference plate to the top of the bonded length, i.e., the unbonded length (feet)
$A =$ Cross-sectional area of the steel (square inches)
$E =$ Young's modulus of steel (typically 29,000 ksi)

Test load No.1 was stopped at 40 Kips (178 kN) with no elastic movement observed because the entire tendon was grouted. Test load No. 2 failed at 23 kips (102 kN). Tests No. 3 and No. 4 did not reach failure criteria up to 56 kips (249 kN) (maximum capacity of the hydraulic jack). This may be due to enlargement of the grout hole, resulting in higher capacity. The design load was 8 kips (35.58 kN) per nail, based on the assumption that the allowable capacity will be 1 kip per linear foot (4.5 kN/0.3 m), which is the minimum allowable capacity per foot.

The following is a summary of the test data.

Test No.	Theoretical Elastic Elongation (inches)	Bonded Length/ Ultimate Capacity (feet / kips)	Movement at 200 % Load (inches)	Ultimate Capacity (kips)	Measured Unit Skin Friction (kips/foot)
1	-	20/40	0.22	>40	> 2
2	0.13	9/18	0.17	Failed at 23	2.56
3	0.25	10/20	0.35	>56	> 5.6
4	0.25	10/20	0.35	>56	> 5.6

1 inch : 25.4 mm
1 kip : 4.448 kN

Test load verses movement results for Tendons 1, 2, and 3 are presented in Figure 3. Test results of test 4 are typical to test 3.

Instrumentation and Monitoring Program

The objective of the instrumentation program was to monitor the performance of both the existing bridge abutment and the new Soil Nail Wall due to the excavation of the soil that existed in front of the abutment foundation. The program consisted of the following:

1) Optical survey to monitor the horizontal and vertical movement of the abutment.

2) Tiltmeters to monitor tilting of the abutment and the wall.

3) Strain gages to monitor the stress/strain in six tendons and to measure the transfer of load from the soil to the tendons.

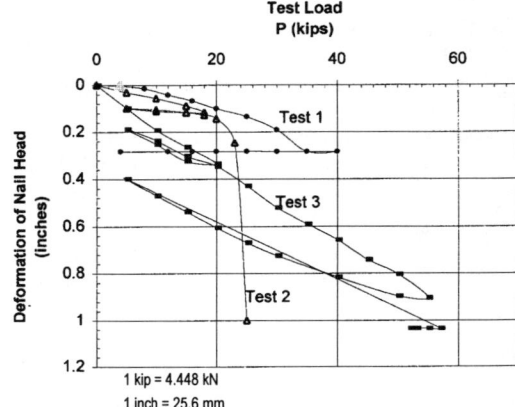

Figure 3 - Load Test Data

The data collection was done once a week for three months, then once a month for twelve months.

Instrument readings for the abutment (optical survey, tiltmeter) started three months before the installation of the soil nails. Strain gage readings were collected right after mounting them on the tendons; another set of readings was taken right after insertion of the tendons in the grouted hole, and one set of readings after the grout had set. The third set of readings was used as the baseline reading as it was considered the zero reading.

Optical Survey Points: To monitor the lateral movement of the face of the existing abutment and to provide additional data to those of the tiltmeters, three lateral displacement reference points were located on the face of the abutment with drilled steel bolts, survey points were located along a line of sight approximately parallel to and as near as practical to the face of the abutment to be monitored.

Six settlement reference points were marked on the surface of the I-80 roadway and top of the abutment footing to monitor the settlement due to Soil Nail Wall construction.

Lateral displacement point offset distances were monitored with Theodolite equipment and a standard Vernia-equipped Philadelphia rod. The offsets were read to the nearest 0.001 foot (0.3 mm). Optical survey readings are plotted in Figure 4. The data indicate that the abutment experienced an outward movement of 0.12 to 0.2 inch (3 to 5 mm).

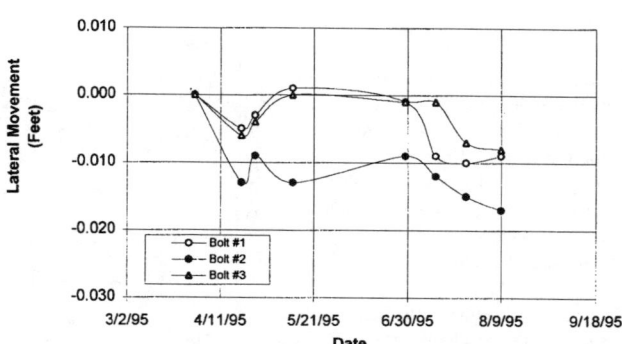

Figure 4 - Lateral Movement of Existing Abutment Optical Survey

1 foot = 0.3048 m

Tiltmeters: Five uniaxial tiltmeters were installed and monitored to get an accurate record of vertical angular movement of the bridge abutment. Three tiltmeters were installed on the new Soil Nail Wall facing after it was constructed. The vertical angular movements are relative measurements, in that all readings are referenced to the base line reading rather than to the absolute vertical. Tilts smaller than 1 microinch per inch (one microradian) are easily readable with this tiltmeter. All readings for each tiltmeter are plotted in Figures 5 and 6. The data indicate that the abutment experienced an outward movement between 0.0 and 0.2 inch (one degree correspond to 0.2 inch per 10 feet). The data for the Soil Nail wall indicate that the wall moved inward about 0.13 inch (3.3 mm)

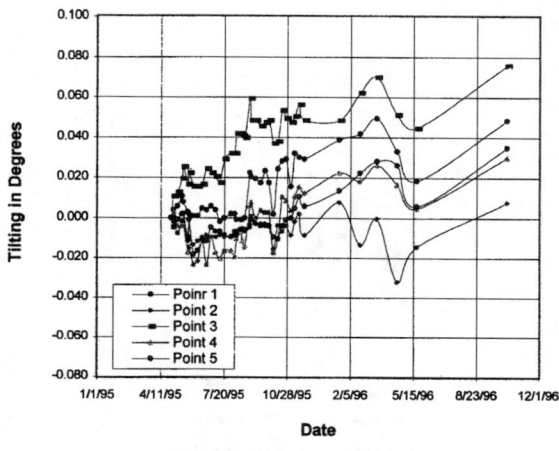

Figure 5 - Existing Abutment - Tiltmeter Data

Strain Gages: A total of six tendons were instrumented with strain gages to measure strain induced by potential ground deformations or wall movement. The selected strain gages were weldable vibrating wire strain gages . Three of six instrumented tendons were installed along the upper row of nails, and three along the lower row.

Each of the instrumented tendons was fitted with a total of six strain gages installed in three pairs. Each pair was located near the quarter, half, and three-quarter points along the bar length. Each pair of strain gage elements was spot welded according to the manufacturer's recommendations at diagonally opposite positions at the top and bottom of the bar. An additional bar centralizer was attached over the strain gage elements and sensors for each pair were added for extra protection during bar insertion into the hole.

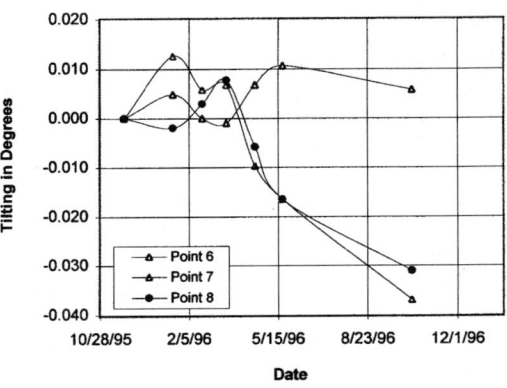

Figure 6 - Soil Nailed Wall - Tiltmeter Data

The strain gages measured the strains along the extreme fibers of the rod. In order to convert the strains into stresses, the following stress/strain formula was used:

Load = Strain x Young's Modulus x Cross Section Area of the Rod

This relation is true if the strain throughout the tendon cross section is equal. But, if there is any bending in the rod, then the strains at the bottom of the rod will be different from the top. To avoid getting bending effect in the rod, an additional centralizer was mounted on the strain gage. In addition, the use of the average reading from the top and the bottom strain gage at each location was designed to offset the bending effect, if any.

Monitoring started right after the tendons were grouted to establish base line information for assessment of the impact of construction on the existing abutment structure and the soil-nailed mass beneath it. Tendon strain gage measurements provided information concerning ground deformations within the soil-nailed soil mass during construction, including information concerning the effectiveness of the tendons in controlling ground deformations. In particular, the variation of axial strain with distance behind the excavated face, from nail to nail, and time during and after excavation were examined to assess the performance of the tendons and the impacts of construction

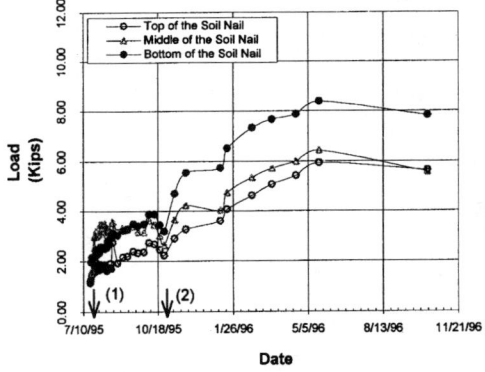

Notes:
1. Bottom row of the soil nails were installed on July 24, 1995.
2. Permanent concrete facing was installed on November 1, 1995.
3. Assumed that soil nail loads were zero before the soil nail grouted.

Figure 7 - Average Load verses Time - The Soil Nail T-38

operations. Data collected from top and bottom nail No. 38 (T38 and B38) are presented in Figures 7 to 10. Figure 11 presents all data from the six instrumented tendons.

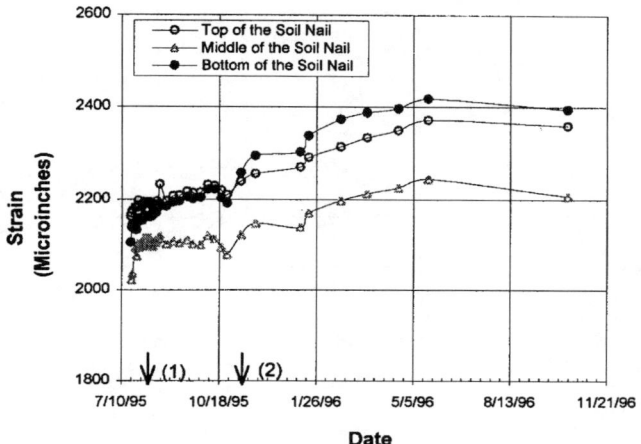

Notes:
1. Bottom row of the soil nails were installed on July 24, 1995.
2. Permanent concrete facing was installed on November 1, 1995.
3. Assumed that soil nail loads were zero before the soil nail grouted.

Figure 8 - Average Strains vs. Time - The Soil Nail T-38

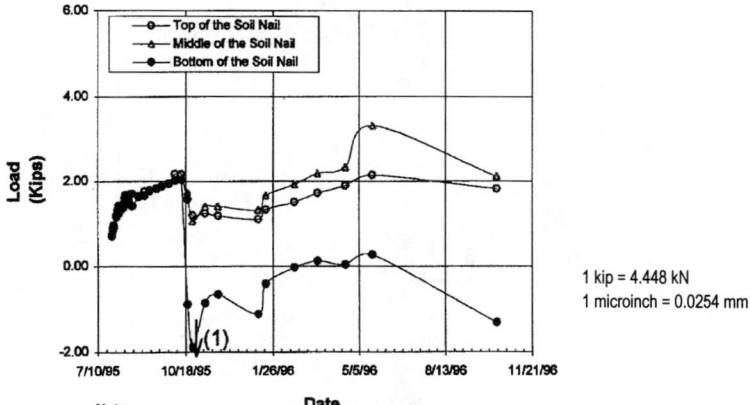

1 kip = 4.448 kN
1 microinch = 0.0254 mm

Notes:
1. Permanent concrete facing was installed November 1, 1995.
2. One Strain Gage at the bottom of the soil nail was not functioning from July 28 to October 13, 1995

Figure 9 - Average Load vs. Time - The Soil Nail B-38

236 SOIL IMPROVEMENT FOR BIG DIGS

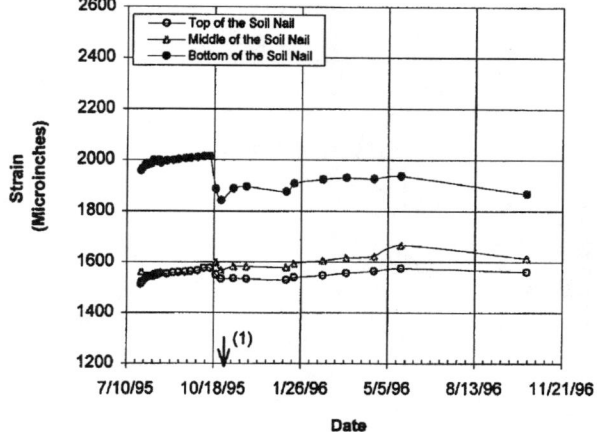

Notes:
1. Permanent concrete facing was installed on November 1, 1995.
2. One Strain Gage at the bottom of the soil nail was not functioning from July 28 to October 13, 1995.

Figure 10 - Average Strains vs. Time - The Soil Nail B-38

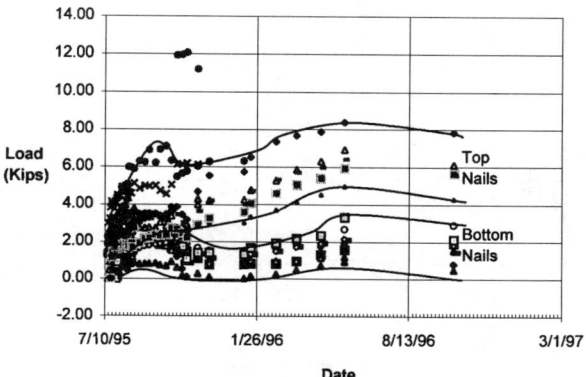

Figure 11 - Range of Loads - Top and Bottom Soil Nails

1 kip = 4.448 kN
1 microinch = 0.0254 mm

DISCUSSION OF THE RESULTS AND CONCLUSION

Load Tests: The load testing program was successful and indicated that the tendons have the required ultimate pull-out capacity of 2 kips (8.89 kN) per linear foot. The creep tests met the design criteria (deflection at 200 percent design capacity at a constant load). Based on the test data, the tendons have the required capacity to carry the designed load. Results indicate that probably enlargement of the grouted hole resulted in a much higher capacity than predicted.

Horizontal Movements and Settlement: The optical survey reading and tiltmeter did not show any significant tilting, rotating, or settlement of the abutment or SNW. Tiltmeter readings ranged from -0.03 to +0.08 degrees on the existing abutment and -0.035 to + 0.015 degrees on SNW, while the optical survey indicated a lateral movement of -0.018 feet (-5.5 mm) and settlement between -0.001 to +0.008 feet (-0.3 to 2.4 mm) on the existing abutment. The optical survey and tiltmeter of the abutment indicate that the wall tilted outward. Some variations in the readings can be attributed to temperature changes. The remainder are within the tolerances of the equipment.

Stress and Strain in the Nails: Most of the strain gages functioned well until the end of the monitoring program except at location T-2. During construction of the concrete facing, the Contractor cut the wire to provide additional length to move the readout system to the top of the wall. Probably the splices were not complete and the T-2 strain gages started malfunctioning, then ceased to operate.

For all locations, the first reading was taken right after the tendon was grouted and was used as the base line. This assumption is correct if the strain gages were measuring zero at the time the hole was being grouted. But since these gages are very sensitive, any small movement or bending in the steel rod will create internal strain in the gage, which corresponds to an initial value more or less than the assumed zero. As a result, some of the gages at the bottom of the tendon have a higher reading than the one at the center of the tendon. This could be the result that the entire tendon was grouted, not the bonded length only.

It was observed that when excavation for the bottom section was completed, the upper soil nail exhibited a sudden increase in loading. Also, when the concrete wall was added, the soil nails exhibited a sudden drop in strains, which increased and stabilized with time.

Generally, the movement appears to have stabilized with no practical increase in loading between May and October of 1996. However, most of the tendons indicate a slight decrease in loading. This is probably due to creep between the tendon and the grout and steel relaxation. The upper nails are carrying 6 to 8 kips (design load 8 kips) and the lower nails are carrying about 2 kips. This loading appears to be in agreement with the geometry of the walls, which indicate that the upper nails carry more than the lower ones.

Monitoring of the strain/load in the instrumented tendons indicates that they are functioning as designed. The measured loads may be on the low side due to the fact that the existing bridge abutment is carrying some of the load.

Overall, the soil nail wall is a viable and economical solution for this project, and could be implemented on other projects with similar conditions.

Acknowledgment

The wall was constructed by Conduit and Foundation Construction Company, Pennsylvania. The tendons were installed by Spencer White & Prentis, New Jersey, as a subcontractor to Conduit. Applied Geomechanics supplied the Tiltmeter and Geokon, Inc. supplied the strain gages. The design criteria was established by Parsons Brinckerhoffn who reviewed the design and undertook the monitoring program and evaluation of the data.

References

1. Carlton L. Ho, et. al. (1989) *Field Performance of Soil Nail System in Loess*, Foundation Engineering Proceedings Congress, ASCE Division, Evanston, Illinois.

2. Elias V. Juran I., (1991) *Soil Nailing for Stabilization of Highway Slopes and Excavations*, FHWA - RD-89-189, Federal Highway Administration.

3. Stocker, M.F., Korber, G.W., Gassler, G., Gudehous, G. (1979) *Soil Nailing*, International Conference on Soil Reinforcement, Paris.

4. Gassler, G., Gudehus, G. (1981) *Soil Nailing—Some Soil Mechanics Aspects of In-Situ Reinforced Earth*, X, ICSMFE, Proceedings Vol. 3, Session 12, Stockholm.

5. Shen, C.K., Bang, S., and Hermann, L.R. (1981) *Ground Movement Analysis of an Earth Support System*, Journal of the Geotechnical Engineering Division, ASCE, Vol. 107.

6. Shen, C.K., Bang, S., Hermann, L.R., and Romstad, J.M. (1978) *An In-Situ Earth Reinforced Lateral Support System*, Department of Civil Engineering, University of California, Davis Report No. 81-03.

7. Shen, C.K., Bang, S., Romstad, J.M., Kulchin, L., and Denatale, J.S. (1981), *Field Measurement of an Earth Support System*, Journal of the Geotechnical Engineering Division, ASCE, Vol. 107.

8. Schlosser, F., (1983) *Analogies et differences dans le Compotement et le Calcul des Ouvrages de Soutenement en Terre Armee et par Clouage du Sol*, Annales de L'Institut Technique du Batiment et des Travaux Publics, No. 418.

9. Schlosser, F. (1982), *Behavior and Design of Soil Nailing*, International Symposium on Recent Development in Ground Improvement Techniques, Bangkok.

10. *Soil Nailing, Field Inspector's Manual*, Publication No. FHWA-SA-93-068.

Seismic Behavior of Micropile Systems

Aomar Benslimane[1], Ilan Juran[2], Sherif Hanna[3], and Serguey Drabkin[4]

ABSTRACT

A series of centrifuge tests were conducted on micropile group and network systems in order to investigate the response to earthquake loading and soil-micropile interaction behavior. Model tests on group and network systems embedded in loose to medium dense dry sand are described. Micropile bending moment, deflection, and acceleration were measured during testing. Dynamic p-y curves were derived from the measurements for low and high levels of shaking and were compared with the backbone p-y curves for sand recommended by API and other published data. Group and network effects were investigated for different configurations and at different levels of loading. For the selected frequency of excitation, the results indicate a positive group effect increasing with the number of piles and the batter angle. This paper describes the experimental procedures used to carry out the centrifugal model tests and summarizes the main preliminary results.

INTRODUCTION

The behavior of micropile group systems under static loading has been investigated and design guidelines have been assessed in a state of practice review (Bruce and Juran, 1997). However, the application of micropiles to seismic retrofitting is facing the need for established and reliable design guidelines.

As reported by Herbst (1994), foundation with root piles in Italy have already survived earthquakes. Micropile is a very flexible pile. Due to its slenderness and its ductile steel core, it can be assumed that it follows best the shock induced displacements in the ground and that it remains integrated with the soil. With a group of micropiles being parallel or battered, a reinforced soil body is created

---
[1,3] Research Fellow, Polytechnic University, Brooklyn NY 11201
[2] Professor and Dept. Head, [4]Research Professor, Polytechnic University, Brooklyn NY 11201

which performs also in a flexible way. However, many issues needs to be addressed at this level and for proper design, still considerable research work needs to be done in order to establish and assess seismic design guidelines for micropile systems.

In order to investigate the seismic behavior of micropile group systems (isolated piles, groups and networks of reticulated micropiles) under axial, lateral and combined loading in selected types of engineering applications, a proposed workplan for laboratory centrifugal and numerical model studies on the seismic behavior of micropile systems has been adopted by the FHWA in conjunction with the FOREVER French program. The prime objective of these model studies, accomplished by the Polytechnic University at New York in cooperation with the Ecole Nationale des Ponts et Chaussees Geotechnical Research Center - CERMES at Paris and the University of Canterburry at New Zealand, is to provide the experimental data base to develop and evaluate seismic design methods for selected engineering applications.

It is anticipated that this cooperative research program will provide the necessary database for the development and evaluation of seismic design methods for micropile groups and networks for infrastructure applications.

A series of centrifuge tests on instrumented micropile group systems were performed at the Rensselaer Polytechnic Institute (RPI) Geotechnical Centrifuge Research Center and several configurations of model micropile group systems were tested in loose to medium dense dry sand. The main objectives of this phase were to investigate the effects of (i) acceleration shaking level, (ii) group and two-dimensional network configuration, (iii) superstructure loading level, and (iv) micropile inclination on the seismic response of selected micropile systems.

This paper describes the experimental procedures used to carry out the centrifugal model tests and summarizes the main preliminary results.

CENTRIFUGE TESTING

The centrifuge tests were performed at the Rensselaer Polytechnic Institute (RPI) Geotechnical Centrifuge Research Center. A detailed description of the RPI centrifuge facility is presented by Elgamal et al. (1991). In the present series of tests, use was made of the rectangular, flexible-wall laminar container built at Rensselaer Polytechnic Institute to closely approximate a continuous shear strain field in the soil during shaking and accommodate possible shear strain concentrations due to boundary conditions.

Pile Properties and Model Layout

The geometrical parameters have been defined according to two main considerations: (i) Similitude requirements in order to model full scale experiments carried out by Plumelle (1993) (L=5 m, D = 10cm, reinforcing steel: tubing steel 50.3 - 40.3 mm), and 1g shaking table tests at University of Canterburry in New Zealand (ii) Size of the laminar box used in the centrifuge tests. However, in the design of the model micropile configurations and the adopted testing procedures, the intention was also to study the mechanism of the seismic response of micropile systems and to generate data for comparison with analytical and numerical simulations.

The structural characteristics of the model piles used in the centrifuge tests were chosen to maximize flexure of the pile during vibration. Interface properties were taken into account by gluing sand particles along the entire pile length, and local compaction around the pile to simulate high grout/ground bound and confining effects (Juran et al, 1998 ; Benslimane et al., 1998, Abdoun, 1996, Vucetic et al., 1993). It was also desirable to have the end of the pile with a sufficient distance from the base of the box to ensure that end bearing of the piles will not be significantly influenced by the laminar box base. For this purpose, the pile tips were about 5 diameters above the container base (Abghari et al., 1995). A pile of sufficient length to ensure that pile tip reactions did not significantly influence the pile head response. Based on this, a scaling factor of 20 has been adopted.

The model piles were constructed of polystyrene with roughened shaft and a Young's modulus of E_m = 2700 Mpa and a length of L=21.3 cm. The selected diameters are 6.5 mm and 9.5 mm achieving slenderness ratios of 33, and 22 respectively. Table 1 summarizes pile model properties.

Table 1. Model micropile properties

Outside Diameter D (mm)	Inertia Modulus I_m (mm^4)	Flexural Rigidity $E_m I_m$ (N.mm^2)	Slenderness Ratio-L/D
9.5	$3.99 \cdot 10^2$	$10.78 \cdot 10^5$	22.4
6.5	$0.87 \cdot 10^2$	$2.36 \cdot 10^5$	32.8

The pile was strain gauged with 6 pairs of foil type strain gauges mounted on the outside of the pile to measure peak bending and axial strains. A mass is generally screwed at the pile head applying an axial loading of 50% and 90% of the failure load determined from 1g static tests in order to simulate the influence of a superstructure. The same procedure has been adopted for pile group tests.

Instrumentation and Data Acquisition

As shown on figure 1, the instrumentation consist of:
- LVDT's : transducers at different locations: lv1 and lv2 to record any lateral soil profile displacement (not expected), lv3 for the pile cap settlements, lv4 for the lateral pile cap displacements, and lv5 to record surface settlements.
- Accelerometers Model 303A03 from PCB Piezotronics were used at different locations for acceleration–time history measurements. The labeled input

242   SOIL IMPROVEMENT FOR BIG DIGS

accelerometer is used to monitor the input base motion, acc1 to acc5 to record the wave propagation along the soil profile and the free field accelerations, and acc6 to measure the pile head accelerations in order to characterize the structural response of soil-micropile system.

• pairs of half bridge circuited strain gauges were installed on the surface along the model piles to monitor bending (sg2, sg4, and sg6) and axial (sg1, sg3, and sg5) strains. The strain gauges were type CEA- 13- 125 UN-120)

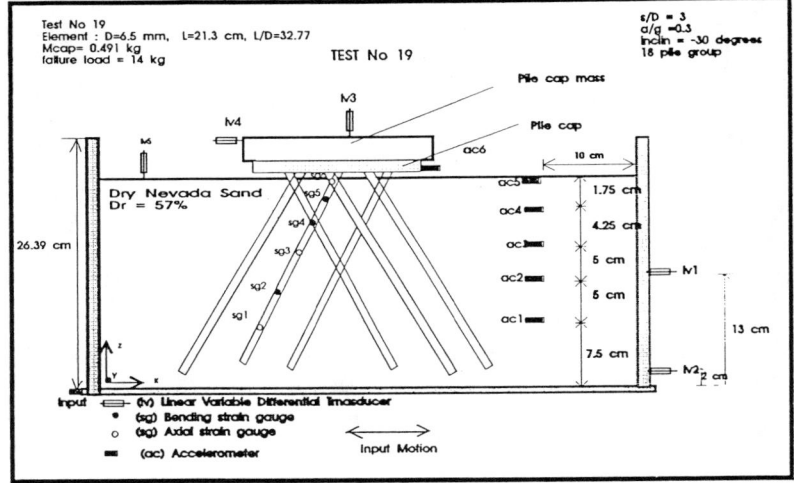

Figure 1.   Typical micropile model system tested by Polytechnic University (From Juran et al., 1998)

Soil Properties

A Nevada sand 120 was used at a relative density of 57 % consistently with the relative density used in the 1g shaking table model tests conducted by Canterburry University in New Zealand. Tests (Arulmoli et al., 1992) conducted on the sand yielded maximum and minimum void ratios of 0.51 and 0.88, specific gravity of 2.67, an average particle size D_{50} of 0.13 mm and a coefficient of uniformity (D_{60}/D_{10}) of 1.6.

The values of a peak friction angle of $\phi=33°$ and $36°$ were obtained from laboratory triaxial tests on Nevada Sand for relative densities of 40% and 60%, respectively. A value of $\phi=35°$ was therefore selected for the testing at a relative density of 57%.

The influence of average grain size (D_{50}) relative to the pile diameter in centrifuge tests has been studied by Oveson (1975) and summarized by Cheney

(1985). It has been concluded that there is no significant influence of the grain size on the load settlement behavior for $30 < D/D_{50} < 180$. The choice of 6.5 and 9.5 mm pile diameter in the Nevada sand corresponds to a ratio of 50-73, which is adequate to minimize the effect of particle size on pile behavior.

Experimental Testing Procedure

The dynamic tests consisted of horizontally shaking the models in flight at 20 g. For each configuration, the horizontal shaking included sequences of 100 uniform cycles of sinusoidal accelerations at 40 Hz. Sinusoidal ground motions were used in order to enable dynamic analysis of basic patterns of model behavior, which are more difficult to perform with more complex input motions. The models were first subjected to a prototype acceleration time history with amplitudes of 0.3 g with cap only, and then under 50% and 90 % of the estimated failure load. The failure load (FL) is defined as the vertical load which causes failure under static loading by occurrence of large deformation and loss of friction. As recommended by Weltman (1980), a limiting displacement of 10 percent of the pile diameter was chosen to define failure in compression . The advantages of this testing procedure are that the effect of the loading level on the seismic behavior of the model micropile configuration being tested can be established. Furthermore, according to the recording of pile cap settlement and ground surface, the overall volume change of soil during shaking was relatively small suggesting that the change in void ratio was insignificant. It is worth noting that this procedure has been extensively used by Zeng (1998) to investigate the seismic response of gravity quay walls and has been proven to be cost effective.

The procedure of model preparation was as follows: first a latex membrane 0.02 cm in thickness, was used to line the inside of the laminar box, prevent leakage of the contents, and to reduce the side friction. During model construction, the external side of the container was sealed and connected to a vacuum pump to remove air from the outer chamber. Dry sand with relative density of 57% was poured in layers of 50 mm thickness or less. The amount of sand needed for each layer to achieve required density was weighed. At appropriate stages, transducers (accelerometers and lvdt's) were placed in the soil.

The main parameters of the experimental program are summarized in table 2. The structural and geometrical properties of the main investigated configurations are reported in Juran et al., (1998) and Benslimane (1998).

ANALYSIS OF THE CENTRIFUGE TEST RESULTS

The interpretation procedure adopted to analyze the centrifuge tests results involved the following:
- Spectral Fourier Analysis of the input ground motion and the recorded accelerations in order to characterize and evaluate the soil-micropile system dynamic response.

- Characterization of the soil pile interaction during base motion excitation through the development and assessment of dynamic p-y curves
- Comparison of the derived p-y curves with the recommended p-y by the American Petroleum Institute (1983) and published data (Gohl, 1991).
- Analytical simulations of the test results using the finite difference computer program LPILE (Reese and Wang, 1989)
- Investigation of the group and network effect on the bending moments, displacement, and axial forces developed in the micropiles under seismic loading.

Table 2. Main parameters of the achieved experimental program.

Microp. Config.	Diam. D (m)	Slend. Ratio L/D	Microp. Inclin. α (deg)	Spac. to Diam. Ratio s/D	Ampl. of Acc. (a/g)	Loading Level
Single Microp.	0.19	32	0	-	0.03 to 0.5	100 cycles- cap only
	0.13	22	0	-		100 cyc. - 0.5 Fl
						100 cyc. - 0.9 Fl
Group of Microp.	0.13	32	0	3 – 5	0.3 until failure	100 cycles- cap only
						100 cyc. - 0.5 Fl
						100 cyc. - 0.9 Fl
(2x1)	0.19	22	0-30	3 – 5	Same	Same
(2x2)	0.13	32	0	3 – 5	Same	Same
3x(2x1)	0.13	32	10 - 30	3	Same	Same
3x3x(2x1)	0.13	32	10 - 30	3	Same	Same

Natural Frequency of the Soil/Micropile System

It is generally agreed that the most meaningful engineering characteristics of ground motion are presented by Fourier and Response Spectra of the recorded ground accelerations. These representations exhibit the recorded acceleration time histories in the frequency domain in order to gain insight on the frequency content and response characteristics. In the case of our tests, Fourier spectra was calculated for each of the accelerometers (input, ac5, ac6) for each event of every test.

The determination of the natural frequency for each investigated configuration is an important parameter for analytical modeling. The general procedure requires to excite the model at different frequencies and from the plot of peak responses to identify the natural frequency of the system as well as to detect any non linearity from the lack of symmetry of the plot of the peak responses. However, due to the experimental constraints, this approach could not be used. Therefore, the following procedure was adopted: Fourier amplitudes computed from the measured pile head accelerations (Aph) have been normalized with respect to the amplitudes computed from the input base accelerations (Ainp). The frequency at which a peak amplitude ratio occurs can be used to characterize the natural frequency of the micropile system. The same procedure is used to obtain the natural frequency of the free field by normalizing the free field accelerations with respect to the input base motion. This procedure has been extensively used by Gohl (1991) for piled systems and Tufenkjian and Vucetic (1993) in the investigation of the seismic behavior of soil nailed retaining structures using centrifuge modeling.

The procedure outlined above was applied to all the configurations during the main events in order to characterize the natural frequency of the micropile system. The results are displayed on figure 2 for the case of single pile (test 2), 2x1 pile group (test 7), 18 pile group system disposed inclined at 10 degrees 3x3x(2x1) (test 18) and 30 degrees (test 19). For all these cases, the micropile system was subjected to the cap loading only.

However, due to use of the laminar box, a lateral inertia correction due to the mass of the rings is needed for the interpretation of the centrifuge test results. As stated by Van Laak et al., (1994), estimates of the lateral forces caused by the rings are important in the interpretation of the measured natural frequencies of the model. The lateral forces can be accounted for by increasing the total unit weight of the soil, γ_t, to a new effective value, γ_{eff}:

$\gamma_{eff} = C \cdot \gamma_t$ [2]

where C is a correction factor given by :

C= 1+ weight of ring / weight of soil = 1 + weight of ring / (L. W. H_r) . γ_t [3]

where the weight of the ring includes that of its bearings. In the equation above, L and W are the length and width of the soil model, and H_r is the height of each ring. Typical values of C= 1.4 are generally used for dry sand

The results clearly show that, for a specified level of shaking (a/g=0.3) : (i) the estimated fundamental frequency of the micropile system is strongly affected by the micropile system configuration. A value of 5.18 Hz can be adopted for the natural frequency of the single micropile (after lateral inertial correction). This value increased to 5.6 Hz for the 2x1 pile vertical group and 10.15 Hz for the 18 micropile network system where the piles are inclined at 10 degrees (Test 18). ii) a greater interaction occurs for higher pile inclination, as a value of 11.9 Hz was observed for an inclination of 30 degrees (Test 19).

To estimate analytically the fundamental natural frequency of the micropile system, the coupled analysis procedure of which basic assumptions can be found in Flores-Berrones and Whitman (1982), Gohl (1991), and Benslimane (1998) was applied.

In a coupled analysis approach, the base motions are harmonic of the form $u_b = u_0 e^{i\omega t}$ and are applied at the pile tip. The pile is surrounded by a homogeneous Winkler medium with the soil reaction, p, given by $p = k_s (y - u)$ where k_s is the

Winkler subgrade modulus, y is the relative lateral pile deflection with respect to the moving base, and u is the relative free field displacement. The undamped pile-superstructure response is then solved assuming that the pile vibrates in its first mode.

Using the single pile structural and geometrical properties (Test 2-330), the coupled analysis procedure leads to an estimated value for the natural frequency of 5.25 Hz which is in good agreement with the corrected natural frequency (5.18 Hz) of the pile system based on Fourier analysis.

Soil-Pile Interaction

To determine the nature of interaction between the soil and the pile during base motion excitation, cyclic p-y curves were derived from the single pile data using procedures described by Ting (1987) and briefly summarized herein.

From simple beam deflection theory, the moment in the pile M is proportional to the recorded flexural strain ε:

$$M = EI \frac{d^2 y}{dz^2} = EI \frac{\varepsilon}{h} \qquad [4]$$

where y is the lateral pile deflection, z is the vertical coordinate along the pile, h is the distance from the strain gauge to the neutral axis of bending, and EI is the flexural stiffness of the pile. The moment may be integrated twice for the deflection y or differentiated twice for the pressure p. Since the strain data are known at discrete locations along the pile, a numerical scheme is necessary to obtain the needed pressures and deflections and dynamic p-y curves. The method developed by Ting (1987) for the full scale data was used with a suitable assumption to take into account the rigidity of the pile cap connection. This method involves the use of a least square fitting procedure for the moment profile with a seventh degree polynomial.

Since the dynamic tests were conducted in sand, the polynomial should be subject to the constraint that the net soil pressure is zero at the ground surface. The other main boundary conditions assume zero moment and shear forces at the pile tip.

The numerical procedure was evaluated by comparing the computed deflections at the top of the pile head mass with measured displacement, a good agreement was observed.

Cyclic p-y curves were computed for the same test at low level vibrations during the shaking cycle where maximum pile deflection occurred and are shown for a range of depths in figure 3. The p-y curves up to about 9 pile diameter are non linear and exhibit hysteresis loops. Up to 6 pile diameter, the secant lateral stiffness of the soil, defined as the slope of the line passing through the end point of the loop gradually increased with the depth. No signs of gapping between the sand and the piles are evident.

Cyclic p-y curves were derived from single pile analysis using the same procedure for the case of a strong shaking (a/g=0.3). Similar behavior was observed except that signs of gapping were observed at shallow depths. Figure 4 compares the computed p-y curves under strong and low level shaking with the cyclic p-y curves recommended by the API (1993) and the p-y curves reported by Gohl (1991) based on centrifuge tests on model piles with prototype bending stiffness of EI = 172 MN.m². The API curves were computed using a peak friction angle of 32 degrees and an n_h value of 6750 kN/m³. The latter defines the initial slope of p-y curve and was based on values recommended for loose dry sand. It can be seen that for low level shaking the secant lateral stiffness corresponds fairly well to the

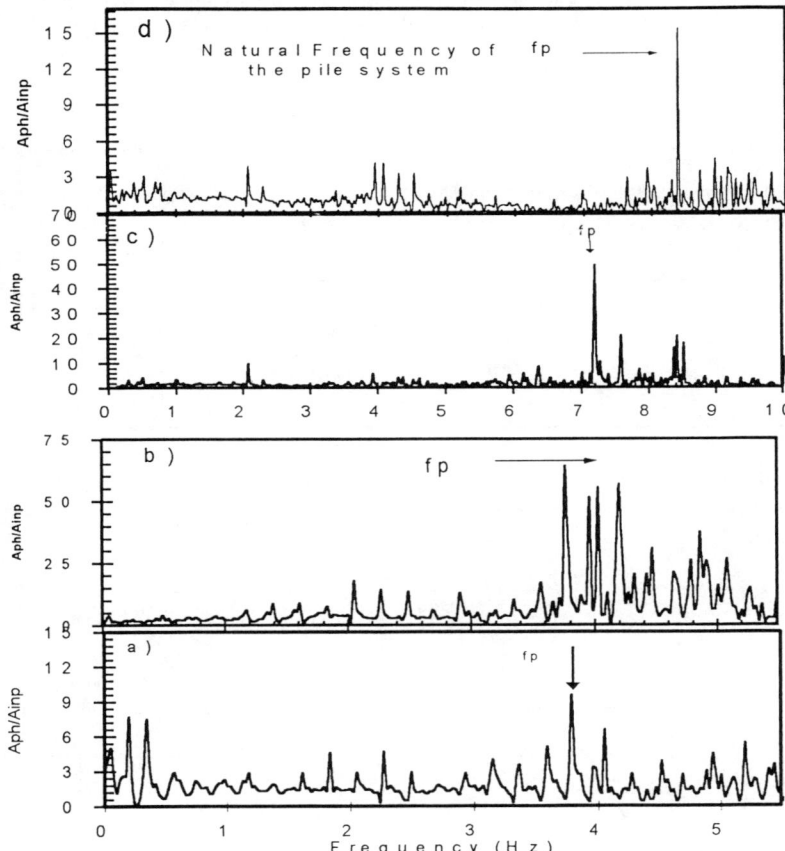

Figure 2. Variation of the Fourier amplitude spectral ratio of the pile head acc. normalized with respect to the free field acc. for : a) single pile, (b) 2x1 vertical pile group, c) 3x3x(2x1) network system (α=10°), d)3x3x(2x1) network system (α=30°) - Cap loading only - s/D= 3 - L/D =32 - a/g=0.3

results obtained by Gohl (1991). For strong shaking, the API and the Gohl (1991) p-y curves are considerably stiffer compared to the experimentally derived p-y curves based on the centrifuge tests on model micropiles.

In order to evaluate the potential use of the LPILE finite difference computer program for the soil-pile interaction analysis, the program has been used with the experimentally derived p-y curves to compute the bending and pile displacement profiles. The numerical results were compared with the experimental measurements. For low level shaking, the flexural response of the model pile was assumed to be dominated by structural inertia rather than by the free field ground motions. The kinematical loading was neglected as the free field nearly follows the motion of the pile (Juran et al., 1998). The inertial loading at the pile head was estimated based on the difference between the recorded pile cap accelerations and the input base motion.

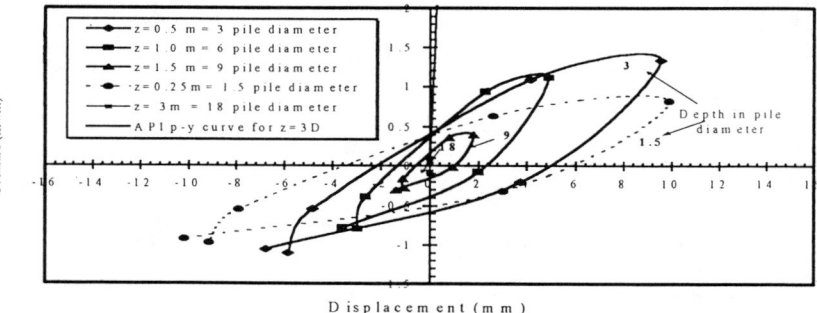

Figure 3. Cyclic p-y curves at various depths during steady state shaking cycle (t=30-30.5 sec) and Comparison with API p-y curve - centrifuge test 1(203). (a/g =0.03)

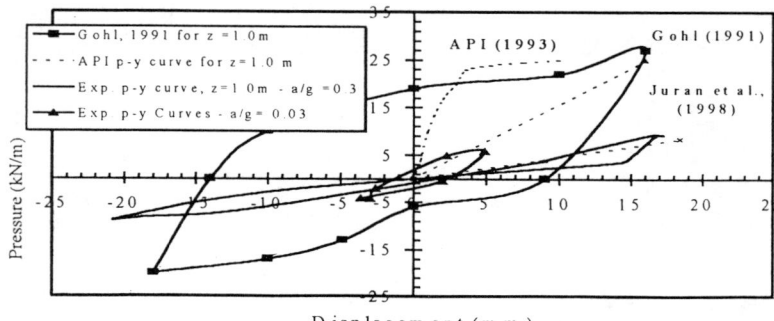

Figure 4. Comparison of the experimentally derived cyclic p-y curves during steady state shaking cycle (t=30-30.5 sec- centrifuge test 1-203 and) with API (1993) p-y curve, and (Gohl, 1991).

Considering test (I-203), several iterations were done to define the fixity conditions at the pile head (moment and shear force) and full fixation was assumed to yield the most appropriate matching between pile displacement computed with LPILE and the experimentally derived pile displacement profile obtained using Ting's procedure. LPILE was used considering both the experimentally derived p-y curves and API p-y curves. Figure 5a and 5b illustrate that the LPILE simulations using the experimentally derived p-y curves agree fairly well with the pile bending and displacement profiles obtained using Ting's procedure.

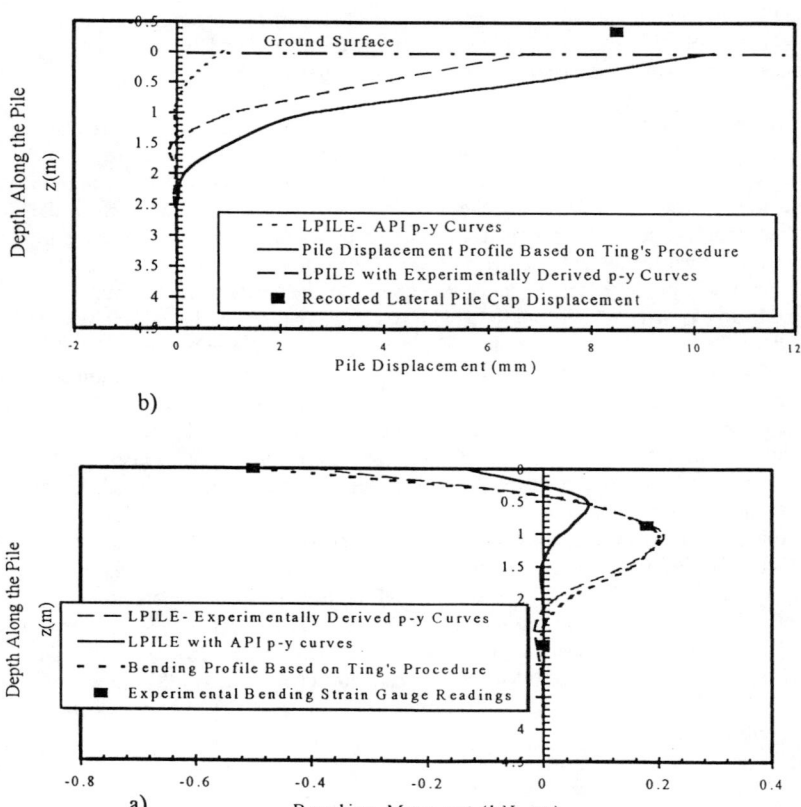

Figure 5. Comparison of a) the bending profile obtained by Ting's procedure with bending profiles obtained through iterations using LPILE based on API and experimentally derived p-y curves. b) corresponding displacement profiles Test 1 (203) - single pile- low level shaking (a/g=0.03)

GROUP EFFECT

Based on the bending strain gauge recordings, bending moment profiles of the instrumented micropile in the various configurations for each event were derived to investigate the group effect. Figure 6(a), and 6(b) display respectively the maximum relative pile displacement with respect to the free field and the associated distributions of the bending moment profiles for (i) single pile (D=0.13m), (ii) 2x1 and 2x(2x1) vertical pile groups with D=0.13m, (iii) (2x1) inclined pile group with a 30 degrees pile inclination and D= 0.13 m, (iv) 3x(2x1) group with a pile inclination of 10 degrees and D=0.13m ; (v) 3x(2x1) network system with a 30 degrees pile inclination and D= 0.13 m, (vi) (2x1) vertical pile group with D=0.19m. All tests configurations were subjected to a vertical loading equivalent to 50 percent of the static failure load (0.5 Fl). Computed lateral pile cap displacements were compared with measured pile cap displacements.

The experimental results illustrate that : (i) bending moment and displacement profiles for single pile and a group of (2x1) vertical piles with s/D=5, under the same equivalent loading conditions, are quite similar illustrating a negligible interaction effect for s/D=5; (ii) the bending stiffness significantly affect the bending moment and displacement magnitude of the pile group. The bending moment obtained for the pile diameter of D=0.19m is about four times greater than that measured for D=0.13m which corresponds to the ratio of to the ratio I_1/I_2 where I_1 and I_2 refer to pile inertia modulus corresponding to the pile diameters 0.19 m and 0.13 m. (iii) For the selected frequency of excitation, the experimental data illustrate a ''positive'' group effect, which results in smaller bending moments and displacements for the 2x(2x1) and 3x(2x1) pile groups with spacing to diameter ratio of 3 as compared with data measured for identical single pile and 2x1 pile group with s/D=5. Similar observations were reported by Ousta (1998) who conducted three dimensional finite element simulations on different configurations of vertical micropile group systems within the FOREVER French research program.. These studies were limited to the kinematical loading and were conducted for different vibration modes. It is worth noting that dynamic pile group analysis for conventional piles in elastic and visco-elastic media are generally associated with a negative group effect. (i.e. Dobry and Gazetas,. 1988 ; Velez et al., 1984). However, the centrifuge test results appear to be consistent with the positive group effect observed by several investigators (i. e. Lizzi, 1982; Bruce and Juran, 1997; O'Neill, 1983) for vertical pile groups subjected to static loading in cohesionless soils.

NETWORK EFFECT

The experimental data for 2x1 and 3x2x1 networks with 30 degrees batter piles displayed on figure 6b and 6c illustrate that the bi-dimensional network effect results in a decrease of the bending moments as compared with vertical pile groups as well as in further reduction in lateral pile cap displacement as compared with 2x1 and 3x2x1 vertical groups.

Figure 7 illustrates the effect of the loading and shaking level on the bending moment profile during Test 7- p3 (2x1 - s/D=3 - alpha=30 degrees). For the case of battered pile with pile inclination of 30 degrees, the measured bending moment for different levels of acceleration are relatively small as compared with vertical pile systems, except at the pile head. These results indicate that bending moment are due mainly to boundary conditions/pile cap restraint rather than lateral

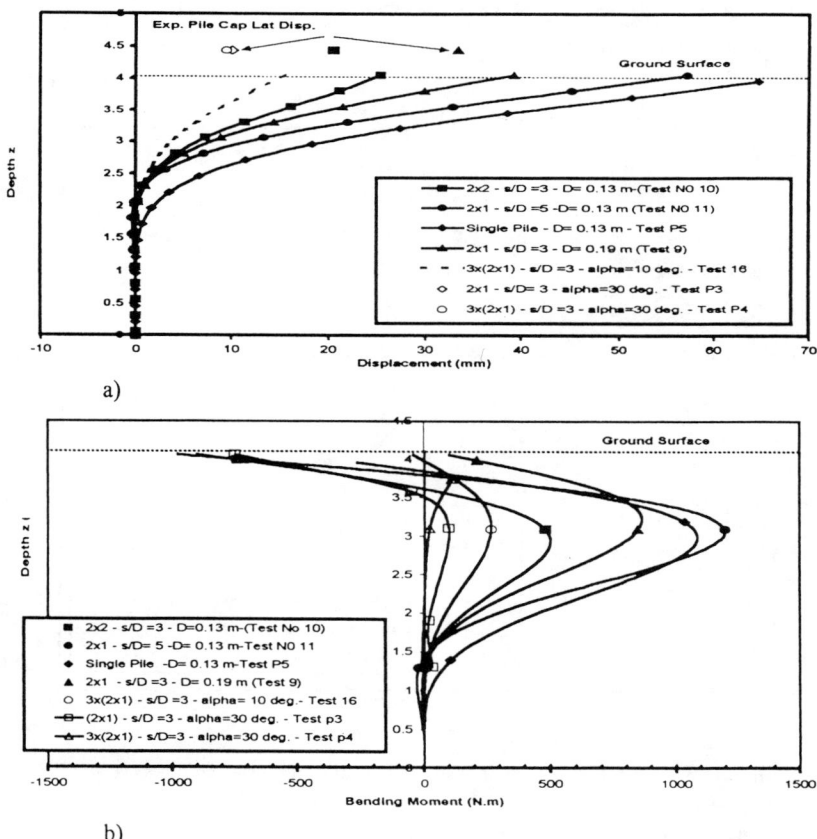

Figure 6. Comparison of: (a) Recorded displacement profiles for single pile, (2x1), (2x2) vertical pile groups and 3x(2x1) network system ($\alpha =10°$) during peak pile lateral displacement (a/g =0.3 - s/D =3 -L/D =32) (b) Corresponding bending moment profiles.

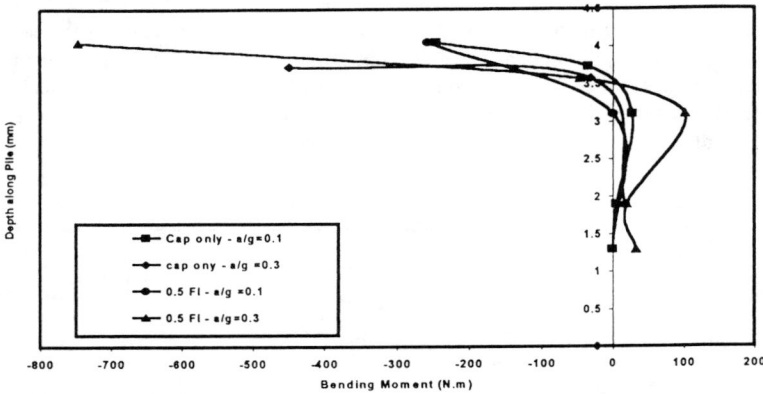

Figure 7. Effect of the loading and shaking level on the bending moment profile at peak pile displacement during Test 7- p3 (2x1- s/D=3 – $\alpha=30°$).

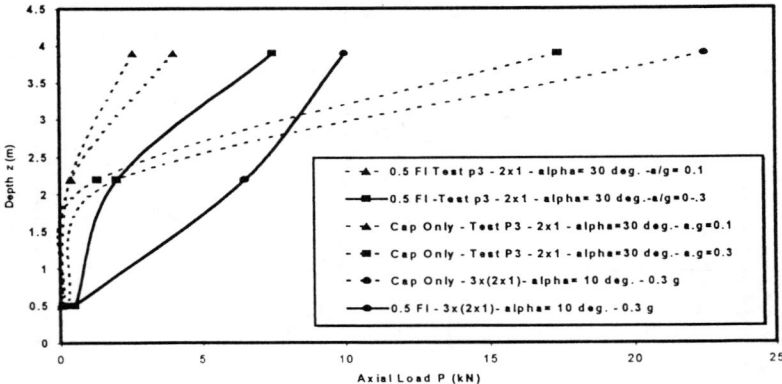

Figure 8. Effect of the loading level, shaking level on the axial load transfer in a 2x1 pile group inclined at $\alpha=30°$ (Test P3) and 3x(2x1) network system (Test 16) with pile inclination of $\alpha=10°$ (L/D= 32-a/g=0.3)

Figure 8 illustrates the effect of the superstructure loading and shaking level on the axial load transfer for 2x1 group of battered micropiles inclined at 30 degrees and 3x(2x1) inclined at 10 degrees. The experimental results indicate that, for the case of cap loading, the maximum axial forces develop at the pile head illustrating the effect of the amplification of the input base motion towards the ground surface.

Measured pile cap accelerations records indicate that for this case, the amplification of the input motion is about 100 %. However, no amplification of the input base motion was observed in the case of a superstructure loading of 0.5 Fl with the measured pile cap accelerations being practically equal to the input base acceleration. Comparing the recorded axial forces for 0.5 Fl and cap loading only, it appears that for the case of the battered piles at 30 degrees, the superstructure loading of 50% Fl results in a decrease of the maximum dynamic axial forces. This reduction can be related to (i) increase of the natural frequency due to pile inclination as illustrated in figure 5 using the spectral Fourier analysis of the recorded acceleration time histories. (ii) Stiffening of the micropile system which is clearly illustrated by the significant decrease of pile cap lateral displacement observed for the case of 0.5 Fl.

These results are consistent with the experimental results reported by Prevost and Scanlan (1983), indicating that the superstructure loading effect results in a stiffening of the system, increase in natural frequency of the soil micropile system and therefore, in a significant decrease of the amplification of the input base motion towards the ground surface. The results obtained for the 3x(2x1) micropile network system with a pile inclination of 30 degrees subjected to 0.3 g are quite similar to those observed for the 2x1 battered micropile system under same acceleration levels.

A preliminary estimate of the dynamic axial forces developed in the piles during simulated seismic motion can be obtained from a pseudo static analysis of the cap–superstructure motion through the seismic events. Figure 9 illustrates the free body diagram of forces acting on (i) 2x1 vertical pile group (Case A), (ii) 2x1 inclined pile system at $\alpha=30°$ (Case B), and (iii) Intermediate Case of 2x1 pile system with $\alpha=10°$ (Case C) under in-line base shaking. For Case A, the equation of moment of equilibrium of the free standing portion of the pile group is given as:

$$m. (a/g). z - (P_1.s/2 + P_2. s/2) = M_1 + M_2 - I_{cg}. \theta_c \quad [5]$$

For Case B and Case C, The equation of moment equilibrium is given as:

$$m. (a/g) z - [P_1.(s/2-l. \tan \alpha) + P_2.(s/2 -l. \tan \alpha)] \cos \alpha = M_1 + M_2 - I_{cg}. \theta_c + [V_1.(s/2 -l. \tan \alpha) + V_2.(s/2 -l. \tan \alpha)] \sin \alpha \quad [6]$$

where m is the pile cap and/or structural mass, a is the relative acceleration at the center of gravity of the mass, g the gravitational acceleration, z is the distance of the center of gravity to the ground surface, l is pile cap stick up above ground surface, s is the pile spacing measured at the level of the pile cap connection, α is the pile inclination with respect to the vertical, I_{cg} the mass moment of inertia with respect to the center of gravity of the structural mass for the direction of shaking, M_1 or M_2 is the average bending moment per pile at the soil surface, P_1 or P_2 is the dynamic axial load in each pile, V_1 or V_2 is the dynamic shear forces, and θ_c is the angular acceleration of the mass.

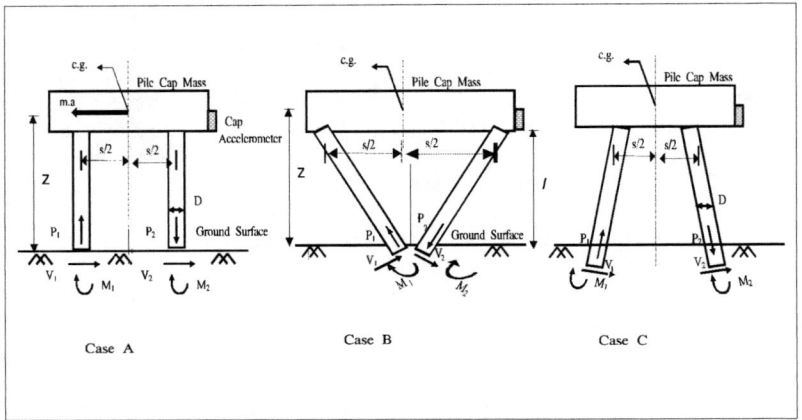

Figure 9. Free body diagram for the free standing portion of 2x1 pile group subjected to an in-line shaking for the considered basic reference cases

Table 3. Comparison of recorded axial forces with the estimated axial forces using the moment equilibrium of the free body diagram for the free standing portion of 2x1 pile group subjected to an in-line shaking for the considered basic reference cases.

Reference Case	a/g	Loading Type (Cap / Superstructure)								Test No.*
		Cap Only				Cap +50 % Failure Load				
		m. a/g (kN)	M_0 (kNm)	Exp. P (kN)	Eq [5-6] P (kN)	m. a/g (kN)	M_0 (kNm)	Exp P (kN)	Eq [5-6] P (kN)	
Case A	0.3					7.2	60	11	8.2	1150
Case B	0.1	0.8	22	4.3	4	0.25	25	4	4.8	P3
Case B	0.3	2.4	45	17.2	12.5	0.75	75	7.5	10	P3
Case C	0.3					10.4	40	10	8.5	1650

Nota m : Pile cap (with or without superstructure) weight
M_0 : Recorded pile bending moment near the ground surface.
$M_0 = (M_1+M_2-I_{cg}\cdot\theta_c)/2$
P : Recorded or calculated axial effort in the pile.
* : For test details, refer to Juran et al., (1998) (FHWA Report)

Table 3 illustrates that, for the considered basic reference cases, the estimated axial forces agree fairly well with the experimental data. This simple procedure, which has been adopted for preliminary test result interpretation, seems to illustrate that the experimental scheme yields a consistent description of the observed centrifugal model behavior.

However, this analysis presents limitations which are related to the simplifying assumptions with regard to the dynamic soil-pile interaction, the effect of the kinematical forces induced by the free field motion, and the assumed distributions of moments and axial forces. Further experimental and numerical studies are presently being conducted for development and evaluation of an analytical approach for the seismic performance evaluation of micropile groups and reticulated network systems.

CONCLUSIONS

The principal conclusions from the preliminary analysis of the centrifuge tests conducted on model micropile configurations can be summarized as follows:

(i) Micropile systems present a flexible behavior. For the selected normalized frequency simulating earthquake excitation, due to the relatively low micropile bending stiffness, pile deformation follows closely free field motion except at shallow depths. This results in a composite seismic response as the micropile system transfers the inertial force of the accelerated superstructure to the soil through soil-structure interaction with relatively low dynamic sresses.

(ii) The estimated fundamental frequency of the micropile system is significantly affected by the micropile system configuration. Based on the spectral Fourier analysis conducted on the selected configurations under cap loading and subjected to a prototype input base motion of 0.3 g amplitude, a value of 5.18 Hz was adopted for the natural frequency of the single micropile. This value increased to 10.15 Hz for the 18 micropile network system where the piles were inclined at 10 degrees. A greater interaction occurred for higher pile inclination, as a value of 11.9 Hz was observed for an inclination of 30 degrees.

(iii) Soil-pile interaction under lateral base shaking was evaluated in terms of dynamic p-y curves. The p-y curves obtained under low level shaking were found to be non-linear and low damping. Similar results were found for strong shaking with signs of gapping at shallow depths.

(iv) The experimental data illustrate a "positive" group effect for the selected frequency of excitation, which results in smaller bending moments and displacements of the 2x(2x1) and 3x(2x1) pile group with spacing to diameter ratio of 3 as compared with data measured for identical single and (2x1) pile group with s/D=5. These observations appear to be consistent with the postive group effect observed by several investigators for vertical pile groups subjected to static loading in cohesionless soils.

(v) Network configuration tends to resist earthquake loading resulting in higher axial stresses compared with vertical groups, while there is a considerable reduction in pile bending moments and cap displacements. The superstructure loading effect results in a stiffening of the system, increasing in natural frequency of the soil micropile system and therefore, in a significant decrease of the amplification of the input base motion towards the ground surface.

The present study has resulted in the creation of a significant database relating to the response of single, group, and networks of micropiles to simulated earthquake excitation. Testing methods and interpretation procedures have been presented. However, these preliminary results need to be further investigated in order to develop seismic design guidelines for micropile systems. In addition, these data are limited in that they are confined to pile response in loose to medium dry sand, where pore water pressure build up during earthquakes has not occurred. It is therefore suggested that further studies be conducted where the effect of pore water pressure build up on free field soil response are taken into account. This will provide an additional data base against which various methods of assessing single pile and pile group response to earthquake excitation can be checked and relevant design guidelines can be assessed for the complex soil-pile-superstructure interaction under seismic loading for micropile groups and network systems.

AKNOWLEDGEMENTS

The authors gratefully acknowledge the support of the Federal Highway Administration. In particular gratitude is expressed to Mr. Al. Dimillio of the FHWA for his effective cooperation, Dr. Ricardo Dobry of the RPI for his advice and help in using the centrifugal research facilities, Dr. Roger Frank of the French laboratory CERMES, and S Perlo for the help in analyzing the centrifuge test results

REFERENCES

Abdoun, T. (1996), "Modeling of seismically induced lateral spreading of multi-layered soil and its effect on pile foundations" Ph.D dissertation, Dept. of Civ. Engrg., Rensselaer Polytechnic Institute.

Abghari, A., and Chai, J., . (1995), " Modeling of Soil Pile Superstructure Interaction for Bridge Foundation". *ASCE Conf. on Performance of Deep Foundations Under Seismic Loading, Geotech. Spec. Publ.* No. 51. 45-59.

Arulmoli, K., Muraleetharan, K. K., Hossain, M. M., and Fruth, L. S., (1992), " VELACS Laboratory Testing Program-Soil Data Report." The Earth Technology Corporation, Project N0 90-0562

American Petroleum Institute (1993), "Recommended Practice for Planning, Design and Construction Fixed Offshore Platforms", Washington, D. C.

Benslimane, A., Juran , I., Hanna, S., Drabkin, S., Perlo, S., Frank, R., (1998). "

Seismic Retrofitting Using Micropile Systems: Centrifugal Model Studies. *IV Int. Conf.* on Case Histories in Geotechnical Engineering, St Louis, Missouri March 8-15.

Benslimane, A., (1998). " Seismic Behavior of Micropile Systems- Centrifugal Model Studies". Ph.D dissertation, Dept. of Civ. Engrg., Polytechnic University.

Bruce, D. A., and I. Juran, *(1997)*. "Drilled and Grouted Micropiles: State-of-Practice Review". U.S. Federal Highway Administration, Publication No. FHWA-RD-96-017.

Cheney, J. A. (1985), " Physical Modeling in Geotechnical Engineering", Proceedings from a Centrifuge Workshop conducted during 12th International Conference on Soil Mechanics and Foundation Engineering, San Francisco.

Dobry, R., and Gazetas, G., (1988) ''Simple methods for dynamic stiffness and damping of floating pile group'' *Geotechnique, 38 No. 4*, 557-574

Elgammal, A. W., Dobry, R., and Van Laak, P., (1991). "Design, Construction and Operation of 100g-ton Centrifuge at RPI " *Int. Conf. Centrifuge 1991*, Boulder, CO, Balkema, Rotterdam, pp. 27-34

Gazetas, G., Fan, K., Tazoh, T., Shimizu, K., Kavvadas, M., and Makris. N. (1992). ''Seismic Pile-Soil-Pile Interaction ''. Proc. of Session Sponsored by the Geotech. Engrg. Div,, ASCE, New York.

Gohl, W. B. (1991), "Response of Pile Foundations to Simulated Earthquake Loading. Experimental and Analytical Results," Ph.D dissertation, University of British Columbia.

Herbst, T. F. (1994). " The GEWI-PILE, A Micropile for Retrofitting, Seismic Upgrading and Difficult Installation". *Int. Conf. on Design and Construction of Deep Foundations* Sponsored by the U.S Federal Highway Administrartion (FHWA), Vol (2), 913-930.

Juran, I., Benslimane, A., Hanna, S., and Drabkin, S., (1998). "Seismic Behavior of Micropile Systems – Centrifuge Test Results " Preliminary Report - FHWA Contract No DTFH61-96-00021

O'Neill, M.N. (1983) "Group action in offshore piles", *ASCE Conf on Geotechnical Practice in Offshore Engineering*, Austin, 25-64

Ousta, R. (1998). " Seismic Behavior of Micropiles". Ph.D dissertation University of Sciences and Technology of Lilles. France.

Oveson, N. K. (1975), " Centrifugal Testing Applied to Bearing Capacity Problems of Footings on Sand", *Geotechnique, Vol. 25, No 2, June, 394-401*

Reese, L. C., W. C. Cox, and S. E. Koop (1974). " Analysis of Laterally Loaded Piles in Sand ", 6th Offshore Technology Conference, Houston, Texas, OTC 2080

Reese, L. C., Wang, S. T. (1989). " Documentation of computer program LPILE. " Ensoft, Inc.

Scott, R. F., (1979), " Cyclic and Static Model Pile Tests in a Centrifuge", OTC 3492, 11th OTC, Houston, Texas, 159-1168

Steedman, R. S. and Zeng, X., 1990. "The influence of Phase on the Calculation of Pseudo-Static Earth Pressure on a Retaining Wall ". *Geotechnique, 40, No 1, 103-112*.

Takemura, J., Kimura, T., Hiro-oka., and Okamura, M., 1991. " Dynamic Stability of a Bridge Foundation in a High Embankment, " *Int. Conf. Centrifuge 1991*, Boulder, Colorado, Balkema, Rotterdam, pp. 449-456.

Ting, J. M. (1987), " Full Scale Cyclic Dynamic Lateral Pile Response, " *J. Geotechn. Engrg. Div.*, ASCE , 113 (1), 30-45.

Tse, F. S., Morse, I. E., and Hinkle, R. T., 1964, *Mechanical Vibrations*, Allyn and Bacon , p . 73.

Vucetic, M., Tufenkjian, M., and Doroudian, M. (1993). "Dynamic Centrifuge Testing of Soil-Nailed Excavations". *J. Geotech. Testing*, ASTM, 16 (2), 12-187.

Weltmann, A. J., (1980), " Pile Load testing Procedure" Report PG7, Construction Industry Research and Information Association (CIRIA), London, 1980, 53 pp.

Zeng, S., (1998). " Seismic Response of Gravity Quay Walls. I: Centrifuge Modeling "*J. Geotech.Engrg,*, ASCE, 124 (5), 406-417.

Recent Developments in Soil-Nailing - Design & Practice

Sherif Hanna[1], Ilan Juran[2], Ofer Levy[3], and Aomar Benslimane[4]

Abstract

Soil nailing is an in-situ soil reinforcement technique that has been effectively used during the past three decades. The most significant construction technology innovations have been the development and use of jet nailing in conjunction with jet piling, and launched soil nails. Construction and retrofitting of bridge abutments, excavation support, as well as seismic retrofitting in earthquake zones have been increasingly used. Available experiments, and observations on full-scale structures have been recently reviewed and documented in an International Knowledge Database for Ground Improvement Technologies (IKDGIT). This paper focuses on the recent technologies, new applications, the development of soil nailing database, and the design aspects of soil nailed structures in seismic zones.

Introduction

Soil nailing is an in-situ soil reinforcement technique that has been effectively used during the past three decades, mainly in France, Germany and in the United States. It presents significant technical advantages over conventional rigid gravity retaining walls or external bracing systems that result in substantial cost savings and reduced construction periods in cut slope retaining systems and slope stabilization.

Soil nailing has been primarily used for temporary retaining structures. This is mainly due to the engineering concerns with regard to durability of metallic inclusions in the ground and shortcomings of facing technology. In recent years,

---
[1,3,4]Research Fellow, Polytechnic University, Brooklyn, NY 11201
[2]Prof. and Head, Polytechnic University, Brooklyn, NY 11201

technological developments have included low-cost corrosion-protected nails, innovative installation techniques such as jet nailing and nail launching as well as prefabricated concrete or steel panels to overcome these limitations. Soil nailing has now become a common construction technique for a wide variety of engineering applications including: stabilization of railroad and highway cut slopes, excavation retaining structures in urban areas for high-rise building and underground facilities, tunnel portals in steep and unstable stratified slopes, construction and retrofitting of bridge abutments with complex boundaries involving interaction with piled foundations, and other civil and industrial projects. A significant research and full scale experiments have been conducted during the past decade to develop and evaluate reliable analysis methods and establish relevant design codes. It is worth noting that in France, a National Research Project CLOUTERRE was conducted from 1986 to 1990 with a total budget of 22M French Francs, which resulted in the development of the CLOUTERRE 1991 Recommendations (translated into English by the Federal Highway Administration, 1991). In the United States, the engineering use of this technology for temporary and permanent structures is currently growing with increasing local experience (Abramson and Hansmire, 1988; AASHTO, 1990; Bruce and Jewell, 1987; Fannin and Bowden, 1991; Nicholson, 1986; Thompson and Miller, 1990), and both federal and state DOTs increasingly recognize the specific advantages of this technology (Chassie, 1994). In particular, the research conducted by the FHWA (Elias and Juran, 1990; Byrne et al, 1996) has resulted in the development of a manual of practice for design, construction, quality control, and monitoring of soil-nailed structures.

Available experiments, and observations on full-scale structures have been recently reviewed and documented (Levy et al, 1998) in an International Knowledge Database for Ground Improvement Technologies (IKDGIT) by the Federal Highway Administration in cooperation with the Technical Committee-17 of the International Society of Soil Mechanics and Geotechnical Engineering. This database, available on the world wide web, is developed to provide a relevant design aid for practitioners, contractors, and governmental agencies.

Of particular interest for earthquake zones is the seismic performance of soil nailed excavations which has been observed during the 1989 Loma Prieta Earthquake in the San Francisco Bay, where several soil nailed structures were subjected to significant levels of shaking (Barar, 1990; Felio et al., 1990). The 1995 Kobe earthquake provided many opportunities for documenting the behavior of soil-reinforced structures along transportation facilities, and through various public and private developments (Tatsuoka et al, 1996). Soil nailed structures are systems that are coherent and flexible, therefore, they present inherent advantages of withstanding larger deformations with high resistance to dynamic loading. Due to these advantages, these systems appear to offer a valuable and cost effective technical solutions for geotechnical construction in seismic zones. The high performance of soil nailed systems to earthquake loading in seismic zones was demonstrated by post earthquake observations (Barar,1990; Felio et al., 1990;

Tatsuoka et al, 1996). Centrifugal model tests have been conducted by Vucetic et al., (1993 and 1996) to evaluate the failure mechanisms of soil nailed structures under seismic loading. However to date, only limited studies have been conducted to evaluate the dynamic response of soil nailed structures and a comprehensive investigation is needed in order to develop and experimentally evaluate reliable seismic design methods.

This paper focuses on the recent construction technologies, new applications, the development of the soil nailing data base, and design aspects of soil-nailed structures in seismic zones.

Construction Technologies

The main components of a soil nailed retaining system are the in-situ ground, the tension-resisting nails, and the facing or the structural retaining element. The economy of the system is predominantly dependent upon the technology used (i.e., structural elements and installation process of inclusions) and the construction rate achieved.

The nails used in soil-nailing retaining structures, are generally steel bars or other metallic elements that can resist tensile stresses, shear stresses, and bending moments. They are generally, either placed in drilled boreholes and grouted along their total length, or driven into the ground. The nails are not prestressed but are closely spaced (e.g., one driven nail per 2.5 ft^2, one grouted nail per 10-50 ft^2) to provide an anisotropic apparent cohesion to the native ground. A variety of proprietary nails, corrosion-protection systems, and installation techniques such as coupling nail driving with jet grouting, driving encapsulated nails, or driving prefabricated nails that consist of prestressed bars in compression tubes, have been developed by specialty French contractors (Intrafor-Cofor; Solrenfor) to be used in permanent structures.

During the past decade, the most significant technological innovations have been the development and use of the jet-grouted nails (Louis, 1986) and the launched soil nails (Ingold and Miles, 1996). A brief description of these nailing systems is outlined below.

Jet-grouted nails are composite inclusions made of a grouted soil with a central steel rod, which can be as thick as 30 to 40 cm. A technique that combines the vibropercussion driving and high-pressure (greater than 20MPa) jet grouting has been developed by Louis (1986). The nails are installed (figure1) using a high frequency (up to 70Hz) vibropercussion hammer, and cement jet grouting is performed during installation. The inner nail is protected against corrosion using a steel tube. The jet-grouting installation technique provides recompaction and improvement of the surrounding ground and increases significantly the effective nail diameter and the pull-out resistance of the composite inclusion providing effective means for constructing soil nailed structures in clayey soils.

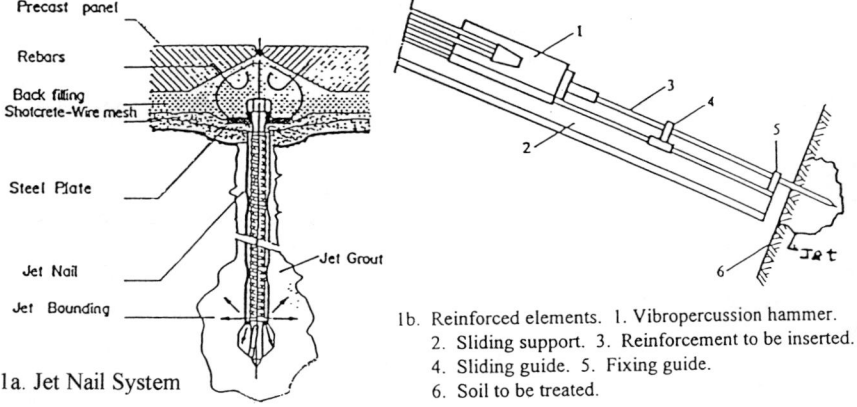

1a. Jet Nail System

1b. Reinforced elements. 1. Vibropercussion hammer. 2. Sliding support. 3. Reinforcement to be inserted. 4. Sliding guide. 5. Fixing guide. 6. Soil to be treated.

Figure 1.   Jet nailing (after Louis, 1986).

Table 1 presents typical grouted nail diameter and ultimate pull out capacity values for different types of soils.

Table 1.   Jet nailing typical grouted nail diameter and ultimate pull-out capacity values for different soil types. (Louis, 1990)

Ground	Gravel	Sand	Silt	Clay
Bulb Diameter. (cm)	60	40	30	20
Ultimate Pull-out (KN/m)	1240	540	205	75

This technique can also be successfully used in conjunction with jet piling using similar type of equipment. Jet piling consists of inserting a reinforced steel element into a jet grouted column prior to the hardening of the jet grouted materials to allow for tension resistance and uplift loading capacity of the jet grouted column. Combination of the two techniques will result in shorter nailing length, larger spacing between the nails, as well as better structural performance and corrosion protection for the jet nailed head. A typical use of jet nails and jet piling is illustrated in figure (2). The figure shows a process of underpinning a retaining wall for the construction of an under ground parking in an urban area. In the underpinning process, jet piling was used to provide vertical support to control settlement, as well as lateral support to reduce the horizontal earth pressure on the parking lot wall. The following step was to jet nail the alluvium soil beneath the retaining wall to control lateral deflections. This technique has been successfully used for the underpinning of historical monuments.

SOIL IMPROVEMENT FOR BIG DIGS 263

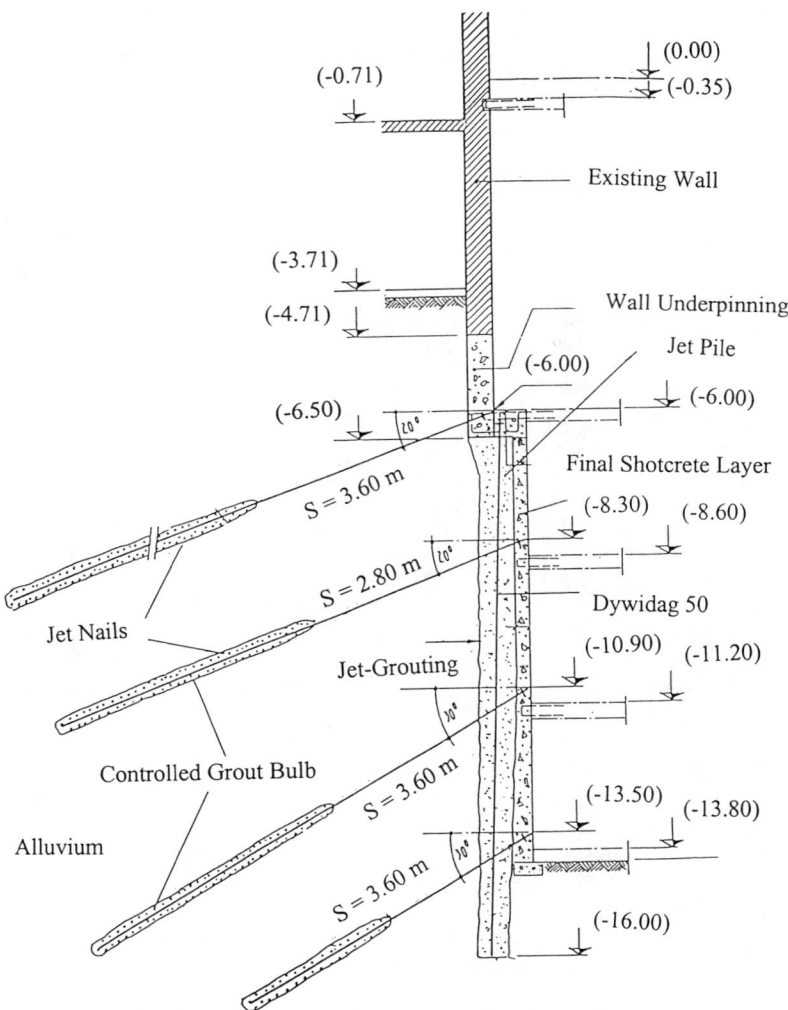

Figure 2. Typical Use of Jet Nailing and Jet Piling (Louis, 1990)

Figure (3) shows the use of jet nailing and jet piling for the excavation of the LOUVRE Pyramid in Paris (Louis, 1990). For this sensitive structure, the jet piles were used in a complex urban soils involving fills and alluvium to prevent excessive settlement and create lateral support, while jet nails were used to anchor the lateral support and control lateral deflection. Table (II) presents typical grouted diameters and ultimate capacity of jet piling for different types of soils.

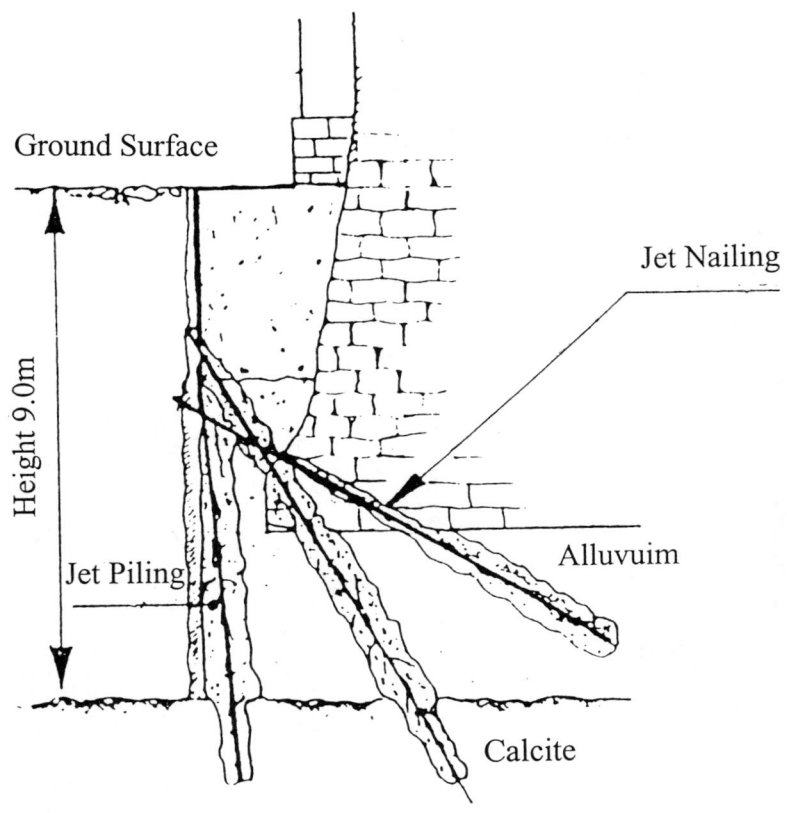

Figure 3. Application of Jet nailing and Jet Piling for the excavation of the LOUVRES Pyramid in Paris (Louis, 1990)

Table II. Jet piling typical grouted pile diameter and ultimate capacity for different soil types.(Louis, 1990)

Ground	Gravel	Sand	Silt	Clay
Column Diameter (cm)	100	80	60	40
Working Load (KN)	2500	250	350	50
Ultimate Pull-out (KN/m)	2200	1100	450	130

As illustrated in figure (4), jet nailing has been successfully used in Paris for retrofitting reinforced earth structures. It also provides a cost effective technical solution for excavation support in clayey soils, as illustrated by Louis, (1990) by jet nailing application for a 30 m deep excavation support in a clayey soil in Lyon. For this case the jet nailing design relies primarily on creating the jet nail column in the clayey soil with a diameter of up to 20 cm and pull-out capacity of up to 75 KN/m

Launched Nails - The nail launching technology (Bridle and Myles, 1991; Ingold and Myles, 1996) consists of firing directly into the ground, using a compressed air launcher, nails of 25mm and 38mm in diameter, made from bright bar (EN3B to BS982) with nail lengths of 6 meters or more. The installation of nails by means of compressed air is rapid, flexible, and economical. The nails are installed at speeds of 200 mph with an energy transfer of up to 100 kJ, which makes it possible to insert up to 15 nails per hour. This installation technique enables an optimization of nail installation with a minimum of site disruption, as compared with conventional technology of drilling and grouting.

As illustrated in Fig. (5), in the compressed air launcher, the force from the air pressure is transferred to the tip of the nail thus placing the nail in tension, assuring straightness and preventing buckling on ground penetration. During penetration, the ground around the nail is displaced and compressed. The annulus of compression developed reduces the surface friction and minimizes damage to protective coatings such as galvanize and epoxy.

The technology is presently used primarily for slope stabilization especially in sensitive soils, such as loose and quick sand formations, as an alternative to conventional methods. It also appears to be cost effective for large slope stabilization, as used by the Forest Department and US Department of Agriculture. Successful applications have also been recorded for retrofitting of retaining systems. However, a rigorous evaluation of the pull-out resistance of launched nails is required prior to their use in retaining structures. Future developments envisage the adoption of different shaped cross sections, increasing the perimeter/area ratio and the cross section stiffness, as well as the potential use of synthetic nails.

Soil-Nailed Systems- Recent Applications

It is worth noting that during the past two decades, soil nailing systems have been increasingly used for the construction and retrofitting of bridge abutments, as well as for excavation support and seismic retrofitting in earthquake zones. Observations on new applications and recent research results on the seismic performance of soil nailed systems are briefly summarized below.

Figure 4.   Use of Jet nailing for Retrofitting of reinforced earth Structure.

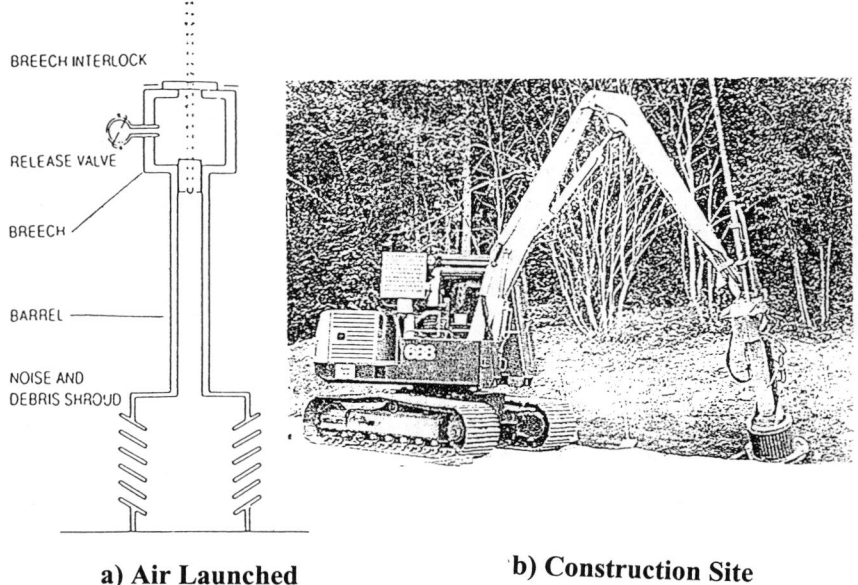

a) Air Launched   b) Construction Site

Figure 5.   Launched Soil Nails (After Ingold and Myles, 1996)

Soil-Nailing for Bridge Abutments

For bridge abutments, as with tiebacks, soil nailing presents the advantage of construction from the top-down in cut situations, where generally a precast or cast-in-place diaphragm wall is constructed from existing grade. The nails are installed during the excavation phase. For more common cases of underpass widening, nails offer the advantage in that their installations does not require soldier piles, which would be difficult to install under restricted head room conditions. Soil nailing presents a significant advantage for construction in urban areas as nail installation can proceed under existing highway facility without practically any disturbance to traffic, and with a minimum right of way acquisition.

A case study of monitoring a soil nailed wall under a piled bridge abutment located under the south end of the Oregon Slough bridge was reported by the Oregon Department Of Transportation and the Federal HighWay Administration in Portland (Kimmerling and Chassie, 1993). The wall 4.2 to 5.7m height shown in figure (6) was designed using the limit equilibrium analysis (Davis Method) to provide for excavation support and retention of the existing 2:1 soil slope beneath the bridge abutment. The bridge abutment is supported on a single row of 360 mm diameter steel pipe piles which are filled with unreinforced concrete, .and are spaced 1.4 m on center. The wall was instrumented to monitor the performance during construction and service state and to evaluate the design assumptions and methodology. Instrumentation involved (i) horizontal movement of the wall face, (ii) horizontal movement and rotation of the pile-cap; (iii) axial strain and load along the soil nail; (iv) nail load at the wall face; (v) bending strain in the nail near shotcrete face; (vi) bending strain in the piles below the pile cap; and (vii) horizontal earth pressures on the shotcrete wall face.

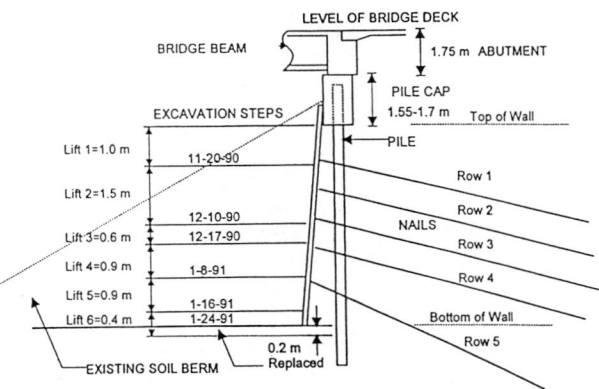

Figure 6. Aerial view and construction sequence (Kimmerling and Chassie, 1993)

Extensive 3-D finite element modeling was carried out by Briaud and Lim (1997), for the analysis of the experimental data. The axial forces versus length distribution in the nails were matched in the calibration processes as mentioned by the authors. For each nail an apparent earth pressure was obtained by calculating the ratio of the maximum nail force over the tributary area of the wall facing for that nail. As reported by Briaud and Lim, (1997), the results shown in figure (7) suggest that, as previously reported by Juran and Elias (1987), the maximum value of the apparent earth pressure and its distribution are close to that proposed by Terzaghi and Peck (1967), for braced excavation. Figure (7) also illustrates that the experimental data agree fairly well with predictions obtained with the KADRENSS working stress analysis code (Juran and Elias, 1991) assuming that the cap and bridge loading is entirely supported by the piles, while the soil nailed wall is subjected to its self weight loading and to the surcharge loading due to the 3.4m thick backfill layer behind the bridge slab.

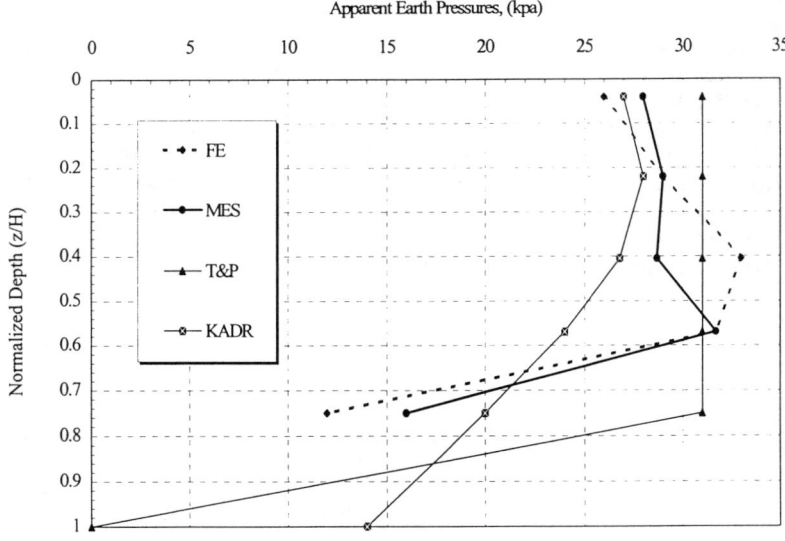

Figure 7. Comparison Between Measured Values (MES) of earth pressure, Finite element Simulations (FE), Terzaghi and Peck empirical distribution (T&P), and values obtained from the KADRENSS (KADR)

It is worth noting that the experimental data was used as reported by Kimmling and Chassie, (1993) to evaluate the limit equilibrium analysis method (Modified Davis Method). As indicated by the authors, comparison of predicted and measured nail forces for instrumented section away from the bridge slab, illustrate that, consistently with results reported by different investigators (Thomson and Miller, 1991; Juran and Elias, 1987), the Davis method over predicts nail forces in the lower nails and is likely to under predict the nail force in the upper nails. Figure (8) illustrates the comparison of nail forces measured at the end of construction in an instrumented wall section under the bridge slab, the modified Davis Method prediction, the Terzaghi and Peck (1967) earth pressure diagram for braced excavation, and the KADRENSS working stress analysis code. It illustrates that the triangular distribution of nail loads, top to bottom of the wall, predicted by the Modified Davis Method is not consistent with the more uniform distribution of nail loads typically measured (Juran et al, 1991) in soil-nailed structures, and results in significant overestimate of the measured nail forces. The Terzaghi and Peck (1967), earth pressure diagram and particularly the KADRENSS code appear to yield reliable predictions of the measured nail forces. As indicated by Thompson and Miller (1990), while limit equilibrium analyses are currently common tools for the design of soil nailed walls, such analyses do not provide good estimates of the magnitude and location of the maximum nail forces.

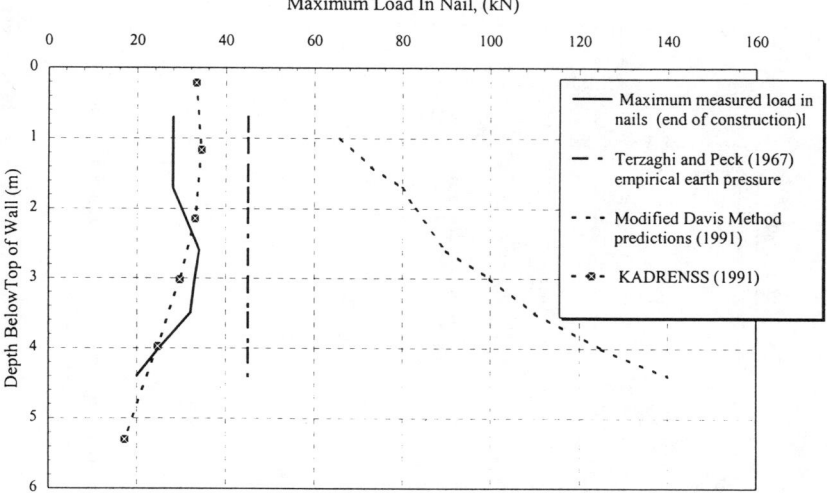

Figure 8.   Comparison between measured nail forces , modified Davis Method Prediction, Terzaghi and Peck empirical earth pressure distribution, and the KADRENSS code predictions.

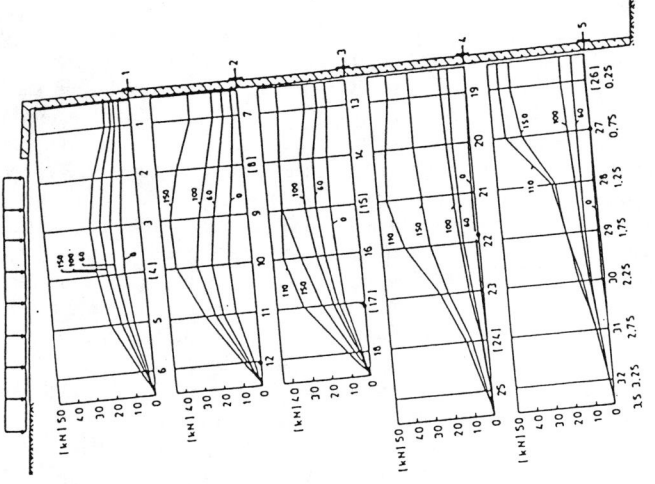
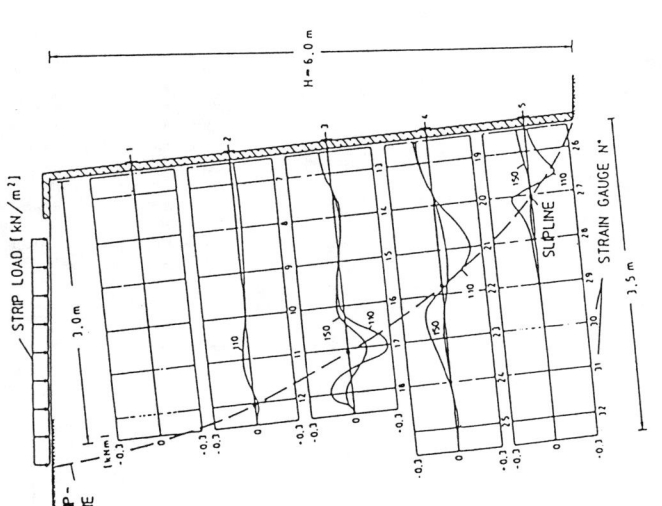

Figure 9. Field test results conducted By Gassler (1993)

a. Bending moments at failure ($p_f = 150$ kN/m^2) and after failure ($p_f = 110$ kN/m^2).

b. Distribution of axial force.

Of particular interest is the experiment conducted by Gassler (1993) on a soil nailed retaining wall subjected to a surcharge loading at the ground surface. The results of this full scale experiment, shown in figure (9), illustrate the nail bending moments measured by Gassler (1993) under surcharge loading. At the state of failure which occurred by nail sliding under the load of $p_f = 150$ kn/m^2, the bending moments remained rather low and did not generate significant shear forces in comparison to the axial nail forces. As a conclusion from several field test walls, Gassler (1993) suggested that under service loading, the shear forces in nails of diameters generally used in practice can be neglected. figure (9) illustrates that the surcharge does not practically affect the location of the maximum tension force in the nails. However, at failure by slippage of the nails, the observed slip line in the soil is located behind the locus of maximum tension forces in the nails. It is worth noting that the measured nail forces shown in figure (10) were found to agree fairly well with the KADRENSS design code predictions for both cases of (i) self weight loading, and (ii) surcharge loading. These results also illustrate that the observed behavior of nailed cut slopes is quite similar to that of braced excavations.

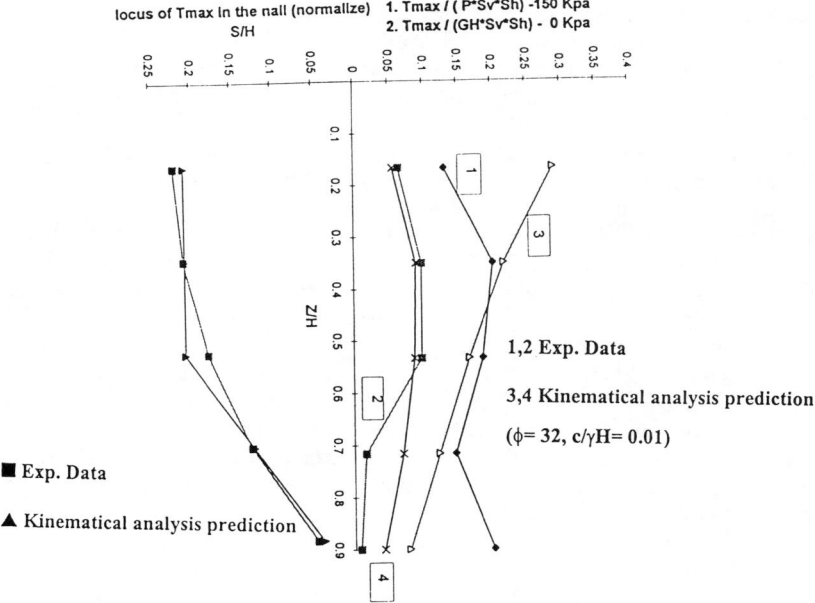

Figure 10   Comparison between measured data and KADRENSS code predictions for (I) self weight. (ii) surcharge loading 150 kpa

The major limitation of the slope stability analysis procedures currently used in design of soil-nailed retaining structures lies in the basic definition of a global factor of safety. Observations on both full-scale structures and reduced-scale laboratory models have illustrated that pull-out failure is a progressive phenomenon that is generally induced by the sliding of the upper inclusions. Therefore, this internal failure mechanism cannot be adequately defined using a "global" value of a unique safety factor for all the inclusions. The local stability (or safety factor) at the level of the sliding inclusion can be significantly more critical than the "global" stability with respect to general sliding in the retaining system or the surrounding ground. Therefore, as suggested by Juran and Elias (1991), for a reliable design of these composite structures, it is essential that the engineer should attempt to rationally evaluate both the working forces in the nails to assess the local internal stability at the level of each inclusion and the "global" structure stability.

Seismic Performance of Soil-Nailed Retaining Structures

Soil nailed structures are systems that are coherent and flexible, offering inherent advantages in withstanding large deformations and, as illustrated by post-earthquake observations (Bara, 1990; Felio et al, 1990; Tatsuoka et al, 1996) they present high resistance to earthquake loading, However, to date, the analysis of the available case histories and dynamic model tests results has been primarily limited to qualitative evaluation of the system performance and failure mechanisms.

The design methods most currently used for seismic stability analysis of soil nailed systems are derived from the pseudo-static Mononobe-Okabe analysis. Two fundamentally different pseudo-static design approaches have been developed: (i) limit equilibrium analysis (Schlosser, 1983; Koga et al, 1988; Calterance; 1990) which yields only a global safety factor with respect to a rotational or transitional failure of the reinforced soil mass and/or the surrounding ground along the potential sliding surface, and (ii) the working stress analysis using empirical correlations (Richardson and Lee, 1975), or numerically derived design assumptions (Seed and Mitchell, 1981; Dhouib, 1987; Segrestin and Bastick, 1988) to evaluate the seismically induced forces in the reinforcements. Displacement methods have also been incorporated in global limit equilibrium analysis (Bathrust and Cai, 1995), extending the sliding-block theory proposed by Newmark (1965) to predict the permanent horizontal displacements that may accumulate at the base of the structure during seismic events.

The limit equilibrium methods provide only a global safety factor with respect to the shear strength characteristics of the soil and/or the pull-out capacity of the reinforcements. They do not allow for an estimate of the seismic loading effect on the maximum tension and shear forces generated in the nails, and therefore cannot be used to evaluate the local seismic stability of the nailed soil at each reinforcement level. The requirement to evaluate the local pull-out stability of soil nailed retaining systems under both static and seismic loading conditions raises the need for the development and experimental evaluation of working stress design methods in order to estimate the seismic loading effect on the forces mobilized at each reinforcement level.

The KADRENSS computer code was extended (Choukeir et al, 1997) to allow for seismic pseudo static stability analysis of soil nailed retaining systems. The basic assumptions considered in this analysis implies that the seismic loading effect can be represented by a pseudo static self-weight inertia force due to the horizontal acceleration of the potentially sliding active zone limited by the locus of maximum tension forces in the nails. This pseudo static inertia force is equivalent to a uniform horizontal earth pressure acting along the potential sliding surface in the soil nailed mass. For a given earthquake acceleration, its magnitude is therefore directly related to the geometry of the active zone determined from the kinematical working stress analysis.

This pseudo static working stress analysis approach was evaluated (Choukeir et al, 1997) through the comparison of pull-out failure simulations with experimental observations on centrifugal soil nailed model walls conducted by Vucetic et al., (1993) and numerical model simulations conducted by Choukeir (1996). Pull-out failure observations on the centrifugal shaking table soil nailed model walls are illustrated in figure (11).

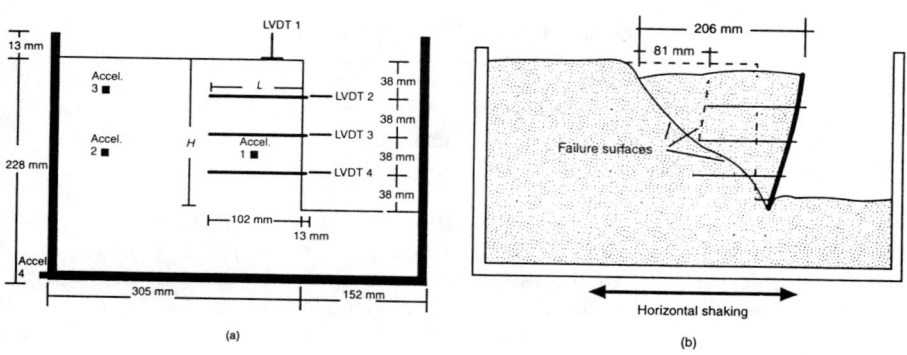

Figure 11   Configuration of centrifuge model wall (Vucetic et al, 1993). (a) Longitudinal view of the model box. (b) Failure surface.

Figure (12) illustrates the comparison of predicted and measured failure geometry in the centrifugal soil-nailed model walls for different acceleration levels of a_m/g = 0.1, 0.28, and 0.43. This comparison illustrates that, the method predictions agree fairly well with the experimental values of L/H and with the numerical test simulations. The low L/H values obtained for a/g=0.1 can be probably related to the effect of the experimental technique (Choukeir et al, 1997).

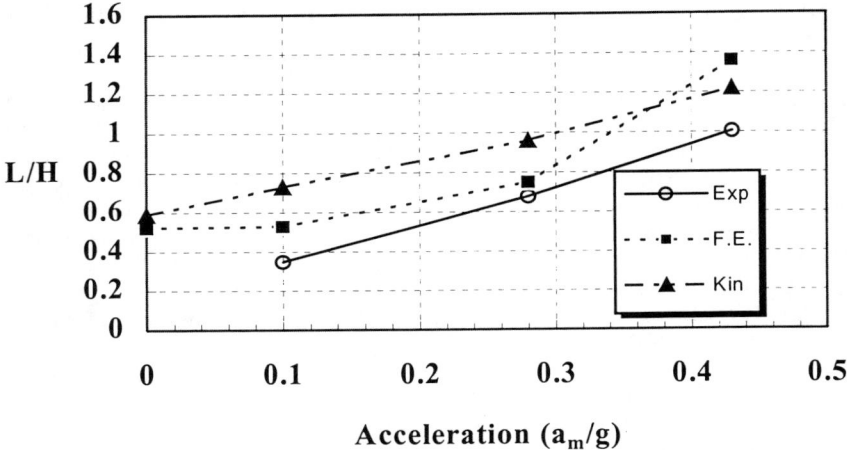

Figure 12 Comparison between experimental results (Exp.) of pull-out failure (L/H values), and values obtained from the finite element simulations (FE) and from KADRENSS code predictions (KADR)

Figure (13) illustrates the comparison between the method predictions, the numerical values and the experimental results obtained for the normalized locus of maximum tensile force S/H for various input base acceleration levels a_m/g= 0.0, 0.1, 0.28, and 0.43. This comparison illustrates that the method predictions agree fairly well with the finite element simulations and under predicts the post-failure measured values for S/H.

The large difference between the observed post-failure S/H values recorded by Vucetic et al, (1993) and the predicted S/H values is most probably due to the basic difference between the assumptions incorporated in the numerical and kinematical analyses and the observed propagation mode of the pull-out failure through the soil-nailed mass under working stress conditions. As shown by Schlosser and Juran (1983), the locus of maximum tension forces under working loading conditions is quite different from that developed through a composite reinforced soil mass during pull-out failure propagation, as the soil-reinforcement interaction is not large enough to prevent the progressive sliding of the reinforcement prior to failure.

However, it should be emphasized that further experimental research and particularly centrifugal and shaking table model testing is necessary in order to establish a statistically significant data base for the seismic performance assessment as well as for the development and experimental evaluation of reliable seismic design methods for the engineering use of soil nailing in earthquake zones.

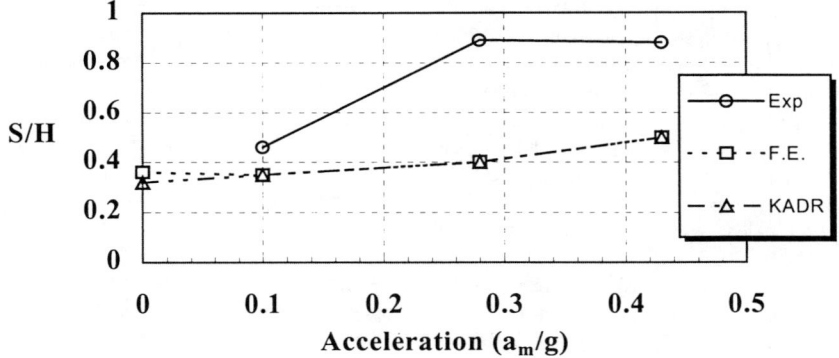

Figure 13. Comparison between experimental results (Exp) of pull-out failure (L/H values), and values obtained from finite element simulations (FE) and from KADRENSS code predictions (KADR)

Soil Nailing Database Development

As soil nailing currently offer effective technical solutions for a variety of geotechnical applications, the inherent value of an international Database, which will efficiently collect, synthesize, analyze, and display case studies of soil nailing has become increasingly evident for Industry, Universities, and Governmental Agencies. In order to effectively foster worldwide technology transfer and know-how exchange, the Technical Committee on Ground Improvement Reinforcement and Grouting of the International Society for Soil Mechanics and Geotechnical Engineering (ISSMGE TC-17), with the support of the Federal Highway Administration has initiated the development of an International Knowledge Database for Ground Improvement Technologies (IKDGIT).

IKDGIT is to serve engineers by responding to specific engineering queries that are critical and frequently posed during various phases of planning, design and construction of ground improvement projects from different countries. Such a system will help an engineer to retrieve information such as possible technologies for the project, similar case histories, problems encountered, possible remedial action schemes, comparative cost data, QA/QC, engineering specifications, applicable codes, etc. The engineer may use the information to supplement his own experiences and allow for comparative analysis to be part in developing the design and construction recommendations for the project. It is developed with an Internet Interface, in order to provide a world-wide user friendly access for data search, feed-in, statistics, retrieval and display and thereby create a bridge for technology transfer among the geotechnical communities. The database presently contain about 70 soil nailing case studies including Full Scale Experiments, Research Projects, and observations on actual structures.

The IKD provides the user with an interactive model to retrieve valuable statistical information with regard to each technology from the available case studies. This model provides the user with a design aid framework for comparative assessment of alternative technical solutions with the current state of practice. For this purpose the statistical analysis model is designed to yield relevant empirical correlations among selected design parameters. The following are some correlations derived from the IKDGIT, using the statistical analysis model which tend to represent the current state of soil nailing design practice. For specific details on the case studies refer to the IKDGIT web site currently at http://TC17.poly.edu.

Figure 14 shows the statistical model interface which allows the user to assess field correlations for a specific Technology, Application, Soil Type, and Country. The following soil nailing design correlations were selected in order to illustrate the current design practice with respect to working nail forces and pull-out evaluation.

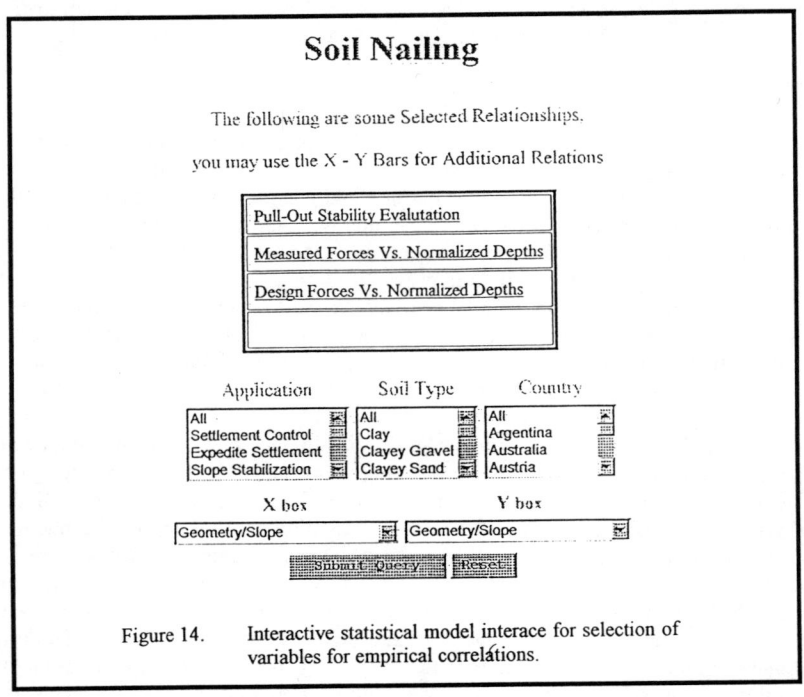

Figure 14. Interactive statistical model interace for selection of variables for empirical correlátions.

Figure (15) shows the variation of the values of maximum normalized forces T_N (Eq. 2) measured in the nail with respect to the structural geometry L/H (where L is the inclusion length and H is the structure height). The results indicate that (i) measured nail forces are independent of the structural geometry, (ii) the measured normalized nail forces, are in good agreement with the values predicted

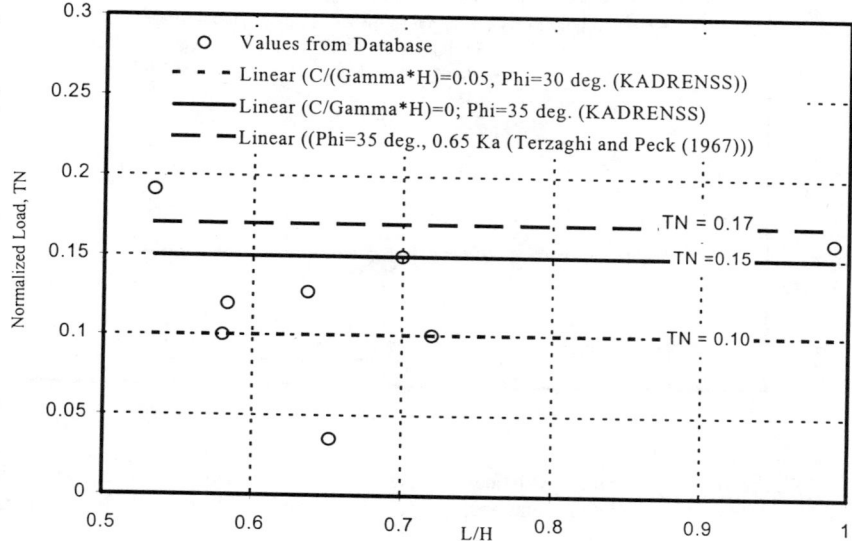

Figure 15. IKDGIT statistical analysis for comparison of measured and predicted relationship of normalized nail forces, TN, versus structural geometry L/H.

according to Terzaghi and Peck lateral earth pressure diagram (1967), and with the working forces predicted with the KADRENSS code (Juran and Elias, 1991) for the specified conditions.

Figure (16) illustrates the current design practice in terms of the relationship between the normalized design nail forces and the structural geometry. The results tend to confirm that (i) in current design practice, the normalized design nail forces are independent of the structure geometry; (ii) the design values (T_N=0.55) are significantly higher then the measured values and the values predicted from the Terzaghi and Peck diagram and from the KADRENSS computer code (T_N=0.17, and 0.1) respectively; (iii) the difference between design and measured values tends to indicate a conservative design practice which would correspond to a Factor of Safety of about 3 to 4 with respect to the normalized nail strength as often used by practitioners.

Figure (17) shows the results from the database with regard to current design practice with respect to pull-out resistance evaluation.

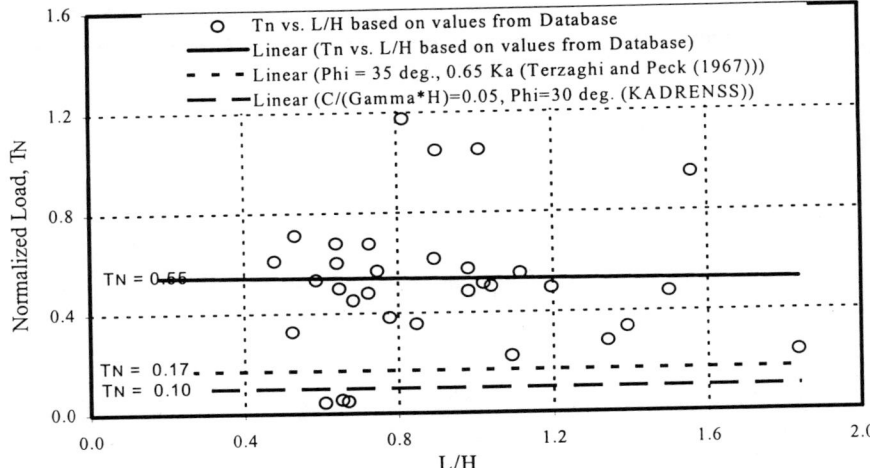

Figure 16. IKDGIT statistical analysis for comparison of designed and predicted relationship of normalized nail forces, TN, versus structural geometry L/H.

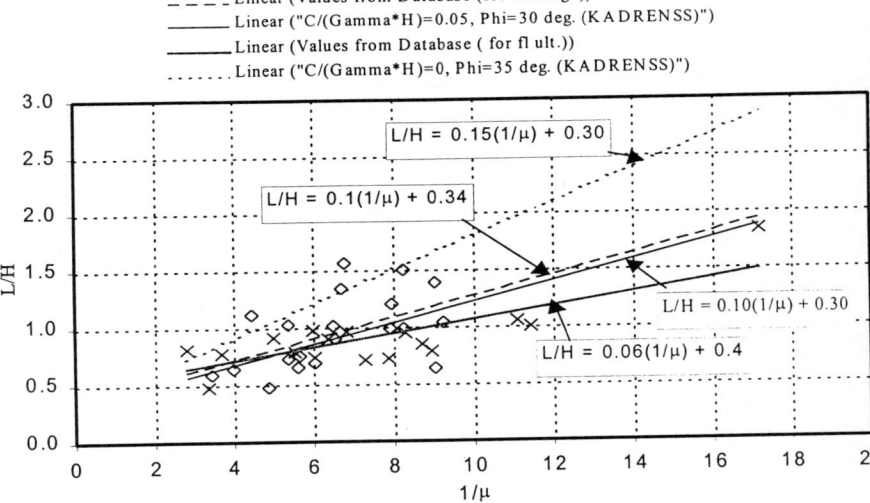

Figure 17. IKDGIT statistical analysis for comparison of measured and predicted relationship of structural geometry L/H versus the pull-out resistance factor.

The results are expressed in terms of the relationship between the structural geometry L/H and the pull-out resistance factor μ (Eq. 3). The linear regression of the experimental data is compared with the analytical expression:

$$\frac{L}{H} = \frac{1}{\mu} \cdot T_N + 0.3 \tag{1}$$

where,

$$T_N = \frac{T_{max}}{\gamma \cdot H \cdot S_H \cdot S_V} \quad (2), \quad \text{and} \quad \mu = \frac{f_l \cdot \pi \cdot D}{\gamma \cdot S_H \cdot S_V} \tag{3}$$

where γ is the soil unit weight, H is the structure height, and S_H, S_V are the horizontal and vertical spacing respectively.

Equation (1) is derived from the equilibrium of forces between the maximum working nail force T_{max} and the pull-out resistance force, T, given by:

$$T = f_l \cdot \pi \cdot D \cdot (L - L_{ac,max}) \tag{4}$$

where $L_{ac,max}$ is the maximum length of the inclusion in the active zone which can be expressed as $L_{ac,max} \approx 0.3$ H, D is the nail diameter, and f_l is the design or ultimate ground-grout bond adhesion (kn/m^2).

The database values are compared with the KADRENSS predictions for the specified conditions (i.e. T_N values of 0.10 and 0.15). The linear regression of the experimental data obtained for the f_l design values tend to agree with the KADRENSS predictions yielding the experimentally derived relationship:

$$\frac{L}{H} = 0.10 \cdot \frac{1}{\mu} + 0.34 \tag{5}$$

The linear regression of the experimental data obtained for the f_l ultimate values yields:

$$\frac{L}{H} = 0.06 \cdot \frac{1}{\mu} + 0.4 \tag{6}$$

Comparison between Eq. 5 and 6 suggests that in soil nailing design practice, the factor of safety with respect to pull-out resistance commonly used by the practitioners is within the range of 2 to 3. It is worth noting that most of the experimental data obtained for f_l design values are within the range of the KADRENSS predictions obtained for T_N values of 0.1 and 0.15 respectively.

It should be emphasized that the practical use of such field correlations is subjected to limitations due to differences in the site conditions, soil types, installation techniques etc. However, in-spite of these inherent differences, the IKD appears to provide relevant information for the assessment of the current state of practice and experimental evaluation of basic design equations.

Technological Development and Research Needs

The increasing use of soil nails in permanent structures is a key parameter in current technological developments. Durability of the inclusions, long-term performance in fine-grained soil, and environmental/architectural requirements for soil-nailed facing have become major design considerations.

The potential application of the technology in more aggressive environments has stimulated specialty contractors to continuously invest in the development of more reliable and cost-effective corrosion-protection schemes. In particular, it is noted that driven soil nails, which are commonly used in parts of Europe, are not corrosion-protected. Their implementation in the construction of permanent structures requires innovative improvements in the manufacturing and/or installation process to provide adequate protective coatings that may resist construction damage. Furthermore, implementation of newly developed corrosion-protection systems raises a substantial requirement for innovative quality control procedures to properly assess the performance of the proposed protection scheme.

During the past decade, technological efforts have been invested by European contractors to develop cost-effective driving installation processes, such as jet nailing and nail launching that may effectively increase construction rate and thereby substantially decrease the project cost while providing effective means for slope stabilization and construction of retaining systems in fine grained soils. However, the engineering use for such innovative technologies in permanent structures and their potential applications in fine-grained soils raises significant research needs pertaining to long-term performance and, in particular, to the effect of the installation process on the creep response and on soil-nail interaction in clayey soils.

As it is often the case in geotechnical engineering, technology and construction practice with soil nailing has generally preceded any fundamental research on the behavior or long-term performance of the soil nailed systems and any experimental or theoretical developments of appropriate design methods. With regard to such considerations, the most significant contribution of the past ten years to the current state of practice has undoubtedly been the full scale experiments and monitoring of in-service structures that have demonstrated the performance of soil nailed structures under a variety of static and seismic loading conditions.

At this stage, synthesis and world-wide dissemination of local experience can play a key role in advancing the state of practice. In particular the IKDGIT developed by the ISSMGE TC-17 and the International Ground Improvement Center at *Polytechnic University* with the support of the Federal HighWay Administration through the cooperation with the national geotechnical societies around the word, appears to provide an effective user friendly tool toward this goal. Its future development is expected to provide the user with an interactive model to retrieve valuable statistical information from the available worldwide experience for comparative assessment of alternative technical solutions and evaluation of selected technologies and engineering design schemes. The ISSMGE-TC 17 is presently working with the national societies on the constitution of regional centers to establish an effective information sharing network to accomplish there objectives.

Acknowledgments

The authors wish to acknowledge TC 17 delegates in the development of IKDGIT. The authors also wish to thank The Federal HighWay Administration for their support and, in particular Mr. Al Dimillio for his effective cooperation in this study.

References

Abramson, L.W. and W.H. Hansmire, (1988), "Three Examples of Innovative Retaining Wall Construction," *Proceedings of the 2nd Int. Conf.. On Case Histories in Geotechnical Engineering*, June, St. Louis, Paper No. 6.11, pp. 1191-1200.

American Association of State Highway and Transportation Officials (AASHTO), 1990, "In-Situ Soil Improvement Techniques," *American Association of State Highway and Transportation Officials (AASHTO) — The Associated General Contractors of America (AGC) — American Road & Transportation Builders Association (ARTBA) Joint Committee Subcommittee on New Highway Materials*, Task Force 27 Report, Jan.

Bayrne,R,, Cotton,D, Porterfield,J., Wolschlag,C., and Ueblacker,G (1996), "Manual for Design and Construction Monitoring of Soil Nail Walls", *FHWA report No. SA-96-069*

Bang, S. (1991), "Contribution to Summary of Design Methods Comparison for Nailed Retaining Walls", *FHWA Demonstration Project 82*.

Briaud, J.L., and Lim, Y., (1997),"Soil-Nailed wall under Piled Bridge Abutment – Simulation and Guidelines" *„Journal of Geotechnical Engineering, ASCE, Vol. 123, No. 11, Nov.*

Bridle, R. and Myles, B. (1991), "A Machine for Soil Nailing Process and Design", *ECOLE Nationale des Ponts et Chaussees, Civil Engineering European Courses in Soil Reinforcement*, Paris.

Bruce, D.A. and R.A. Jewell (1987), "Soil Nailing: Application and Practice — 2 Parts," *Ground Engineering*, Vol. 20, No 1, Jan., pp. 21-28.

Byrne, J. (1991), Contribution to Summary of Design Methods Comparison for Nailed Retaining Walls, *FHWA Demonstration Project 82*.

CALTRANS (1991), "A User's Manual for the SNAIL Program, Version 2.02 — Updated PC Version," Obtained from Mr. K. A. Jackson, *California*

Department of Transportation, Division of New Technology, Material & Research, Office of Geotechnical Engineering, Sacramento, Oct.

Chassie, R.G., 1994, "FHWA "Ground Nailing Demonstration Project, Guideline Manual and Workshop," *US Federal Highway Administration Publication*.

Choukeir, M.H. (1996) "Seismic Analysis of Reinforced Earth and Soil Nailed Structures", *PhD Dissertation, Polytechnic Univeristy, New York*.

Choukeir,M., Juran, I, and Hanna, S., (1997) "Seismic Design of Reinforced-Earth and Soil-Nailed Structures" *Journal of Ground Improvement, Vol. I, No.4, Oct, 97*

Elias V. and Juran, I. (1991), "Soil Nailing for Stabilization of Highway Slopes and Excavations," *United States Federal Highway Administration, Publication No. FHWA-RD-89-193*, June.

Fannin, R. J. and R. K. Bowden, (1991), "Soil Nailing: An In-Situ Reinforcement Technique," *Geotechnical News*, June, pp. 32-34.

Felio, G.Y., Vucetic, M., Hudson, M., Barar, O., and Chapman, R. (1990), Performance of Soil Nailed Walls during the October 17, 1989 Loma Prieta Earthquake. *Proceedings, Forty-third Canadian geotechnical Conference*, Quebec, pp. 165-173.

French National Research Project CLOUTERRE (1993) "Recommendations CLOUTERRE 1991 - Soil Nailing Recommendations 1991" *Presses de l'Ecole Nationale des Ponts et Claussees* English Translation, July. Gassler, G. and

Gudehus, G. (1981), Soil nailing: Some mechanical aspects of in situ reinforced earth, *Proceedings of the 10th International Conference on Soil Mechanics and Foundation Engineering*, Stockholm, Sweden, **3**, pp. 665-670.

Gassler, G. (1993), The first two field tests in the history of soil nailing on nailed walls pushed to failure *"Soil reinforcement: full scale experiments of the 80's" Presses de l'ecole Nationale des Ponts et Chaussees*, CEEC, pp. 7-34.

Ingold, T.S. and Myles, B. (1996), Ballistic soil nailing, Earth Reinforcement, H. Ochiai, N. Yasufuku, and K. Omine, eds., *Proceedings of the International Symposium on Earth Reinforcement*, Japan.

Juran, I. (1977) "Dimensionement Interne, Ovrages en Terre Armee," These de Docteur-Ingenieur, ENPC, Paris.

Juran, I. (1987), Soil Retaining Structures: Design and Practice, Trans. Research Record 1119, Trans. Research Board, Washington DC, pp. 139-150.

Juran, I., and Elias, V. (1987), Soil nailed retaining structures: Analysis of case histories, Soil Improvement - A Ten Year Update, *ASCE Special Geotechnical Publication* No. 12, Ed. JP Welsh, pp. 232-244.

Juran, I., Baudrand, G., Farrag, K. and Elias, V. (1988) "Kinematical limit analysis for design of nailed structures", *Journal of Geotechnical Engineering, ASCE, Vol. 116, No. 1, pp. 54-72*.

Juran, I. and Elias, V. (1991) Ground Anchors and Soil Nails in Retaining Structures, *Foundation Engineering Handbook*, 2nd Edition, pp. 868-905.

Kimmerling, and Chassie (1993), Performance monitoring of a soil nail wall used in a bridge abatement application, "Soil reinforcement: full scale experiments of the 80's" Presses de l'ecole Nationale des Ponts et Chaussees, CEEC, pp.

63-92.

Levy,O., Juran,I., Dimillio,Al (1998),"Development of International Knowledge Data Base for Ground Improvement", sub. to *Journal of Ground Improvement* July,98.

Long, J.H., Chow, E., Cording, E.T., and Sieczkowski, W.J. (1990) Stability Analysis for Soil Nailed Walls, Geotechnical Special Publication No. 25, ASCE, pp. 676-691.

Louis, C. (1986), "Theory and practice in soil nailing temporary or permanent works", *ASCE Annual Conference*, Boston.

Louis, C. (1990),"Soil Nailing System – Slope Stabilization and Protection of Underground Excavation", *Procedes Specieux Geotechnique Cenie Civil Mines Energine, Paris*.

Matsui, T., et al, (1993) Case histories on reinforced cut slopes due to root piles in Osaka, "Soil reinforcement full scale experiments of the 80's" pp. 133-167.

Mitchell, J.K., et al. (1987), Reinforcement of earth slopes and embankments, *National Cooperative Highway Research Program Report* No. 290, Transportation Research Board, June.

Nicholson, P.J., (1986), "In Situ Earth Reinforcement at Cumberland Gap, U.S. 25E," ASCE and Pennsylvania Department of Transportation Joint Conference, Harrisburg, Pa., April, 21 pp.

Plumelle, C. Schlosser, F., Dolage, P. and Knochenmus, G. (1990), French National Research Project on Soil Nailing: CLOUTERRE, Geotechnical Special Publication No. 25, ASCE, pp. 660-675.

Plumelle, C. (1993), Full scale test of a soil nailed wall: design, construction and experimentation, "Soil reinforcement: full scale experiments of the 80's" Presses de l'ecole Nationale des Ponts et Chaussees, CEEC, pp 35-62.

Plumelle, C. (1993), Conception, execution et experimentation d'un massif en sol cloue en vraie grandeur, C.R. Symposium International sur le Renforcement des Sols: Experimentations des annees 1980, ENPC, Paris, pp. 35-62.

Sabhahit, N., Madhav, M.R. and Basudhar, P.K. (1996), Seismic analysis of nailed soil slopes -- A pseudo-dynamic approach, H. Ochiai, N. Yasufuku, and K. Omine, eds., *Proceedings of the International Symposium on Earth Reinforcement*, Japan.

Schlosser, F. and Segrestin, P. (1979), Dimensionnement des Ouvrages en Terre Armee par la Methode de l'Equilibre Local, *International Conference on Soil Reinforcement: Reinforced Earth and Other Techniques*, Paris.

Schlosser, F., Unterreiner, P., and Plumelle, C. (1993) Validation des methodes de calcul de clouage par les experimentations du Projet National CLOUTERRE. Revue Francaise de Geotechnique, 64.

Shen, C.K., Herrmann, L.R., Romstand, K.M., Bang, S., Kim, Y.S., and Denatale, J.S. (1981a), *In-situ Earth Reinforcement Lateral Support System*, Report No. 81-03, Department of Civil Engineering, University of California, Davis.

Shen, C.K., Bang, S. Romstad, J.M., Kulchin, L., and Denatale, J.S. (1981b), Field measurements of an earth support system, *Journal of the Geotechnical Engineering Division, ASCE*, **107**, No. GT-12.

Shen, C.K., (1991), Contribution to Summary of Design Methods Comparison for

Nailed Retaining Walls, FHWA Demonstration Project 82.

Stocker, M.F., Korber, G.W., Gassler, G., and Gudehus, G. (1979), Soil nailing, *International Conference on Soil Reinforcement*, Paris, **2**, pp. 463-474.

Stocker, M.F. and Riedinger (1990), The bearing behavior of nailed retaining structures. "Design and Performance of Earth Retaining Structures, P.C. Lambe and L.A. Hansen, Eds., ASCE Geotechnical Special Publication No. 25, New York, pp. 612-628.

Tatsuoka,F., Tateyama,M., Koseki,J., (1996), "Performance of Soil Retaining Walls for Railway Embankments", Soils and Foundations, Special Issue on Geotechnical Aspects of the Jan.,1995 Hyogoken-Nanbu Earthquake, pp 311-324

Terzaghi, K. and Peck, R.B., (1948 & 1967), *Soil Mechanics in Engineering Practice*, John Wiley and Sons, Inc., New York, N.Y.

Thompson, S.R. and I.R. Miller, (1990), "Design, Construction and Performance of a Soil Nailed Wall in Seattle, Washington," Design and Performance of Earth Retaining Structures, P.C. Lambe and L.A. Hansen, Eds., ASCE Geotechnical Special Publication No. 25, New York, pp. 629-643.

Unterreiner, P., Benhamida, B., and Schlosser, F. (1997), Finite element modelling of the construction of a full-scale experimental soil-nailed wall. French National Research Project CLOUTERRE, *Journal of Ground Improvement*, Vol. 1, No. 1: 1-8.

Vucetic, M., Tufenkjian, M., and Doroudian, M. (1993), Dynamic Centrifuge Testing of Soil-Nailed Excavations. *ASTM Geotechnical Testing Journal*, Vol. 16, No. 2: 172-187.

Vucetic, M., Iskandar, V.E., Doroudian, M. and Luccioni, L. (1996), "Dynamic Failure of Soil-Nailed Excavation in Centrifuge", *Earth Reinforcement, Balkama, Inc., pp. 829-834.*

Welsh, J.P. et al. (1987), "Soil improvement — A ten year update", *ASCE Geotechnical Special Publication* No. 12.

Xanthakos, P., Abramson, L. W., and Bruce, D. A. (1994), "In situ ground reinforcement", *Ground Control and Improvement*, pp. 331-402.

CHEMICAL STABILIZATION OF KAOLINITE BY ELECTROCHEMICAL INJECTION

by

Senda Ozkan, M.S.*[1]

Robert J. Gale, Ph.D.[2]

Roger K. Seals, Ph.D., P.E.[1]

* Corresponding Author
Phone: (504) 388-8512
Fax: (504) 388-4945
e-mail: sozkan@unix1.sncc.lsu.edu

---

[1] Graduate Research Assistant, and Professor respectively; Civil and Environmental Engineering Department, Louisiana State University, Baton Rouge, LA 70803.

[2] Professor; Chemistry Department, Louisiana State University, Baton Rouge, LA 70803.

CHEMICAL STABILIZATION OF KAOLINITE BY ELECTROCHEMICAL INJECTION

Abstract

Cementation of kaolinite by the electrochemical injection of aluminum and phosphate ions has been investigated. Electrodes were placed across a kaolinite bed, 35.5 cm length, 13 cm width, and 16 cm depth. A current of 90 mA is applied across the electrodes. At completion of the tests, changes in the kaolinite properties such as undrained shear strength, water content, and Atterberg limits were analyzed. An average increase in undrained shear strength of 600% was observed for the phosphoric acid treatment of the kaolinite, whereas an average increase of 500% in undrained shear strength is observed for aluminum sulfate/phosphoric acid treatment. The formation of insoluble salts in the kaolinite specimens can lead to increase in shear strength by ion exchange and/or precipitation mechanisms.

Introduction

Stabilization of fine-grained soils by injecting chemicals under electrical fields is a possible stabilization technique. Methods of introduction of stabilizing agents have always been a problem for *in-situ* soil stabilization. To date, *in-situ* injection of chemicals has been achieved chiefly through hydraulic techniques. Such chemicals often follow the high permeability deposits resulting in nonhomogeneous stabilization. In addition, it is difficult to inject chemical agents into clays with low permeability. A comprehensive understanding of species transport mechanisms in porous media under electrical fields can provide a means for effective and efficient injection of chemical species into soil deposits. Understanding of multi-species transport in a soil medium under electrical fields was advanced through the development of the electrokinetic remediation technique (e.g.1-6). The primary emphasis of this study is to investigate injection of ionic species to achieve stabilization by homogenous precipitation of the species, pore fluid reconstitution and appropriate ion exchange mechanisms.

Background

Electroosmotic technology has been used since the 1930's for extracting water from clays, silts and fine sands. Casagrande (7) introduced this technique as a method of soil stabilization and there are a number of successful field applications of consolidation by electroosmosis as a means for soil stabilization (8, 9). Bjerrum used electroosmosis for the consolidation of a quick clay in Norway. He observed a 10-fold increase in shear strength near the anode, a gradual decrease in strength towards the cathode, and almost no change in shear strength near the cathode. The theory of consolidation by electroosmosis was developed and well-established by Gray and Mitchell (10), Esrig (11), Wan and Mitchell (12), and Mitchell (13). When electroosmosis is used to move stabilizing agents into the soil, the technique is called "Electrokinetic Stabilization". Esrig and Gemeinhardt (11) investigated electrokinetic stabilization of an illitic clay by introducing calcium ions. They observed advantageous changes in plasticity and strength as a result of the electrokinetic treatment. The strength change was explained to be the result of ion exchange reactions, formation of precipitates, and structural rearrangement of the clay fabric. Madshus and Janbu (14) reported a field study by electrokinetic soil stabilization of a quick clay deposit. The electrolyte concentration was increased in the pore fluid by placing potassium chloride at the anode. The soils with an initial shear strength of about 20-25 kPa strengthened about 100% due to the increase in the electrolyte concentration in the pore fluid and also due to the increase in temperature under the electric fields (15).

Chemical stabilization of soils is chemical cementing of soils, in other words strengthening soils by bonding them with a cementing agent produced by chemical reactions. Mitchell (16) listed the most studied soil stabilization chemicals as phosphoric acid, salts, resins, sodium silicate, lignin, organic cations, and hydroxides. Silicates are the most popular chemical grouts. Van Impe (17) described formation of insoluble calcium silicate and binding of soil particles. The final undrained shear strength of soils, stabilized with the addition of 7-8% of lime, can be 10 to 50 times higher than that of untreated soils. Acids can react with carbonates and silicates to dissolve them and then precipitate salts as binders between the grains. Kelly and Kinter (18) studied the plasticity, moisture-density relations, volume change and unconfined compressive strength of several fine-grained plastic soils stabilized by phosphoric acid. They showed that the addition of phosphoric acid causes moisture absorption, volume changes at satisfactory levels, and increases in unconfined compressive strengths by varying degrees. Aluminum sulfate addition to improve stability of high plasticity clays has also been studied. It was found that even 0.1% was enough to increase the strength in certain cases (19).

Experimental Program

A variety of batch tests were conducted to determine the effects on the strength of kaolinite by mixing it with acidic, basic, aluminum, calcium and phosphate formulations. Objective criteria were established and used to select the best candidate formulations from the batch tests for the electrokinetic tests. The first test studied the effects of phosphate ions under an acidic environment. Phosphoric acid (1 M) was used at the anode and at the cathode compartment for this purpose. For the second test, it was decided to inject aluminum at the anode compartment and phosphate at the cathode compartment to study simultaneous injection of ions and possible precipitation of them throughout the soil medium. Aluminum sulfate solution (0.5 M) at the anode compartment and phosphoric acid solution (0.5 M) at the cathode compartment were used for this purpose.

Batch Tests

The batch tests consisted of mixing chemicals with certain amounts of kaolinite. Soils were mixed in bowls manually and placed into plastic jars. The specimens were prepared approximately at the same water content to be used in the bench-scale electrokinetic tests with the appropriate added weights of chemicals, 0.1%, 0.5%, 1.0%, 2.0%, 3.5%, and 5.0% by weight of product precipitant. Shear strength values were measured using a hand vane tester, after a curing period of 1 day and 5 days. During the curing period, the plastic jars containing the specimens were kept closed to avoid any change in water content. Measured shear strength values were compared to the shear strength values of untreated kaolinite at corresponding water contents.

Experimental Apparatus and Materials

The test container for the electrokinetic tests was made of 18 mm thick acrylic of width 13 cm, length 50 cm, and depth of 16 cm. Tungsten wires were used as voltage probes to determine the electrical gradient across the soil bed. The electrode compartments are located at the two ends of the container and the soil was placed between them. Titanium oxide/titanium mesh electrodes were used. Deionized distilled water, or a particular electrolyte, was used in the electrode compartments. The soil and the electrode compartments were separated from each other by acrylic frames. Each electrode chamber was connected to conditioning reservoirs which are a means of selecting the particular ions to be injected. The reservoirs maintained a constant water level across the specimen so the development of any external hydraulic gradient was prevented, and also supplied fluid that may be lost in the electrode compartments due to electroosmotic flow. Chemical solutions in the containers were circulated into the system by peristaltic pumps. A constant current of 10 mA was applied to the system. An acrylic lid was placed at the top of the box to minimize evaporation of moisture from the soil. A schematic diagram of the test setup is shown in Figure 1. The characteristics of experimented soil, EPK Kaolin are presented in Table 1.

Test Program

Two electrokinetic tests were conducted. For each test, 10 kgs of kaolinite were mixed with 4.3 l of distilled deionized water to achieve an initial water content of 44 %, midway between liquid and plastic limit values which are 56% and 30%. The moist soil was cured for one day in a covered container before being placed in the apparatus. The shear strength of the soil was measured prior to placing it into the acrylic box by a hand vane tester. The initial shear strength of the soil specimens was 7 kPa.

The first experiment (Test 1) was conducted by circulating 1 M phosphoric acid as the anolyte and catholyte. The test was conducted over a period of 14 days. For the second experiment (Test 2), 0.5 M aluminum sulfate solution (anolyte) and 0.5 M phosphoric acid solution (catholyte) were used. The similar constant current (10 mA) was applied to the system. Test 2 was conducted for a period of 21 days. Voltage readings and the pH of the electrolyte solutions were taken during both of the experiments. To assess possible thixotropic effects, a duplicate soil mixture was placed in another acrylic box. The electrode compartments were filled with deionized distilled water but no electric current was applied. Shear strength measurements were taken at the beginning of the test and at the end of 21 days. Changes in the undrained shear strength due to age hardening were measured in the duplicate soil specimen so that the effects of the electrokinetic phenomena could be differentiated.

The test specimens were sampled after electrokinetic processing. Water content, pH, and Atterberg limits were conducted on defined portions of the test specimens. Test specimens were divided into 2 horizontal layers, and 5 transverse sections. *In-situ* undisturbed, undrained shear strength values were measured after the completion of the electrokinetic tests by the hand vane tester. The measurements were taken at 24 points, 12 of them at 5 cm depth, and 12 of them at 10 cm depth. For water content measurements, soil samples were taken from the points where shear strength values were measured. The samples for the pH measurements of the soils were prepared using 1:10 ratio mixture by weight of soil and distilled water, respectively.

Results

Batch tests

Batch tests were utilized to assess the strength behavior of kaolinite mixed with different varieties and amounts of calcium, aluminum and phosphate formations. The effects of the chemicals were analyzed by comparing the vane shear test results with similar results for untreated kaolinite. Figure 2 shows the effects of some selected chemicals on the strength of kaolinite. These chemicals produced a peak effect as a function of weight (0-5%). The results are consistent with data reported in the literature. Excess amounts of phosphoric acid have an adverse effect and tend to reduce the internal angle of friction (20). Deflocculation of sulfate systems was reported by Wendelbo and Rosenqvist (21) due to edge adsorption effects of sulfate systems. Based on the results of the batch tests, phosphoric acid, a common soil stabilizer was selected for the first test (Test 1) and phosphoric acid/aluminum sulfate was selected for the second test (Test 2).

Test 1:

The pH value plays a very important role most in chemical equilibria in dissolution /precipitation aqueous phase reactions and in sorption reactions in the pore fluid as well as effects upon the fabric and engineering characteristics of the soil. If there is only distilled water at the electrode compartments, electrolysis reactions at the electrodes cause oxidation of the molecular water in the anolyte, decreasing its pH, and reduction of molecular water in the catholyte, increasing its pH. Test 1 was performed by 1 M H_3PO_4 solution, which resulted a pH range of 1.3-1.6 for both anolyte and catholyte solutions. The final pH distribution across the soil medium of Test 1 is shown in Figure 3. The entire specimen was acidified as a results of the electrochemical treatment. The soil pH changed from an initial value of 5.8 to values ranging from 2.7 to 4.0. Acidic soil conditions cause ionization of kaolinite, dissolve basic precipitates, desorb surface cation species into the pore fluid, and lead to a flocculated soil structure.

Undrained shear strength values were measured at various locations throughout the Test 1 specimen. There are significant differences in strengths within the top and bottom layers of the specimen (Figure 4a). The shear strength increased, on average, from 7 kPa to 30 kPa at the 5 cm depth of the specimen. It increased from 7 kPa to 70 kPa at the 10 cm depth of the specimen. The liquid and plastic limit values of the soil increased significantly throughout the specimen (Figure 5a). The liquid limit of the soil increased from 56% to 75% whereas the plastic limit increased from 30% to 40%. The plasticity index increased from 26 to 35. Water content distributions through the specimen are shown in Figure 6a. At the top portion of soil specimen, there is little change in water content, but swelling occurred near the anode and near the cathode regions. At the 10 cm depth of the specimen, the average water content decrease is greater. The changes in water content through the soil are an indication of changes in pore water pressure and effective stresses.

Test 2:

An anolyte pH in the range of 1.0 to 2.0 and a catholyte pH in the range of 1.5 to 2.0 were observed during the test. Figure 3 shows the pH distribution of the specimen in

Test 2. The pH values in the specimen vary from 2.5 to 3.5 after treatment, while it was initially 5.8. The entire specimen became acidic during the electrochemical treatment. In general, the soil near the anode is more acidic than that near the cathode. The higher acidity of the anolyte and transport of the hydrogen ions are the likely causes. Ionization and flocculation of kaolinite particles are expected due to the acidic environment.

The shear strength distributions through the soil are shown in Figure 4b. The strengths change from 10 kPa near the anode region to 70 kPa near the cathode region. The low shear strength near the anode may be the result of a high concentration of sulfate ions. Deflocculation of sulfate systems was reported (22) due to edge adsorption of sulfate ions. The liquid and the plastic limit values of the soil increased after the experiment (Figure 5b). The liquid limit varies in the soil mass from 52% at the anode to 80% in the cathode half of the cell, while the plastic limit increased from 30% to 40% respectively. The plasticity index increased, on average, from 26 to 35. The initial and final distributions of water content of the test specimen are shown in Figure 6b. These data indicate swelling of the soil throughout the top layer, especially near the anode and the cathode. This swelling effect may be one reason for the low shear strength near the anode and the difference in strength between the top and the bottom of the soil near the cathode. Electroosmotic consolidation is apparent only at the bottom layer of the soil. The effect of consolidation decreased from the anode to the cathode region.

Shear Strength Components of Tests 1 and 2

The strength increases achieved may not be only from the electrochemical treatment, but may also be due to water content decreases and thixotropic effects. To investigate thixotropic effects, a third test setup was prepared for comparison purposes. The kaolinite specimen was prepared in exactly the same manner as those for Tests 1 and 2 but left in the acrylic box without applying any voltage. The initial shear strength of soil medium was 6.4±0.78 kPa. After 14 days, the strength was 6.7±0.13 kPa. This result showed that when untreated kaolinite is mixed and left for the duration of experiments, no major strength increase will be observed due to time effects. The observed small strength increase is probably due to the water content changes. The shear strength increases due to water content change are estimated on the basis of water content/shear strength relation of untreated kaolinite. The results of these calculations indicated that the water content change observed has very little effect on the increases found experimentally in Test 1 and 2. The estimated effect of water content change on shear strength is 0-17 % and 0-22 % for Tests 1 and 2, respectively. Thus, the strength increase was achieved mainly by electrochemical treatment.

Summary and Conclusions

Electrochemical strengthening of kaolinite has been investigated by electrochemical injection of chemical species, namely phosphate and aluminum ions. A variety of batch tests demonstrated the strength behavior of kaolinite mixed with a selection of chemicals (22). Although an average shear strength increase of 500% - 600% was achieved, both tests above showed nonhomogeneous strength distributions through the specimen. The tests showed an average increase of 30% in Atterberg limits. Phosphoric acid injection showed a constant increase in Atterberg limits through the specimen. In Test 2, increase in Atterberg limits showed similar behavior with the shear strength, the limits increased towards to the cathode end. A change in the water content was observed as a result of the electrokinetic effects. Swelling occurred near the anode and the cathode, where there was direct contact with fluids and in the top region of the specimens, where unrestrained expansion was possible. Water content changes did not appear to significantly contribute to strength changes in either test.

This study aimed to achieve strength increase of kaolinite by appropriate ion exchange mechanisms and homogenous precipitation of injected ions through the specimen. In phosphoric acid treatment, the strength increase in kaolinite is mainly contributed by ion exchange reactions, ion adsorption and fabric changes. In aluminum sulfate and phosphoric acid treatment, besides the reactions that occurred in Test 1, the effects of the formation of insoluble phosphate compounds, especially aluminum phosphate, on the strength of kaolinite are expected. The results of these electrochemical tests are encouraging and demonstrate the possibility of soil stabilization by electrokinetic injection of selected cationic and anionic species into clay matrixes to obtain cementitious products.

Acknowledgements

This project was supported by the Louisiana Department of Transportation and Development Louisiana Transportation Research Center Grant No. 95-9GT. Electrokinetic Inc., of Baton Rouge provided experimental supplies. The writers especially acknowledge the contributions of the late Dr. Yalcin B. Acar, who conceived and initiated this project.

References

(1) Acar, Y.B. and Gale, R.J. (1992) "Electrochemical Decontamination of Soils and Slurries," US Patent No. 5,137,608, Washington, D.C., August 11,1992.
(2) Pamukcu, S. and Whittle, J.K. (1992) "Electrokinetic Removal of Selected Heavy Metals from Soil," Environmental Progress, V. 11, No. 3, pp. 241-250.
(3) Acar, Y.B. and Alshawabkeh, A.N. (1993) "Principles of Electrokinetic Remediation," Environmental Science and Technology , Vol. 27, No. 13, pp. 2683-2647.
(4) Lageman, R. (1993) "Electro-Reclamation," Environmental Science and Technology, Vol.27, No.13.
(5) Probstein, R.F. and Hicks, R.E. (1993) "Removal of Contaminants from Soils by Electrical Fields," Science, 1993, 260, pp 498-504.
(6) Shapiro, A.P. and Probstein, R.F. (1993) "Removal of Contaminants from Saturated Clay by Electroosmosis," Environmental Science and Technology, Vol. 27. No. 2, pp. 283-291.
(7) Casagrande, L. (1949) "Electroosmosis in Soils," Geotechnique, 1, No. 3, pp. 159-177.
(8) Casagrande, L. (1953) "Method of Hardening of Clayey Soils," Soil Mechanics Series, No.45, Harvard University, Cambridge, Mass., p. 2.
(9) Bjerrum, L., Moum, J., and Eide, O. (1967) "Application of Electroosmosis on a Foundation Problem in a Norwegian Quick Clay," Geotechnique, London, England, 17(3), 214-235.

(10) Gray, D.H. and Mitchell, J.K. (1967) "Fundamental Aspects of Electroosmosis in Soils," Journal of Soil Mechanics and Foundations Division, Proceedings of ASCE, Vol. 93, No. SM6, pp. 875-879.
(11) Esrig, M.I. and Gemenhardt, J.P. (1967) "Electrokinetic Stabilization of an Illitic Clay," Journal of Soil Mechanics and Foundation Division, ASCE, 93(3), pp. 109-128.
(12) Wan, T. and Mitchell, J.K. (1976) " Electroosmotic Consolidation of Soils," ASCE, Journal of Geotechnical Engineering Division, vol. 102, No.GT5, pp. 473-491.
(13) Mitchell, J.K. (1981), "Soil Improvement State-of-the-Art Report." Proceedings of the Tenth International Conference on Soil Mechanics and Foundation Engineering, Stockholm, v. 4, pp. 509-565.
(14) Madshus, P.A. and Janbu, N. (1984) "Improvement of Quick Clay by Electrolysis" Scandinavian Geotechnical Meeting, Sweden, Bulletin 17, Department of Geotechnical Engineering, The Norwegian Institute of Technology.
(15) Senneset K. and Acar, Y.B. (1995) "Electrokinetic Soil Improvement : A Glimpse at Past/Present Experience and Future Potential," Broms Memorial Volume, AIT, Singapure.
(16) Mitchell, J.K. and Klainer, E. (1987) "Chemical Stabilization of Landslides," Research Report, UCB-ITS-RR-87-16, Institute of Transportation Studies, University of California, Berkeley, December 1987.
(17) Van Impe (1989) "Soil Improvement Techniques and Their Evolution," Ch.4, Balkema, Netherland.
(18) Kelley, J.A. and Kinter, E.B. (1962) "Evaluation of Phosphoric Acid in Stabilization of Fine-Grained Plastic Soils," Bulletin 318, HRB, National Research Council, Washington, D.C., pp. 52-56.
(19) Sowers, G.B. and Sowers, G.F. (1970) "Introductory Soil Mechanics and Foundations," Collier Macmillan, 556 p.
(20) Demirel, T., and Davidson, D.T.(1962) "Reaction of Phosphoric Acid with Clay Minerals," Bulletin 318, HRB, National Research Council, Washington, D.C., pp. 64-71.
(21) Wendelbo, R. and Rosenqvist, I.T. (1987), "Effects of Anion Adsorption on Mechanical Properties of Clay-Water Systems," Proceedings of International Clay Conference, Denver, pp 422-426.
(22) Ozkan, S (1996), "Electrochemical Soil Stabilization," MS Thesis, Louisiana State University, Baton Rouge, LA.

Table 1. Characteristics of EPK Kaolin

Mineralogical Composition (% by weight)	
Kaolinite	97
Illite	3
Index Properties (ASTM D 4318)	
Liquid Limit (%)	56
Plastic Limit (%)	30
Activity	0.28
Cation Exchange Capacity	4.5
(meq/ 100 gr dry clay)	
Initial pH of Soil (10% Solids-Wt.)	5.8
Free Moisture (%)	1.0-3.0
Specific Surface Area (m^2/g)	22.1

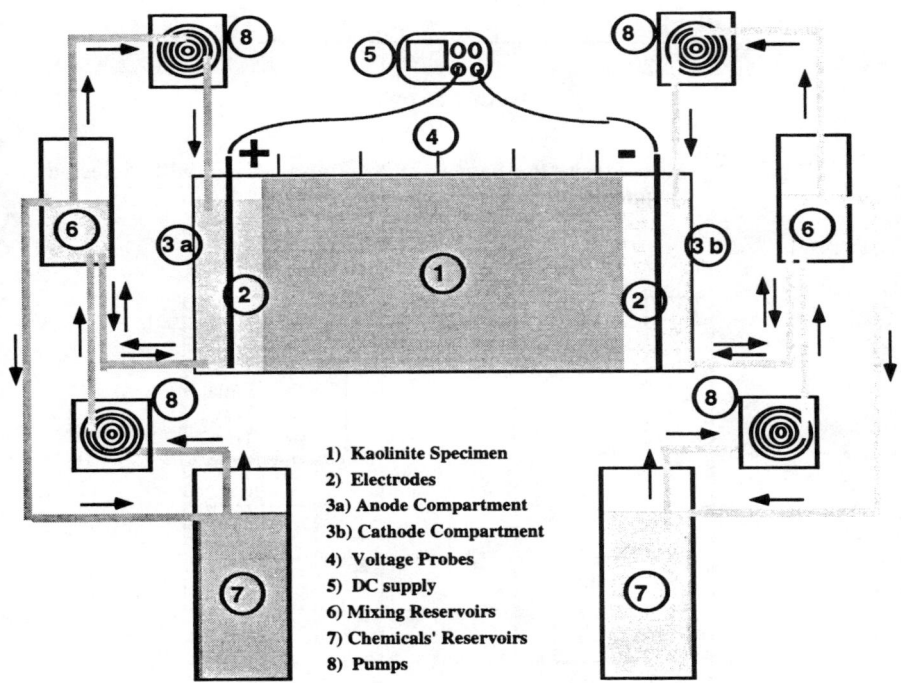

1) Kaolinite Specimen
2) Electrodes
3a) Anode Compartment
3b) Cathode Compartment
4) Voltage Probes
5) DC supply
6) Mixing Reservoirs
7) Chemicals' Reservoirs
8) Pumps

Figure 1. A schematic diagram of the test setup

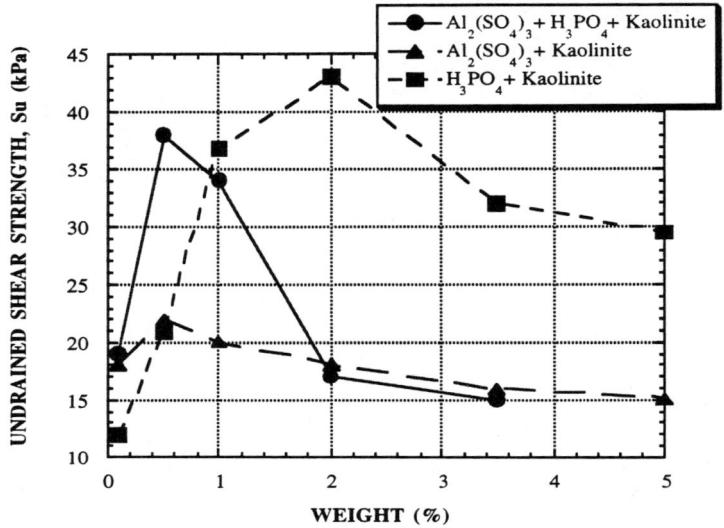

Figure 2. Shear Strength Changes of Kaolinite Mixed with Selected Chemicals after 5 Days of Treatment

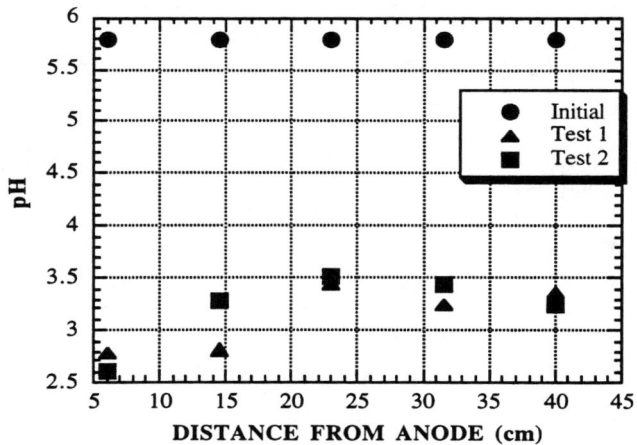

Figure 3. pH Distributions Across the Specimens in Tests 1 and 2.

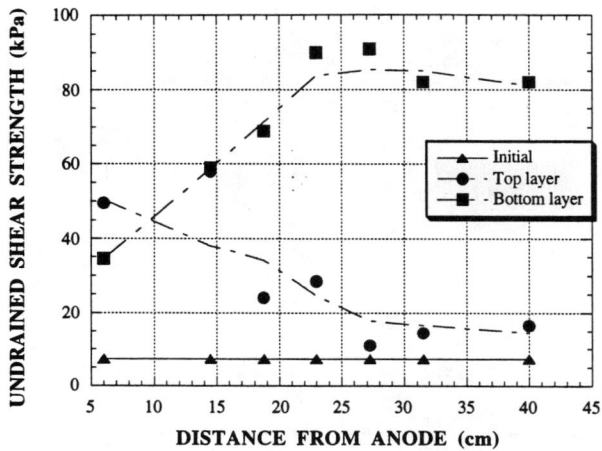

Figure 4a. Undrained Shear Strength Distributions Across the Specimen in Test 1.

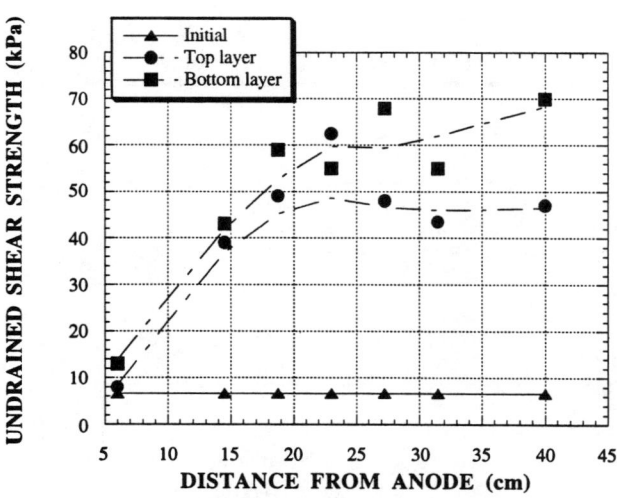

Figure 4b. Undrained Shear Strength Distributions Across the Specimen in Test 2.

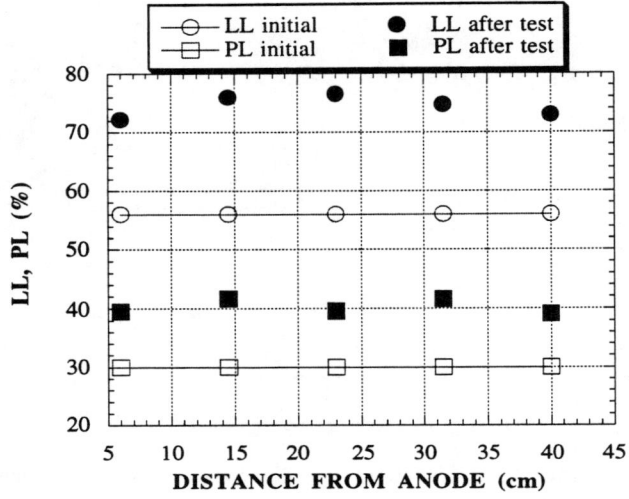

Figure 5a. The Liquid and Plastic Limit Distributions of the Specimen in Test 1.

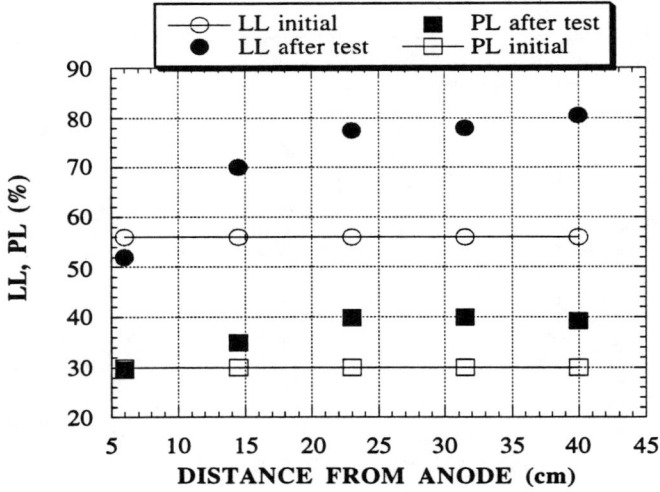

Figure 5b. The Liquid and Plastic Limit Distributions of the Specimen in Test 2.

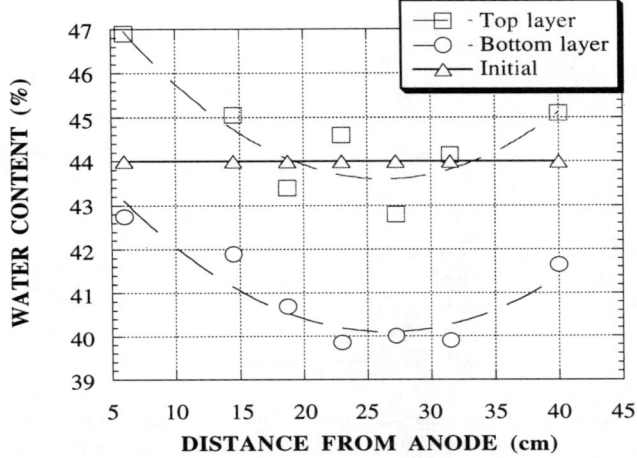

Figure 6a. Distributions of Water Content Across the Specimen in Test 1

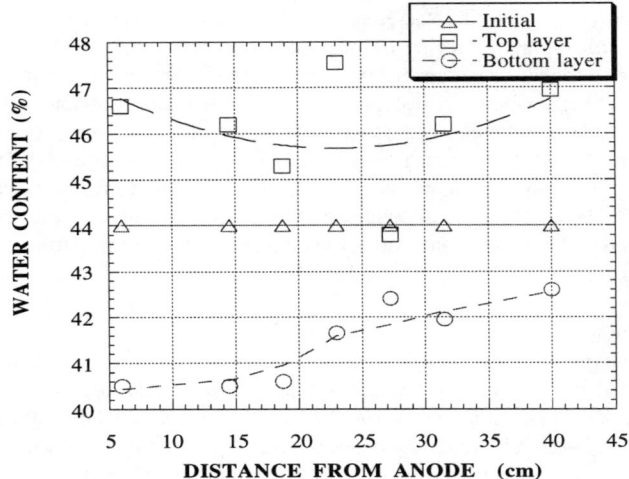

Figure 6b. Distributions of Water Content Across the Specimen in Test 2

SURCHARGE OF PHOSPHATIC WASTE CLAY WITH STRIP DRAINS

Wing Heung[1], Member, ASCE
Ching L. Kuo[2], Associate Member, ASCE
John Roberts[3], Member, ASCE

Abstract: This paper describes a case history of design, construction, monitoring, and data reduction for a surcharge program at Section 5 of Polk Parkway in Lakeland, Florida. Soft phosphatic waste clay with liquid limit up to 285 percent, which is a by-product of the mining activities, underlain the proposed roadway embankments. In order to accelerate the construction schedule, strip (wick) drains were used in the surcharge program to reduce the anticipated surcharge period from approximately 3 years to 18 months. Special consolidation tests performed by the State Materials Office of Florida Department of Transportation and centrifugal consolidation tests performed by the University of Florida studied the feasible use of strip drains for susceptible clay particle intrusion and clogging of strip drain geotextile jackets. Strip drains were installed in a triangular pattern at 1.52-meter (5-foot) spacing. A monitoring program consisted of piezometers, vertical inclinometers, settlement plates, and settlement cells was carried out. A spreadsheet program was prepared to superimpose the observed settlement data with theoretical curves such that parametric studies could be made conveniently to enhance engineering judgement and to evaluate the appropriate time for surcharge release. Based on the analyses of observed data, the surcharge program was terminated in 11 months after surcharge embankments were constructed. Available post-construction monitoring data indicated that adequate improvement was achieved.

Introduction
Polk Parkway is a new 39-kilometer (24.5-mile) limited access toll facility being constructed by the Turnpike District of the Florida Department of Transportation (FDOT). Its alignment traverses near the southern and eastern boundaries of the City of Lakeland in central Florida and connects with Interstate I-4 at both ends (See Figure 1). In Section 5 of the parkway, located at the southeastern side of Lakeland, geotechnical investigation revealed the presence of generally soft to very soft phosphatic

---

[1] Turnpike District Geotechnical Engineer, Parsons Brinckerhoff, Pompano Beach, Florida
[2] Chief Engineer, PSI, Tampa, Florida
[3] CEMC Program Director, Parsons Brinckerhoff Construction Services, Pompano Beach, Florida

waste clay, which is one of the by-products of phosphate mining activities in the past. Three alternatives were used to support the highway over the mine reclaimed areas to prevent excessive post-construction settlement: 1) bridge construction, 2) removal and replacement, and 3) improvement of waste clay through surcharging. Soil improvement by surcharging (or preloading) was selected in three areas. Strip (wick) drains were installed in two of these areas, including a total of approximately 1040 meters (3400 feet) of roadway embankment. This project is the first recorded attempt to improve the phosphatic waste clay through the use of surcharge and strip drains. The purpose of this paper is to summarize the results of the geotechnical investigation, surcharge design, construction, instrumentation, recorded data, and evaluation for the two surcharge areas where strip drains were installed.

Site Conditions and Geotechnical Investigation
During the geotechnical investigation, test borings performed in the surcharge areas revealed highly variable subsurface conditions. Phosphatic waste clay with varying amounts of fine sand was encountered. The maximum thickness of waste clay was approximately 8 meters. Waste clay was not encountered at some boring locations. Soil cover consisting of loose silty or clayey fine sand or sandy clay, typically 0.3 to 1.5 meters thick, overlain the waste clay in most areas. The groundwater table was generally encountered within two meters below the original ground surface.

Laboratory index tests performed on the phosphatic waste clay indicated that the material varies from sandy clay (CL) to clay (CH). According to AASHTO Soil Classification System, the waste clay includes A-6 and A-7 materials. The index soil properties fall within a fairly wide range and are summarized below.

Liquid Limit = 51 to 285 percent
Plasticity Index = 26 to 185 percent
In-situ moisture content = 53 to 183 percent
In-situ void ratio = 5.5 to 7.7
In-situ saturated unit weight = 12.7 to 16.1 kN/m^3 (81 to 103 lb/ft^3)
In-situ dry unit weight = 3.1 to 7.1 kN/m^3 (20 to 45 lb/ft^3)
Specific gravity = 2.69 to 2.93
Fine fraction (smaller than 0.075 mm) = 48 to 100 percent
Clay fraction (smaller than 0.002 mm) = 78 to 99 percent *
Activity = 0.92 to 2.06

* *Tests performed on A-7-5 materials only with reported fine fractions of 92 to 100 percent.*

Due to the very soft consistency of the waste clay, it was difficult to obtain good quality "undisturbed" samples. As a result, only three one-dimension incremental consolidation tests were performed on the waste clay during the design phase. Nine months after the surcharge construction was completed, additional Shelby tube samples were obtained. The primary purpose was to investigate the secondary compression ratio (C-alpha/(1+e_o)) of the waste clay such that the results would confirm the assumptions made for the surcharge release calculations. Eight additional one-dimension incremental consolidation tests were performed on the additional Shelby tube samples. Two of these tests were performed on "undisturbed" samples and were designed to imitate the actual loading conditions in the field. Two other tests were performed on samples remolded at in-situ moisture content. The e-log p curves of the remaining four additional "undisturbed" tests and the three original tests are presented together in Figure 2.

Consolidation tests performed on the samples collected during the construction phase indicated preconsolidation pressures were approximately 67 to 120 kPa (1400 to 2500 psf) higher than the overburden pressure before the construction. The secondary compression ratio ranged from 0.0004 to 0.015, and averaged 0.005. Generally, the consolidation test results indicated that secondary compression ratio decreases with the overconsolidation ratio (preconsolidation pressure of the sample divided by the applied load during the consolidation tests), as shown on Figure 3.

In addition, all the consolidation tests indicated that the compression ratio ($Cc/(1+e_o)$) varied from 0.32 and 0.42 and the recompression ratio ($Cr/(1+e_o)$) varied from 0.016 to 0.042. The compression indexes, Cc were corrected to account for the effect of sample disturbance using the procedure of Peck, Hanson, and Thornburn (1974). The Cr/Cc ratio ranged from 0.07 to 0.16. The vertical coefficient of consolidation, Cv ranged from 1.1×10^{-8} to 2.3×10^{-7} m^2/s (0.01 to 0.21 ft^2/day), which generally increases with the overconsolidation ratio as depicted on Figure 4. The coefficient of permeability, k, calculated from the consolidation tests, ranged from 8.1×10^{-9} to 5.3×10^{-12} m/s (2.3×10^{-3} to 1.5×10^{-6} ft/day).

Standard penetration tests performed in the waste clay provided results ranging from weight of rod over 450 mm to 14 blows per 300 mm, typically less than 4 blows per 300 mm. Results of field vane shear tests performed prior to surcharging indicated that the undrained shear strength of the waste clay ranged from 4 to 22 kPa (80 to 470 psf), averaged 17 kPa (350 psf). These were roughly in agreement with the undrained shear strength of 5 to 28 kPa (100 to 580 psf), with an average of 13 kPa (270 psf), obtained from the laboratory unconsolidated undrained triaxial tests. Consolidated undrained triaxial tests with pore pressure measurements were also performed. Effective stress plot showed a friction angle of 25 degrees with an apparent cohesion intercept of 12 kPa (250 psf). Total stress plot showed a friction angle of 15 degrees with no cohesion intercept. Undrained shear strengths ranging from 19 to 53 kPa (400 to 1100 psf), averaged 37 kPa (780 psf), were measured on the Shelby tube samples, recovered 9 months after surcharge completion using a Torvane shear device, indicating a significant increase from the initial tests.

Initial Design and Surcharge Program
Between Stations 1728+00 and 1747+50 (Surcharge Area 1) and Stations 1768+00 and 1782+90 (Surcharge Area 2), the maximum embankment height is 4.6 meters. The surcharge embankments typically have side slopes of 3H:1V and the top elevations are 4.6 meters above the top of roadway embankments (See Figure 10). Geosynthetic

reinforcements were required to improve the global stability of the surcharge embankments due to the very low shear strength of the waste clay. Strip drains were not included in the initial design due the concern that long-term clogging of the drains might develop. Accordingly, a surcharge waiting period of 36 months was estimated for both surcharge areas with an anticipated settlement of 0.3 to 1.8 meters.

Conical Load Test
Due to the uncertain consolidation characteristics of waste clay, two conical load tests were performed at Station 1774+00 (Site A - Surcharge Area 2) and Station 1745+00 (Site B - Surcharge Area 1). The purpose was to evaluate the compressibility and rate of consolidation of the waste clay in the field. At each site, a conical mound of soil approximately 12 meters in diameter and 4.3 meters in height was constructed. Three layers of uniaxial geogrid were placed criss-crossing at 60 degree intervals below each conical mound. Settlements were measured near the center of the mounds at five selected depths using a magnetic extensometer. The extensometer casing was installed inside a borehole. Settlement movements of installed magnetic "spiders" or plates were monitored in reference to a datum magnet at the casing bottom which was embedded in soil unsusceptible to ground settlement. The magnetic plates were installed at the original ground surface outside the casings. Four magnetic "spiders" were installed inside each extendable casing at 1.5-meter intervals below the plate. In addition, four vibrating strip piezometers were installed near the center of each mound at depth intervals of 1.5 meters (See Figure 5). The conical mounds were constructed in about one day at each site. Monitoring of the instrumentation continued for 47 and 49 days, at Sites A and B, respectively.

Settlement measurements of the conical load test indicated that the average compression ratio ($Cc/(1+e_o)$) of waste clay ranged from 0.28 to 0.48, with a corresponding vertical coefficient of consolidation, Cv of 3.8×10^{-8} and 4.1×10^{-8} m^2/s (0.035 and 0.038 ft^2/day) at Sites A and B, respectively (See Figure 7). Based on the rapid rate of pore pressure dissipation measured by the piezometers as shown on Figure 6, vertical coefficients of consolidation were estimated to be 5.7×10^{-7} and 5.3×10^{-7} m^2/s (0.53 and 0.49 ft^2/day) at Sites A and B, respectively. The reason for the difference between the coefficient of consolidation using the two approaches is not certain. However it is believed that the coefficients of consolidation back-calculated from the settlement measurements are more accurate. It is suspected that the piezometers were not properly installed, leading to a leak in the excess pore water pressure at the measurement locations.

The average compression ratio obtained from the conical load test, ranging from 0.28 to 0.48, are roughly comparable with the values determined from the incremental one-dimension consolidation tests, which ranged from 0.32 to 0.42. The vertical coefficient of consolidation, Cv obtained from consolidation tests ranged between 1.1×10^{-8} and 4.3×10^{-8} m^2/s (0.01 and 0.04 ft^2/day) with an average of 1.7×10^{-8} m^2/s (0.016 ft^2/day) at normally consolidated condition (See Figure 4). These laboratory values are lower than the field results of 3.8×10^{-8} to 4.1×10^{-8} m²/s backcalculated from the conical load tests.

Laboratory Tests for Strip Drain Clogging Evaluation
Due to the potential benefit of accelerating the surcharge waiting period utilizing strip drains, research projects were conducted by the FDOT State Materials Office Laboratory in Gainesville, Florida and by the University of Florida. The goal was to evaluate the potential intrusion of clay particles into the strip drain core through the geotextile jacket which may clog the core and disable its function as an effective drainage element.

At the University of Florida, the clogging potential of the waste clay into the strip drains was studied using the centrifugal tests. Remolded waste clay specimens of approximately 160 mm in diameter and approximately 70 to 120 mm in height were subjected to centrifugal tests at 70 g and 110 g levels. One series of tests included a model strip drain at the center of clay specimens and another series of tests did not include a strip drain. Deformation measurements with time on each specimen were made such that the contrast of consolidation settlement between the two series of tests could be observed. Due to the limitation of modeling the field performance using a strip drain specimen of 100 mm in width in the centrifugal tests, computer modeling using a finite element program, PlasFEM was performed. The results of the computer program predicted that despite the installation of strip drains, the rate of waste clay consolidation in the field would not differ significantly. Details of these tests and finite element modeling were described by Andersen (1997).

In the State Materials Office Laboratory, one-dimension incremental consolidation tests with single drainage were performed with waste clay, such that drainage can only occur through the strip drain geotextile located below the clay specimens. In order to allow free drainage of the effluent and to collect the effluent in a cylinder for examination, a hole was cut at the bottom of the strip drain specimens. Four strip drain products were tested. The stress range applied in these tests ranged from 14 to 110 kPa (2 to 16 psi). The effluent was collected in a cylinder for visual inspection as to the degree of cloudiness in all tests. Also, when the tests were completed, the strip drain specimens were examined for the amount of fines retained inside. Test results indicated that the coefficient of consolidation varied from 5.4×10^{-9} to 2.5×10^{-8} m²/s (0.005 to 0.023 ft²/day), which are slightly lower than the other consolidation tests performed. In addition, visual inspection of the effluent cloudiness indicated that minor waste clay intrusion occurred at low stress levels when the moisture contents of the specimens were relatively high (approximately 160 to 190 percent). The effluent became clearer as the test continued at higher load increments. Examinations of the strip drain specimens after the test revealed that only minor clay intrusion had occurred. These test results were a positive indication that waste clay intrusion into these strip drains would not pose a major problem. The geotextiles of the tested strip drains had apparent opening sizes (AOS) ranging between sieve numbers 140 and 170.

Despite the contrasting results from the two research projects, the Turnpike District decided to modify the surcharge design to install strip drains at a 1.52-meter (5-foot) triangular pattern in Surcharge Areas 1 and 2. Due to the benefit of accelerated drainage through the strip drains, the estimated surcharge period was modified from 36 to 18 months. The strip drains installed in the field have an apparent opening size of sieve number 140, a filter permeability of 1.7×10^{-4} m/s, and a permittivity of 0.45 s^{-1}. Also, the discharge capacity of the strips drains at normal pressures of 10 kPa and 300 kPa are 1.8×10^{-4} and 1.4×10^{-4} m^3/s, respectively.

Geotechnical Instrumentation Program
The geotechnical instrumentation program consisting of 33 settlement plates, 11 piezometers, and 14 vertical inclinometers was planned in the Surcharge Areas 1 and 2 to evaluate the field performance during the surcharge duration. The instrumentation locations are depicted on Figures 8 and 9.

Vibrating wire piezometers with a 340 kPa (50 psi) maximum range were used. The pore pressure transducers were encased with a low air entry filter at the conical tip. During the installation, 100 mm-diameter cased boreholes were drilled and cleaned to approximately 1.5 meters above the design elevations. Pore pressure readings were measured as the piezometers were slowly pushed to their design elevations, such that the maximum allowable limit of 340 kPa was not exceeded.

Vertical inclinometer casings with 85 mm (3.34 inches) I.D. with telescoping joints were used due to expected large settlement. They were installed in 150 mm diameter boreholes and were grouted in place with a mix of 4:1 bentonite to grout ratio. Most vertical inclinometers were installed at the southern side of the surcharge embankments. Test borings indicated that the northern project right-of-way was roughly the limit of the mined area and global stability in that direction was not a concern.

After the strip drain installation, 24 pneumatic settlement cells were added to the instrumentation program. Generally, they were installed in areas where thick deposits of waste clay were known to exist according to the relatively deep strip drain penetrations and where settlement plates were not positioned to monitor those areas.

Embankment and Surcharge Construction
After clearing and grubbing at the site, the ground in the surcharge areas was leveled to create a working platform for construction equipment. Strip drains were then installed at 1.52-meter triangular spacing using a converted excavator. In some areas, very soft waste clay was observed to squeeze to the surface around the mandrel and foam was observed flowing from some strip drains. The maximum penetration depth of the strip drains was 10.7 meters (35 feet). After strip drain installation, a drainage blanket consisting of a one foot thick layer of A-3 clean sand (AASHTO Soil Classification System) was placed. The geotechnical instrumentation was then installed. It was followed by the construction of geosynthetic reinforcements. The reinforcement design consisted of one to two layers of high strength geotextiles in various areas with

allowable tensile strengths ranging from 130 to 390 kN/m (9,000 to 27,000 lb/ft) at 5% strain. The geotextile panels were sewed together along the cross-machine direction in order to enhance geotextile installation. Heavy construction equipment was not allowed to operate above the geotextile until 0.3 meter of sand cover was placed to prevent excessive rutting and deformation of the geotextiles.

In general, the embankment was initially constructed at a rate of 0.3 to 0.6 meter (1 to 2 feet) per week. However, as construction continued, observations from vertical inclinometers in various areas (VI-3, VI-5, VI-6, VI-11, and VI-12) indicated unstable conditions. Most of these vertical inclinometers showed noticeable lateral spreading (See Figure 11). The casings failed after accumulated lateral movements of 200 to 430 mm (8 to 17 inches) were reached. New vertical inclinometers were installed to replace the failed ones. The contractor was directed many times to avoid any earthwork above the unstable areas until the excess pore water pressure in the waste clay dissipated sufficiently and the waste clay gained in shear strength to support additional fill. These resting periods ranging from 1 to 14 days were applied in the unstable areas only when excessive lateral movements and/or excessive settlements, as defined by 13 mm (0.5 inch) per day were measured by the instrumentation. As construction progressed, the lateral movements continued and more vertical inclinometer casings failed. New inclinometers were installed as quickly as possible in order to continue the monitoring. By the end of the surcharge construction, vertical inclinometers VI-5, VI-6, VI-11, and VI-12 had failed two to three times. Due to the lateral movements, some tension cracks were noticed on the embankment during construction.

Methodology of Settlement Evaluation

The approach to evaluate measured settlement data in this project is to compare the observed settlement with theoretical prediction such that the degree of consolidation could be estimated. Establishing the theoretical predictions with a constant value of assumed compression ratio and coefficient of consolidation for all the waste clay on this project would not be appropriate because of the highly variable soil conditions. Instead, it was decided to find the best fit theoretical curve through trial and error of input parameters, including primary settlement, coefficients of consolidation, and Cr/Cc ratio at each settlement plate or settlement cell location. This approach allowed the average soil parameters to be back-calculated and permitted target settlements to be estimated for surcharge release purposes.

Due to the numerous settlement data available for this project and the required trial and error comparisons, a spreadsheet program named "SPANA" (Settlement Plate ANAlyses) was prepared for the repetitious calculations. The observed settlement data was input into the program. This program utilizes the methodology of R.E. Olson (1977) to establish the theoretical settlement curves and can handle up to five ramp loads (linear increments of embankment load with time). The use of the ramp load feature allowed the effect of the slight preconsolidation pressures observed in the settlement data to be included for theoretical calculations and surcharge release decisions. Graphic plots of settlement versus time and settlement versus logarithmic of time were generated by the program for comparisons of observed and theoretical settlement curves. These plots served as useful tools to exercise engineering judgement. In addition, average of difference and average of square of difference between measured and theoretical settlement curves were calculated by the program.

The procedure used to analyze the settlement data involved the identification of any preconsolidation pressure above the existing overburden pressure. This was identified based on an abrupt increase of settlement corresponding to an increase in surcharge height on the settlement versus time plot. From the same plot, SPANA also depicts the fill height versus time. Up to five ramp loads were then selected on the fill height versus time plot. For settlement data which depicted a preconsolidation pressure higher than the existing overburden pressure, at least one ramp load was used to define the recompression range on the fill height versus time plot. In this way, the recompression ratio $(Cr/(1+e_o))$ could be used for the settlement calculation of that ramp load. After that, a total primary settlement was assumed for trial and error calculation. The conventional consolidation equation was used to separate the primary settlements of each ramp load from the total primary settlement as data input for SPANA. Information including applied load, initial waste clay thickness, groundwater level, boundary drainage condition, and an assumed value of Cr/Cc ratio was needed for the calculation. By performing trial and error with the coefficients of consolidation for each ramp loading using SPANA, the best fit curve was established for the assumed total primary settlement.

This procedure was then repeated for a different total primary settlement and Cr/Cc ratio until the best fit match was found. Generally, the coefficients of consolidation selected for back-calculation were assumed to decrease or remain the same, as the load increased. Since it was expected that the effect of vertical drainage would be insignificant compared with the horizontal drainage in the thick waste clay deposits, a constant ratio of 1.5 was assumed between the horizontal and vertical coefficients of consolidation in all calculations for simplicity.

The Olson's theoretical approach adopted by SPANA included the use of the traditional Terzaghi's (1943) theory of consolidation for vertical drainage and Barron's equal strain consolidation solution with no smear (1948) for horizontal drainage. The combined effect was included using the method of Carrillo (1942). These methods are part of the most classical soil consolidation theories available and the authors agreed that better consolidation theories exist. However, from a practical standpoint, due to the highly variable soil conditions, any of the more vigorous theories must also go through a trial and error approach on the input soil parameters and does not necessarily provide a better conclusion. The simplicity of SPANA allows it to be programmed in a relatively short time and permits a convenient trial and error approach through the graphical and statistical functions of the spreadsheet program. This approach assumed a constant compression index value which was found to be reasonable according to the available one-dimension consolidation test results within the applicable stress of less than 190 kPa (4000 psf).

The total target settlement which was the minimum required settlement for surcharge release consisted of primary and secondary target settlements. The primary target settlement was assumed to be the primary settlement of the roadway embankment if the surcharge program was not implemented. It was calculated using the best fit values of compression and re-compression ratios ($Cc/(1+e_o)$ and $Cr/(1+e_o)$) obtained from the calculations of SPANA. The secondary target settlement was calculated as the anticipated secondary settlement for 20 years of design life.

Instrumentation Data and Evaluation

Total settlements ranging from 0.1 to 1.67 meters (4 to 66 inches) were measured under the surcharge. Using the methodology described above, compression ratios ($Cc/(1+e_o)$) ranged from 0.16 to 0.56 and averaged 0.36 were calculated. In addition, most settlement plate and settlement cell data indicated that the preconsolidation pressures are on the average 18 kPa (370 psf) above the existing overburden pressures. These pseudo-preconsolidation pressures are considered to be a result of secondary consolidation of waste clay since its original placement at the site and possibly aging effect.

The calculated results also indicated that the horizontal coefficient of consolidation, Ch was found to be higher at the eastern portion of Surcharge Area 1, which typically varied from 1.1×10^{-7} to 1.1×10^{-6} m^2/s (0.1 to 1.0 ft^2/day) in the normally consolidated condition. Maximum settlement reached 0.94 meter (37 inches) in that area. It was probably a result of the more sandy nature of the waste clay in that region. For the remaining surcharged areas, the horizontal coefficients of consolidation varied from 4.3×10^{-8} to 2.2×10^{-7} m^2/s (0.04 to 0.20 ft^2/day), and averaged 1.2×10^{-7} m^2/s (0.11 ft^2/day) at effective stress levels slightly above the preconsolidation pressure and decreased to 2.2×10^{-8} to 1.1×10^{-7} m^2/s (0.02 to 0.10 ft^2/day) and averaged 5.4×10^{-8} m^2/s (0.05 ft^2/day) near the end of surcharge period.

After 11 months of surcharge, the total primary and secondary target settlements were attained in all settlement plate and settlement cell locations, except in settlement cell SC-23, which was 25 mm (1 inch) below the total target settlement. These results were generally in agreement with the piezometer data. Piezometer readings typically indicated a clear asymptotic decrease starting at the completion of surcharge construction as shown in Figure 12. Some of those piezometers reached the static water pressure within 11 months and the remaining piezometers indicated excess pore water pressures of up to 28 kPa (4 psi). In view of this information, the two surcharge areas were released within 11 months. In the settlement cell SC-23 location, an additional 40 mm (1.5 inches) of settlement was measured between the release and the actual removal of the surcharge fill. As a result, total target settlement requirements were achieved at all settlement monitoring locations.

Immediately after the surcharge was released, 9 settlement cells and 10 piezometers were still operational. The other instrumentations either experienced mechanical failures or were damaged by construction activities. The monitoring of settlement cells and piezometers continued as the surcharge fill was being removed. Typically, as the surcharge was excavated, negative excessive pore water pressures, up to 41 kPa (6 psi) below the static groundwater pressure, was measured. This negative pore water pressure increased gradually to reach the static water pressure typically within 50 to 80 days (See Figure 12), and a maximum of 130 days at piezometer PT-7. Among the 9 settlement cells that survived throughout the surcharge period, five cells experienced mechanical failures as the surcharge was removed. However, settlement cell SC-5 measured a rebound of approximately 130 mm (5 inches) and is depicted in Figure 13. Settlement cell SC-23, which was located in the marginally accepted area, measured an additional settlement of approximately 5 mm (0.2 inch) in 2 months after surcharge removal, which is considered negligible. Approximately 3 months after all the surcharges were released, most remaining instrumentation either developed mechanical failures or were damaged by excavation activities at the site and the instrumentation monitoring program was terminated.

Conclusion

Phosphatic waste clay was successfully improved through surcharging with strip drains in the Polk Parkway project. Construction above the mine reclaimed areas experienced significant subsurface lateral spreading despite the use of high strength geotextiles as soil reinforcement. The geotechnical instrumentation program provided useful information to assist in determining the consolidation process of the waste clay and the appropriate time for terminating the surcharge waiting period. Due to the highly variable soil conditions at the site, a spreadsheet program was used to analyze the settlement data. The program adopted the methodology of R.E. Olson to search for a best fit theoretical match for the measured settlement data. Using this procedure, soil parameters were established to evaluate surcharge release. The surcharge program was terminated after 11 months of waiting period, which was seven months sooner than originally estimated. Limited post surcharge instrumentation readings revealed that the waste clay is sufficiently improved and will not experience significant post-construction settlement.

Acknowledgment

The authors express their appreciation to all the geotechnical engineers, including Jay Casper of PSI, Paul Engeling and Richard A. Hawkins of Consulting Foundation Engineers, and Frank Tejidor of Post, Buckley, Schuh & Jernigan, Inc., who were very involved in the project, and had significant contributions to its success. The research works related to the clogging studies of strip drains by Robert Ho and David Horhota of State Materials Office at Florida's Department of Transportation and Michael McVay and P.G. Andersen of the University of Florida are greatly appreciated. The authors also appreciate the construction management staffs including Charles Wegman and Jim V. Moulton Jr., of the Florida's Turnpike, Neal Penny of Parsons Brinckerhoff Construction Services, Murray Yates of Metric Engineering, for their dedicated efforts on this project. In addition, thanks are also given to Jim Kiesel and his staffs at Atlanta Engineering and Testings for their contributions, especially related to the installation and monitoring of the geotechnical instrumentation.

References

P.G. Andersen "Centrifugal and Numerical Modeling of the Consolidation Behavior of Phosphatic Waste Clays", Thesis for Degree of Master of Engineering, University of Florida, Gainesville, Florida, May 1997

R.A. Barron "Consolidation of Fine Grained Soils by Drain Wells", Transactions ASCE, Vol. 113

N. Carrillo "Simple Two- and Three-Dimensional Cases in the Theory of Consolidation of Soils", Journal of Mathematics and Physics, Vol. 21, 1942, pp 1-5

R.E. Olson "Consolidation Under Time Dependent Loading" Technical Note, ASCE Journal of the Geotechnical Division, January 1977

R.B. Peck, W.E. Hanson, T.H. Thornburn "Foundation Engineering", 2nd edition, John Wiley & Sons, Inc., 1974

K. Terzaghi "Theoretical Soil Mechanics", John Wiley and Sons, New York, 1943

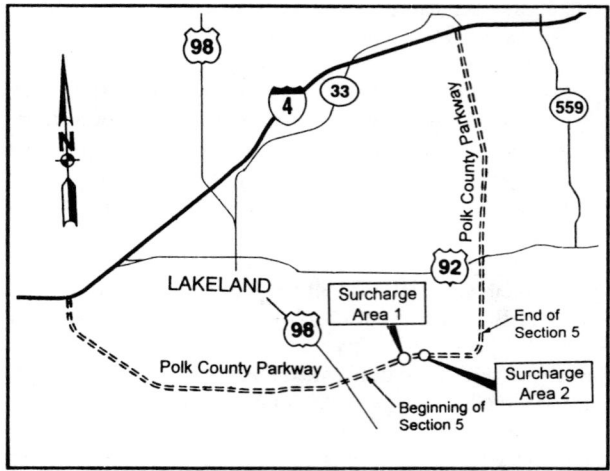

Figure 1: Site Location

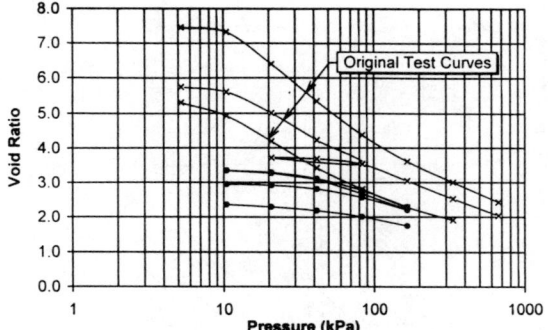

Figure 2: End of primary e - log p Curves

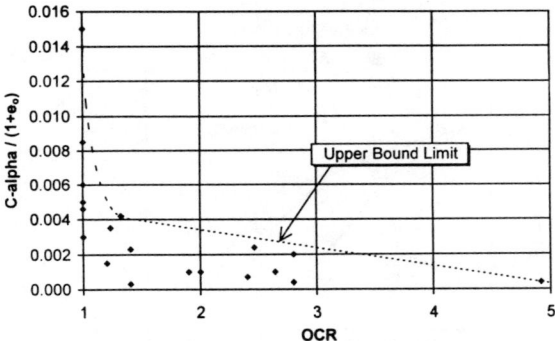

Figure 3: Overconsolidation Ratio versus Secondary Compression Ratio

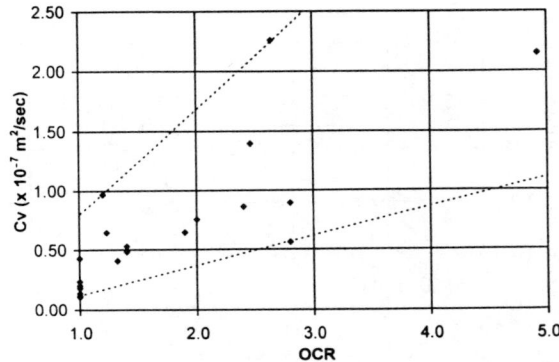

Figure 4: Overconsolidation Ratio versus Vertical Coefficient of Consolidation

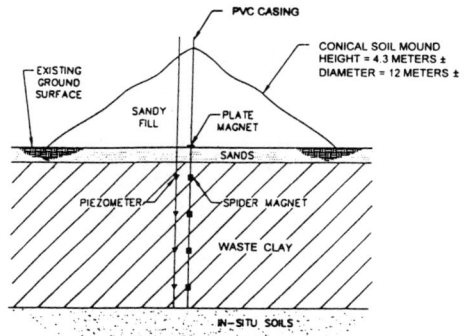

Figure 5: Conical Load Test Setup

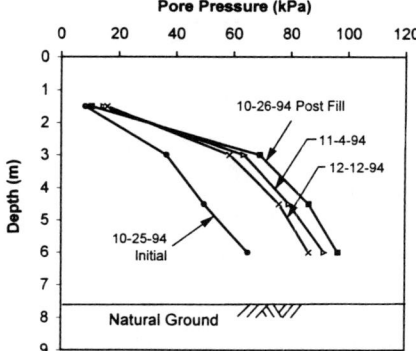

Figure 6: Conical Load Test - Pore Pressure at Site B

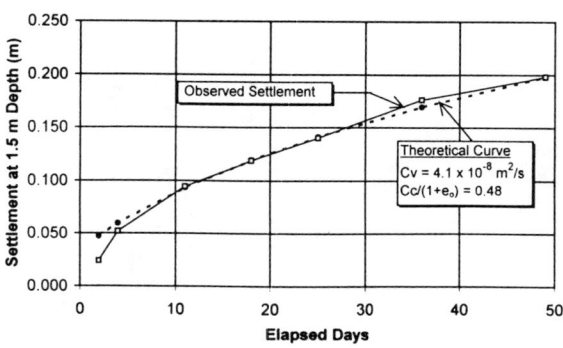

Figure 7: Conical Load Test - Settlement at Site B

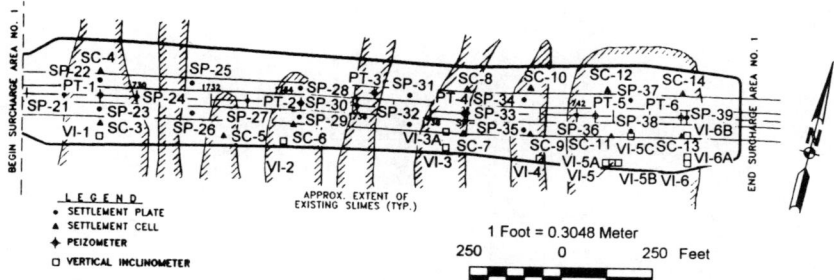

Figure 8: Surcharge Area 1 Instrumentation Plan

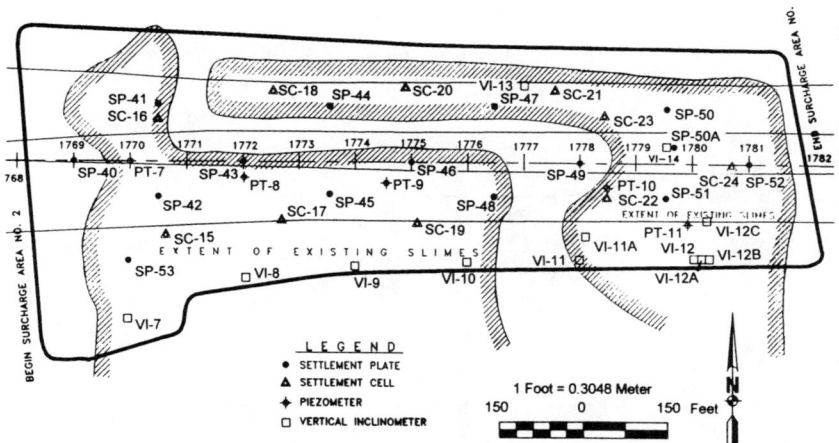

Figure 9: Surcharge Area 2 Instrumentation Plan

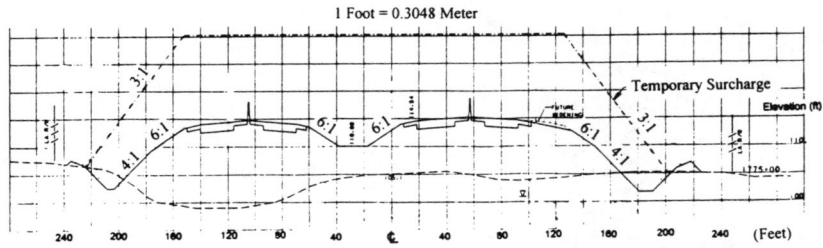

Figure 10: Typical Cross Section of Surcharge Embankment

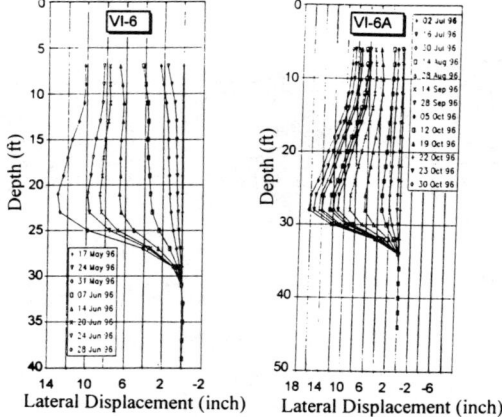

Figure 11: Lateral Spreading of Waste Clays During Construction

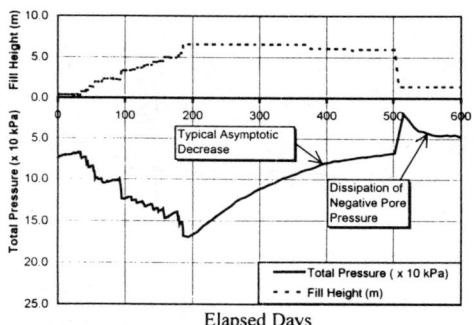

Figure 12: Typical Piezometer Results

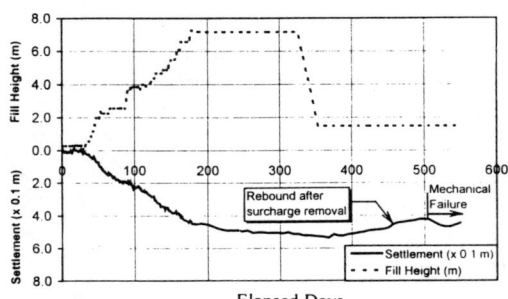

Figure 13: Settlement Rebound at Settlement Cell SC-5

SEISMIC RETROFIT OF FOURTH STREET & RIVERSIDE VIADUCTS WITH MICROPILES

Gary E. Taylor[1], Francis B. Gularte[2], R.G.E., and Greg G. Gularte[3]

Abstract

This paper summarizes design, installation, instrumentation, and performance testing results of high capacity, small diameter tension/compression micropiles in Los Angeles, California.

Fourth Street Viaduct Project (1995)

The Fourth Street Viaduct directs traffic from the Harbor Freeway (I-110), Hope, Flower, and Figueroa Streets, to downtown Los Angeles. The City of Los Angeles selected the Fourth Street Viaduct as the first of several bridge projects to receive seismic retrofit. The project included specialty contractor design and installation of small diameter tie-down piles (micropiles) designed to support 1100 kN to 2000 kN (250 to 450 kip) design loads. Micropiles were chosen over conventional deep foundations because of limited site access, low headroom beneath the viaduct, and ease of attachment to existing footings.

The Fourth Street Viaduct project consists of six structures (see Figure 1a); the Fourth Street Viaduct (V-line), Hope Street Bridge (H-line), Fourth St. Viaduct (south) to Hope St. (D-line), Figueroa St. to Hope St. Ramp (C-line), Hope St. to Fourth St. (south) Ramp (B-line), and Hope St. to Figueroa St. Ramp (A-line). The original structural design specified 198 production micropiles, and one non production performance test pile at each of the six structures. Value engineering reduced the number of production piles by 20% and performance test piles by 50%. The performance specifications required a maximum of 6.3 cm (2.5 inches) of axial deflection at design load. The design engineers specified a design load equal to the ultimate load in both tension and compression.

---

[1] Senior Project Manager / Geologist, Hayward Baker Inc., Santa Paula, California.
[2] Vice President / Regional Manager, Hayward Baker Inc., Santa Paula, California.
[3] Project Engineer, Anderson Consulting Group, Roseville, California.

314 SOIL IMPROVEMENT FOR BIG DIGS

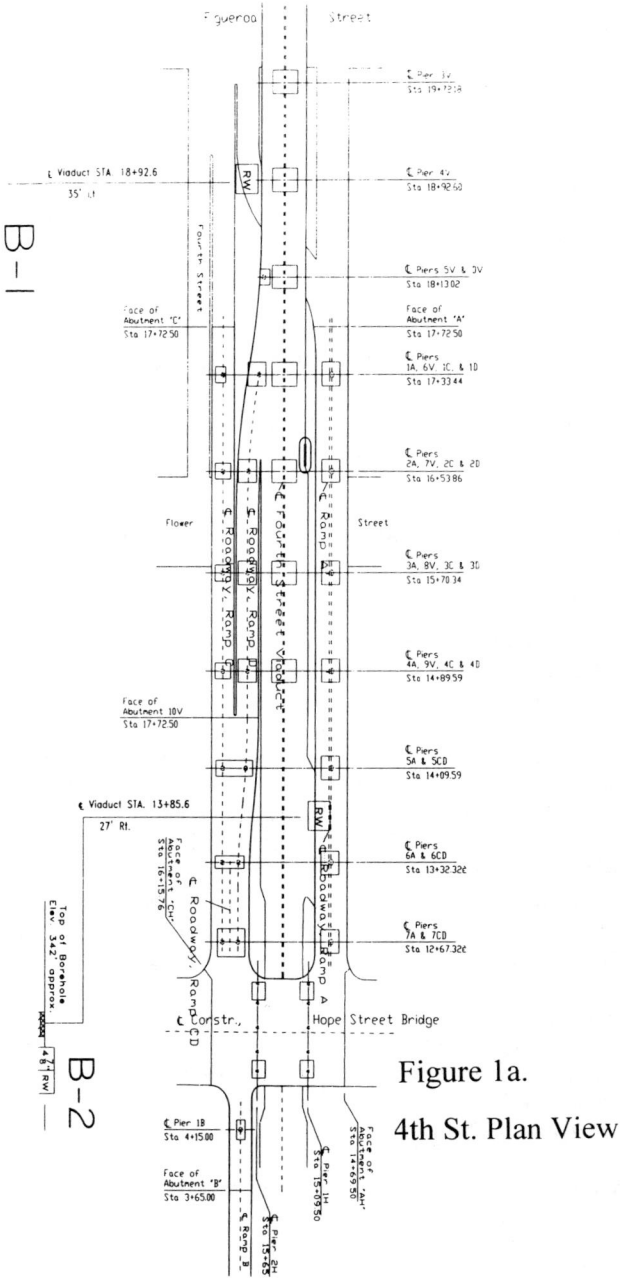

Figure 1a.
4th St. Plan View

Originally, the specifications required proof testing of a randomly selected production pile in tension and compression to design load at each footing. The specialty contractor recommended proof testing in tension only, thus eliminating the need for installation of reaction tendons for compression load tests of production piles.

Performance testing consisted of loading three non-production piles in both tension and compression to 120 percent, 2400 kN (540 kips), of design load. The specialty contractor performance tested one pile to 150 percent (3000 kN, or 675 kips) with satisfactory results.

Site Geology

Site geology consisted of mixed alluvium, fill, and bedrock. Alluvium and fill soil consisted primarily of moist to wet silty clay to clayey silt (CL-ML), lesser amounts of clayey sand (SC), and rare occurrences of gravelly sands (SP). Bedrock consisted of poor to moderately indurated, moist to wet mudstone (Puente Formation, see Figure 1b for the boring log). The contact between the alluvium/fill and bedrock ranged from the ground surface to approximately 12.2 meters (40 feet) below the footings. Often, the highly weathered bedrock near the surface was indistinguishable from overlying alluvium due to weathering caused by high ground water conditions. The weathered nature of the bedrock behaved like a hard clay, rather than exhibiting typical rock characteristics.

Alluvium shear strength varied from 35 to 103 kPa (5 to 15 psi). Bedrock shear strength ranged from 240 to 515 kPa (35 to 75 psi) based on laboratory tested samples obtained from 6 to 30.5 meters (20 to 100 foot) depths, respectively.

Pile Design

Although bedrock shear strengths were relatively high, achieving design capacity required bond lengths within the competent bedrock horizons at depths of approximately 25 meters (80 feet). To increase pile capacity and decrease pile lengths in difficult drilling conditions, the specialty contractor chose post grouting as a cost effective solution to improve alluvium/fill strengths in the upper 18 meters (60 feet). Within the bond length, multiple post grout injections of high strength, neat-cement slurry grout provided the necessary ground modification/reinforcement.

Since pre-stressing the micropiles was not necessary, the selected piling material consisted of 17.8 cm (7-inch) O.D. by 1 cm (0.4-inch) wall N-80 oil country tubing. Piling was shop fabricated in 1.5 to 3 meter (5 to 10 foot) lengths and joined by conventional threaded couplings. External couplings were chosen over flush joint connections to maintain pile capacity through the joint, minimizing reduction of wall thickness and increasing soil contact.

Figure 1b. 4th St. Boring Log, B-1

The specialty contractor chose 30.5 cm (12-inch) diameter borings and incorporated the following parameters for performance test pile installations:

- Design Load = Ultimate Load, 2000 kN max. (450 kips)
- For estimate of pile capacities, design for 1.2 X Design Load or 2400 kN (540 kips)
- Use design shear strength for soils/highly weathered bedrock requiring post grouting (worst case) and competent bedrock to be approximately 206 to 275 kPa (30 to 40 psi).
- Use bond lengths of 12.2 meters (40 feet) for low to moderately weathered bedrock, and 15.2 meters (50 feet) for soil/highly weathered bedrock.

The specialty contractor determined the required amount of post grouting by pile capacity, type of soil/rock encountered, and feedback from grouting operations. For example, if the design load of a pile was 1560 kN (350 kips) and was founded in competent mudstone, then the bond length only required a primary post grouting pass at 3 meter (ten foot) intervals. However, a 2000 kN (450 kip) capacity pile founded in wet, soft clay, required primary, secondary, and possibly tertiary post grouting depending upon grout takes (volumes) and pressures of previous post grout passes.

Pile Installation

Approximately five percent of the production piles could be reached with conventional diesel-hydraulic crawler drills. The remaining 95 percent required specialized, restricted access drilling equipment with sufficient torque (a healthy 13,567 N-m, 10,000 ft.-lbs.) to auger 30.5 cm (12-inch) diameter holes in cohesive or cemented soil to depths up to 18 meters (60 feet). The viaduct generally provided less than 5.5 meters (18 feet) of head room. However, most foundations required excavating and shoring to about 6 meters (20 feet) below existing grade to expose the footings, and allow access for the custom built, site specific portable drills which could be set on the foundations.

Thirty centimeter (12-inch) diameter augers constructed in 1.5 meter (5 foot) lengths, were designed for slurry or wet drilling operations. If the hole became unstable, this system allowed introduction of high strength neat cement grout to stabilize the hole as the augers were withdrawn. The contractor utilized a 30.5 cm (12-inch) diameter down-the-hole hammer to penetrate several unforeseen obstructions and existing belled caissons below the footings.

After drilling the hole to design depth, the pile was assembled in 1.5 to 3.0 meter (5 to 10 foot) sections. The contractor attached centralizers at 3 meter (10 foot) intervals while lowering the pile into the hole. The specialty contractor attached sleeve port grout pipe to the pile, with ports at 1.5 meter (5 foot) intervals. The sleeve port grout pipe provided the ability to perform multiple post grout injections at individual locations along the bond length of the pile.

Neat cement grout consisted of a 0.45:1 water:cement ratio with additives to improve pumpability during post grouting and retard grout on hot days. A straddle packer introduced cement slurry through individual sleeve ports. Prescribed volumes of grout, dependent on soil or rock conditions encountered, were injected under pressures up to 2070 kPa (300 psi). The contractor typically staggered primary and secondary grouting sequences along the bond length at 3 meter (10-foot) intervals.

Performance Testing

Three performance test piles were installed along the alignment of the viaduct. The eastern test pile (near Hope St. footing 7A) was founded primarily in poorly cemented mudstone containing rare layers (up to 1 meter thick, 3 ft.) of well cemented mudstone. The central (near Flower St. footing 6V) and western (near Figueroa St. footing 3V) piles were founded primarily in wet silty clay and poorly cemented, wet mudstone. All test piles were tested with conventional test equipment consisting of three dial indicators placed equidistant around the top of the pile and a 3115 kN (350-ton) hydraulic jack with calibrated pressure gauge.

Digital recording instrumentation, designed by the contractor, was also used on all performance test piles. The digital system consisted of a hydraulic pressure transducer, load cell, and two linear potentiometers placed equidistant around the top of the pile. In addition, the eastern and western performance test piles were instrumented with pairs (opposite sides of the pile) of strain gauges located at 3 meter (10 foot) intervals along the axis of the pile within the bond length. The data from the individual sensors was captured with a programmable data logger and downloaded to a laptop for analysis.

Tension and compression performance testing was performed in two cycles. The first cycle loaded the pile to design load in ten 5-minute stages each sustaining 10% of the design load. After sustaining the design load, the pile was de-stressed in twenty-five percent stages to the seating load. The second cycle was then increased to 120 percent of the design load in stages of ten percent.

The most competent soil and bedrock occurred in the vicinity of Hope Street, which was at a higher elevation than the Flower and Figueroa Street areas. The contractor installed the performance test pile to 18 meters (60 feet) and only post grouted once along the 12 meter (40 foot) bond length. Axial deflection at the 2000 kN (450 kip) design load in compression and tension was approximately 1.3 and 1.8 cm (0.5 and 0.7 inches) respectively (Figure 2a and 2b). Deflection at 120 percent of the design load (2400 kN, 540 kips) was approximately 2.3 cm (0.9 inches) in tension and 1.8 cm (0.7 inches) in compression. Residual deformation after 120 percent loading in compression and tension was approximately 0.5 cm and 0.8 cm (0.2 and 0.3 inches), respectively.

The least competent soil and bedrock occurred in the vicinity of Flower and Figueroa Streets. The alluvial soils and weathered bedrock were soft and saturated from 1.5 to 15 meters, making drilling difficult. The specialty contractor envisioned bond lengths up to 15 meters (50 feet) and/or extensive post grouting.

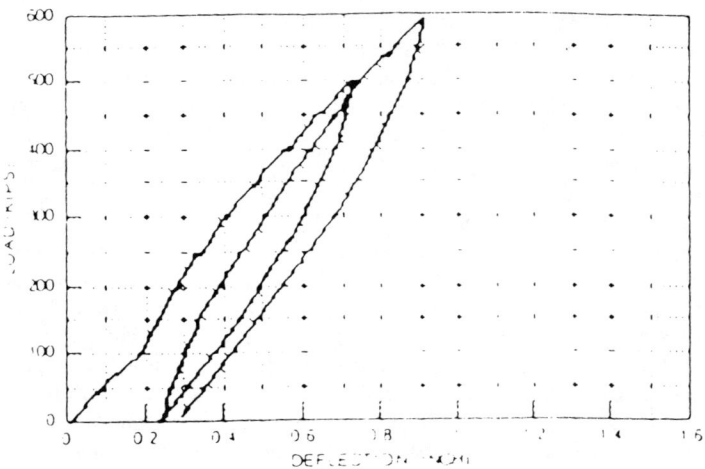

Figure 2a. Hope Street Tension Test

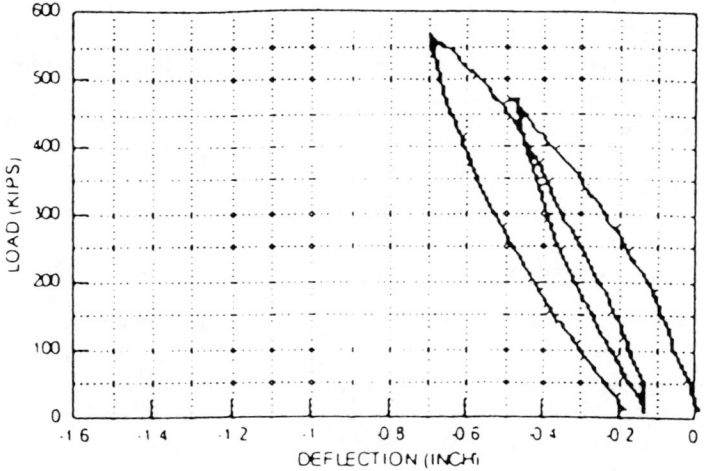

Figure 2b. Hope Street Compression Test

The Flower Street performance test pile was installed to a depth of 18 meters (60 feet) and received a primary and secondary sequence of post grouting along a 15 meter (50 foot) bond length. The contractor grouted from 0.06 to 0.14 cubic meters (2 to 5 cubic feet) of high-strength cement grout (relatively large volumes) at 1.5 meter (5 foot) intervals along the bond length in the primary and secondary passes. Secondary post grouting required 10 to 20 percent higher pressures than primary injections, but large quantities of grout were readily accepted by the soil and bedrock.

Surprisingly, testing of the Flower Street pile yielded exceptional results as shown in Figure 3a and 3b. Compression testing indicated deflection at design load of approximately 0.9 cm (0.35 inches) at 2000 kN (450 kips) and 1.3 cm (0.5 inches) at 120 percent (2400 kN, 540 kips). The contractor attempted to fail the micropile following the 120 percent tensile loading; the pile was loaded to 3000 kN (675 kips) at 25 percent increments and held at 5 minutes each. Deflection in tension was less than 3.8 cm (1.5 inches). A tensile creep test was also conducted on the pile. The creep test consisted of bringing the pile to design load in 25 percent increments, held for approximately 5 minutes, and than increasing the load in 10 percent increments and holding for 30 minutes each. The creep rates at each stage above the design load ranged from 0.03 cm (0.011 in) at 110 percent (2200 kN, 495 kips) to 0.09 cm (0.035 in) at 150 percent (3000 kN, 675 kips) of design load during the 30 minute interval.

The performance test pile near Figueroa Street was also installed in soft, saturated soil with highly weathered bedrock. The pile was installed to a depth of 15 meters (50 feet) and utilized a 12 meter (40 foot) bond length. Primary and secondary grouting was performed under similar conditions as the Flower Street Pile. Deflection at design load (2000 kN, 450 kips) was approximately 1.4 cm (0.55-inch) under compression and 1.9 cm (0.75-inch) under tension. See figures 4a and 4b. At 120 percent of design load (2400 kN, 540 kips) the deflection was 1.9 cm (0.75-inch) and 3.8 cm (1.5 inches) under compression and tension loading respectively. The 3.8 cm (1.5-inch) deflection was well within the specified 6.3 cm (2.5-inch) deflection at design load; the specialty contractor used this design for most piles on the site.

The contractor monitored strain gauge data on a real time basis during testing of the Hope and Figueroa Street piles. This data showed the load distribution along the pile. Pairs of vibrating wire strain gauges were welded to the outside of the piling at 3 meter (10 foot) intervals approximately 180 degrees apart. Figure 5 shows the average of the paired strain gauges response with respect to loading during the second cycle of a typical tensile test.

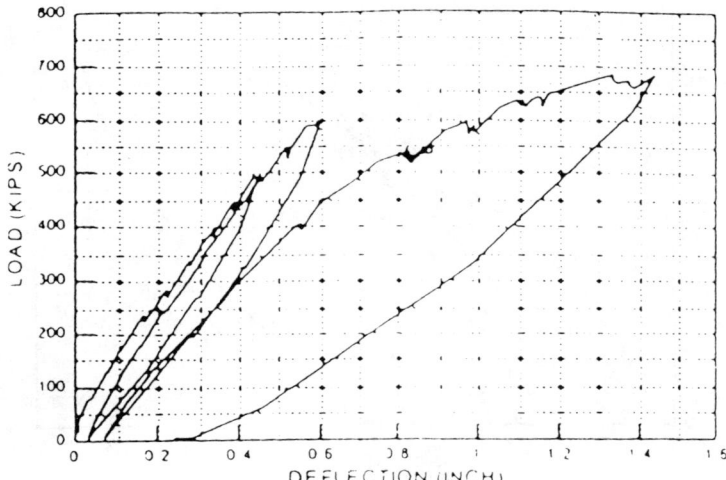

Figure 3a. Flower St. Tension Test

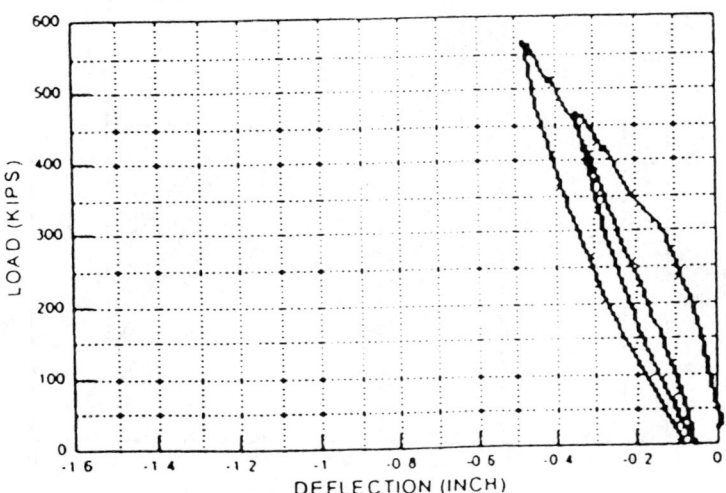

Figure 3b. Flower St. Compression Test

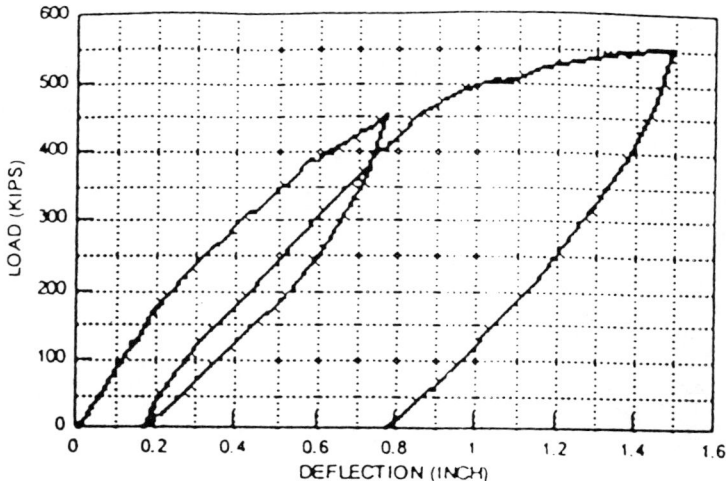

Figure 4a. Figureroa St. Tension Test

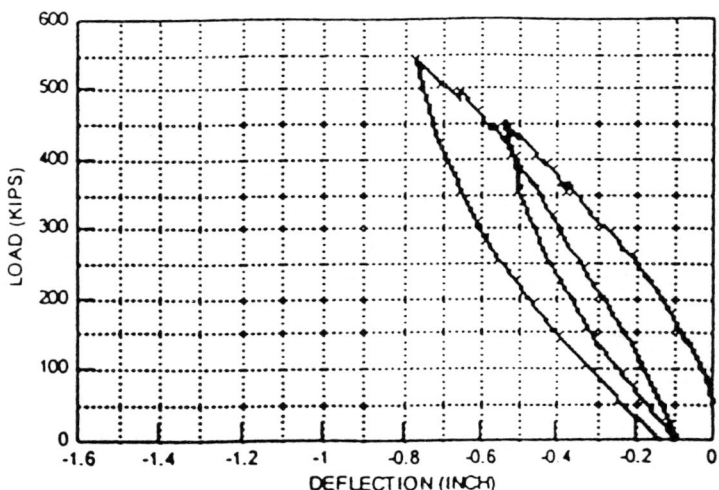

Figure 4b. Figureroa St. Compression Test

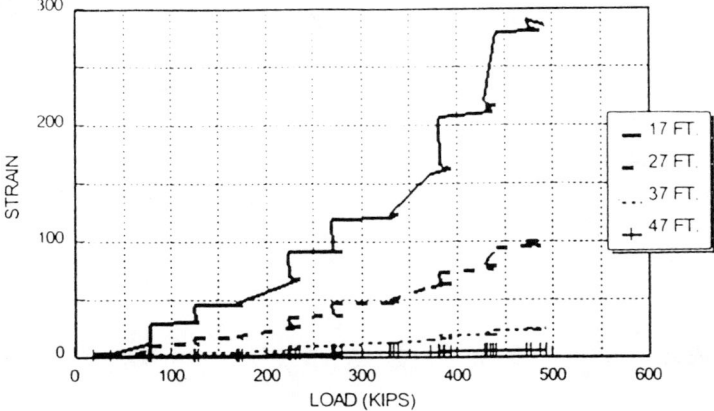

Figure 5. Typical strain gauge response during second cycle (540 kip) tensile loading.

Riverside Drive Project (1995)

The Riverside Drive project was also part of the City of Los Angeles seismic retrofit program. This project required 14 tension/compression piles to hold down/support 3 bents underneath Riverside Drive near the intersection of Interstate 5 and Interstate 10. The specialty contractor installed the 18 cm (7-inch) diameter, approximately 18 meter (60 feet) piles in a similar fashion to the 4th Street Viaduct project above. One non-production pile was performance tested in both tension and compression to 200 percent of the design load 2000 kN (450 kips). Then, one randomly chosen production pile from each bent was tested to 100% of design load (1000 kN, 225 kips) in tension. Specifications required 6.3 cm (2.5 inches) maximum displacement.

Pile Installation

Site geology consisted of loose to medium dense silty sands with layers of gravel and small cobbles. As a result of the cohesionless soil, polymer drilling fluid was used to keep the hole open. However, some of the borings encountered wood and other existing construction debris. To resolve this problem, the contractor drilled the holes with 30.5 cm (12-inch) hollow stem auger, cement grouted the holes, and re-drilled the hole the following day. This resolved the majority of problem borings, except for one boring which required casing.

324   SOIL IMPROVEMENT FOR BIG DIGS

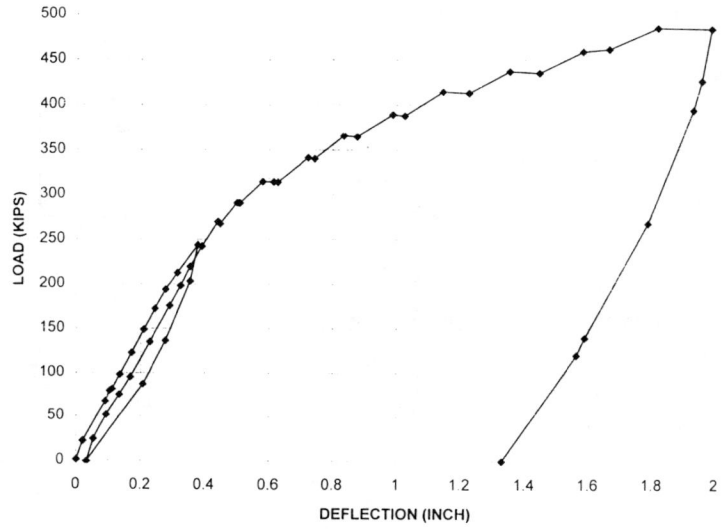

Figure 6a. Riverside Drive Tension Test

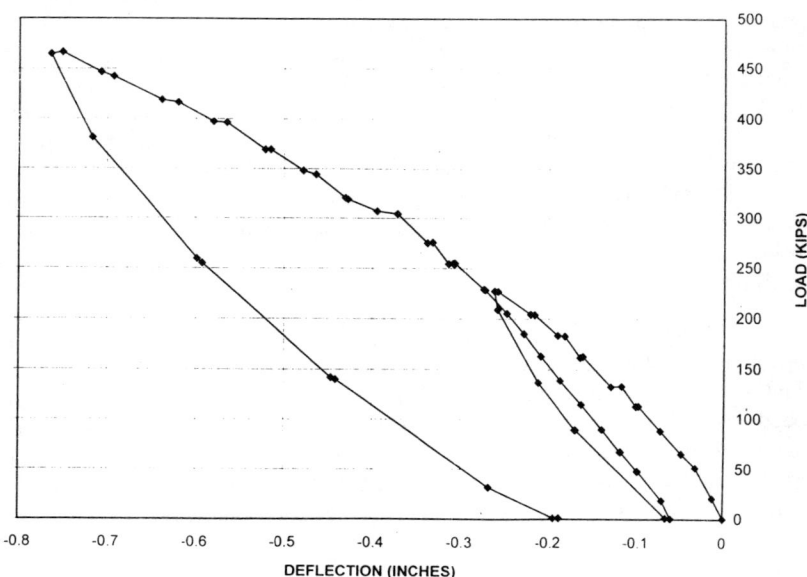

Figure 6b. Riverside Drive Compression Test

Performance Testing

The non-production performance test was performed to 200% of design load, in 5 minute intervals with 5% load increments for tension and compression. Real time instrumentation included vibrating wire strain gauges at 12, 15, and 18 meters (40, 50, and 60 feet), a hydraulic pressure transducer for the jack, and 2 linear potentiometers to measure vertical displacement. The tension test indicated 1 cm (0.38-inch) displacement at 1000 kN (225 kips), and 4.2 cm (1.65 inches) at 2000 kN (450 kips), see Figure 6a. Strain gauges at 2000 kN (450 kips) indicated 1069, 296, and 45 cm/cm for depths of 12, 15, and 18 meters (40, 50, and 60 feet) respectively.

The compression test indicated 0.7 cm (0.26-inch) displacement at 1000 kN (225 kips), and 1.9 cm (0.75 inches) at 2000 kN (450 kips), see Figure 6b. Strain gauges indicated 940, 261, and 13 cm/cm for depths of 12, 15, and 18 meters (40, 50, and 60 feet) respectively. It is reasonable to note that strains in tension were higher than that in compression (especially at depth) due to the lack of end bearing resistance.

One randomly chosen pile from each of the three bents was performance tested to 100% of design load (1000 kN, 225 kips) in tension. Three dial gauges were placed at 120 degree angles on the top of the pile. The average of the three gauges resulted in displacements of 1.0, 1.1, and 1.2 cm (0.38, 0.45, and 0.47 inches) for the three bents, well within the required maximum deflection.

Conclusions

Micropiles provided a cost effective means to resist large tension and compression loads for seismic retrofit of existing freeway foundations. Deflection of production piles at 100% of design load was approximately 20% of the maximum allowed. Various drilling methods including down the hole hammer, hollow stem auger, polymer drilling fluid, and grouting and re-drilling resolved difficult drilling conditions consisting of clays, cohesionless silty sands with gravel and cobbles, and existing belled caissons. Performance testing in tension only reduced project costs.

Acknowledgments

Geotechnical Engineering was performed by Earth Mechanics, Inc. (Ignatius Po Lam), Fountain Valley, California.

Design and Construction of the
Runway 13-31 Overrun Area
at LaGuardia Airport

By: Raymond E. Sandiford [1], M.ASCE, Arnold Aronowitz [2], F.ASCE,
and Stephen Law [3], M.ASCE

ABSTRACT

An earthen embankment was constructed to provide an overrun area for Runway 13-31 at LaGuardia Airport in Queens County, New York. This embankment was constructed on a twenty-three meter thick layer of normally consolidated clay, using high-strength, geotextile reinforcement and vertical strip drains for soil improvement. The stability of the embankment was of paramount concern due to its proximity to an active federal shipping channel and environmentally sensitive areas. Instruments were installed in the embankment and soft subsoil to assess the progression of subsoil improvement and deformation. The instrumentation included settlement plates, slope inclinometers, spider magnets, piezometers and survey range markers. These instruments were analyzed at various stages of the construction to assess the stability of the embankment. The project was completed successfully and on schedule with embankment settlement of approximately five meters.

---

1 Chief Geotechnical Engineer, Port Authority of N.Y. and N.J.
2 Chief Geotechnical Engineer (Retired), Port Authority of N.Y. and N.J.
3 Geotechnical Engineer, Port Authority of N.Y. and N. J.

INTRODUCTION

This paper presents the results of the field performance of an embankment constructed over a thick stratum of highly compressible soil. The embankment was constructed to serve as a new overrun area at the east end of Runway 13-31 at New York's LaGuardia Airport (Figure 1). The methods used were selected to meet unique constraints imposed by the site. They included an adjacent shipping channel, an aircraft runway, a topography consisting of an environmentally sensitive inter-tidal mudflat, and very soft soils that could not support conventional construction equipment. These constraints were mitigated by placing sand hydraulically on to a high-strength geotextile in controlled lifts and installing vertical strip drains to accelerate the consolidation of the underlying compressible soils. These procedures made it possible to fill at all times, even when the runway was in operation and during high tide.

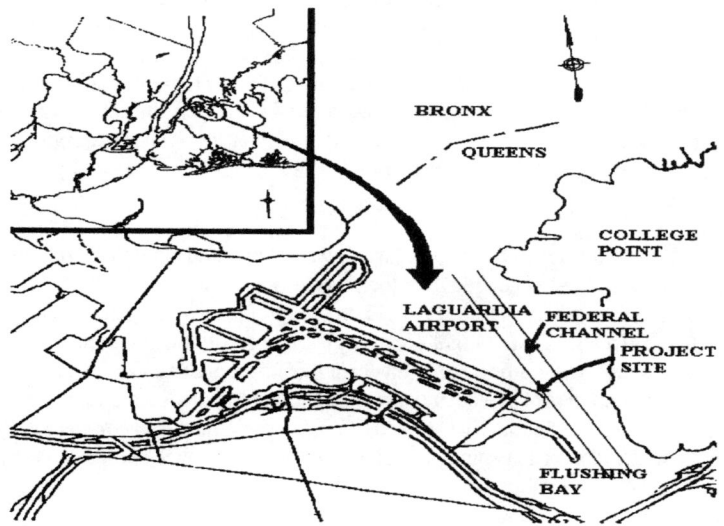

Figure 1 - Location Plan

The overrun construction progressed in stages such that the embankment height corresponded to the appropriate level of strength gain in the underlying soils. The embankment construction staging was initially designed using semi-conventional consolidation and stability calculations. In addition, finite difference models (FDM), performed with FLAC© were used to estimate soil stresses and patterns of soil deformations for selected construction stages and to aid in the selection and placement of instrumentation. One FDM model was used to model the geotextile reinforced soil and to predict the soil strains and creeps. Another FDM model was used to predict pore-pressure dissipation and the resulting consolidation settlement.

Extensive instrumentation was used to assess the various stages of construction. Initial filling was monitored using simple range markers and slope inclinometers. Subsequent consolidation and filling processes were monitored using arrays of pore pressure transducers, spider magnet strain devices and surface settlement plates.

SUBSURFACE CONDITIONS

Site History

LaGuardia Airport is situated on the northern boundary of Queens County between Flushing and Bowery Bays. Prior to the airport development, the area was a shallow bay underlain by soft clay soils. The original airport was constructed as a Works Program Administration (WPA) program. Land area was created by end-dumping primarily incinerated refuse ash on to the bay bottom. The airport has been expanded gradually over its 60 year history. Except for two structurally supported runway extensions, it has grown by extensive filling.

The subsoil conditions of the airport have always presented challenges for foundation engineers. They relate to the thickness and extent of the soft clays and the problems inherent with them. Even though the airport was originally constructed at a significant height above high water level, it has settled in areas to elevations below the storm surge levels in the adjacent bays. To prevent flooding, earthen dikes have been constructed and are maintained along the airport shorelines. The dikes have been releveled periodically to compensate for ongoing settlement. The general settlement of the airport exceeds four meters.

Stratigraphy

The overrun area is located at the east end of Runway 13-31 and extends approximately 150 meters into Flushing Bay. Figure 2 presents a plan of the overrun area and the locations of the soil borings. The site is underlain by a 23 to 25 meter thick stratum of soft, normally consolidated clays. The clay stratum overlies an equally thick sequence of glacial deposits, which consist of dense sands and over-consolidated clays and silts. Bedrock at this site is at a depth in excess of 60 meters (Figure 3).

SOIL IMPROVEMENT FOR BIG DIGS

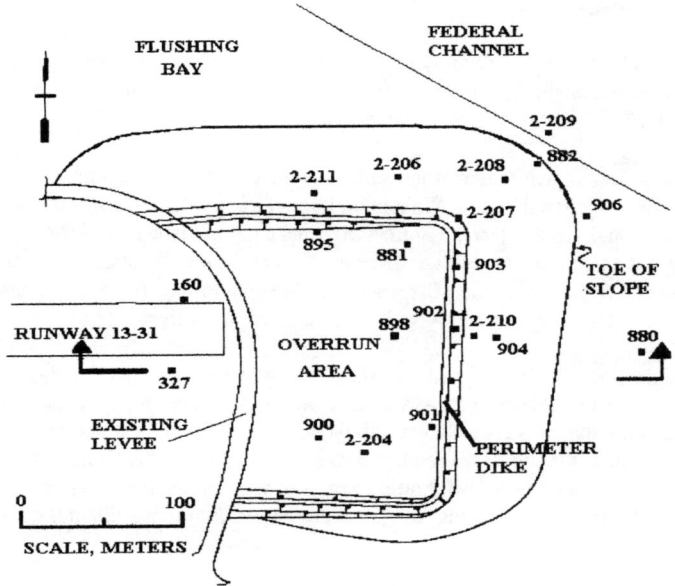

Figure 2 - Boring Location Plan

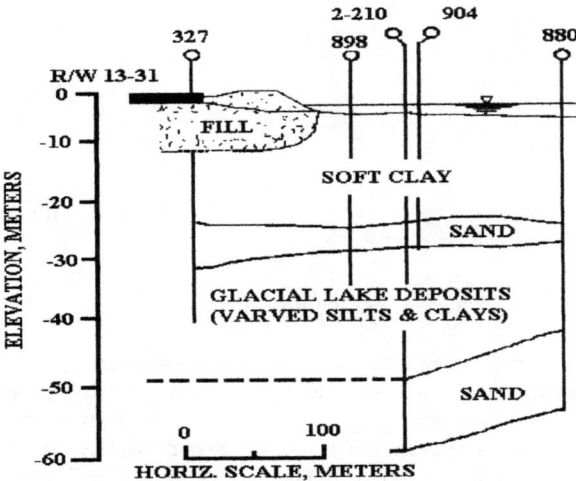

Figure 3 - Geologic Section

Soil Properties

The soft clays were the primary focus of the design effort. Fortunately, the clays have well established engineering properties that have been correlated with laboratory and observed behavior over decades.

The strength characteristics of the soft clays were established from laboratory tests, which included unconsolidated undrained triaxial tests (UU), laboratory vane tests and torvane tests. Figure 4 presents a comparison of UU shear strength data and preconsolidation stress. This relationship suggests a strength profile where $Cu = 1.2$ kPa $+ 0.25 Po$ (with Po equal to the effective overburden pressure). This strength relationship was adopted for design.

The consolidation characteristics of the soft clay were determined from laboratory consolidation tests as well as from field observations. A regional relationship, that was previously established, confirmed the test parameters determined at this site. This relationship indicates that C_c, the Compression Index, is a function of the void ratio of the soil and is related to the stress history. Similar relationships have been developed for other soils in the Port District (Ref. 4).

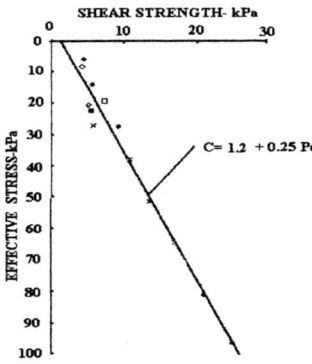

Figure 4 - Strength Profile

In addition to the aforementioned tests, two series of anisotropically consolidated undrained direct simple shear tests [CKoU DSS] were performed previously on these soils for another project [Ref. 1]. The specimens were consolidated to twice the overburden in-situ pressure to remove effects of preconsolidation and sampling disturbance. One series of tests [Fig. 5] were performed at a constant rate of strain. The test results indicated that the normalized shear strength ranged from 0.24 to 0.30 σ'_{vo}.

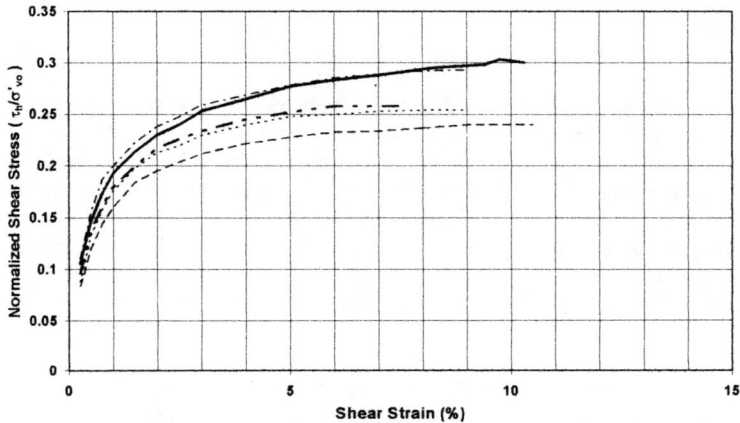

Figure 5 - Normalized Constant Rate Tests [CKoU DSS]

The other series of tests were stress controlled. Specific load increments were applied and maintained for three days to determine the creep properties of the clay. The test results indicated a range in the creep coefficient (e.g. strain per log cycle) of 0.0025 at a stress ratio (S/Sult) of approximately 45% to a value of 0.0215 at a stress ratio of approximately 85%. At stress ratios in excess of 65%, the clay started to creep at accelerating rates (Figure 6).

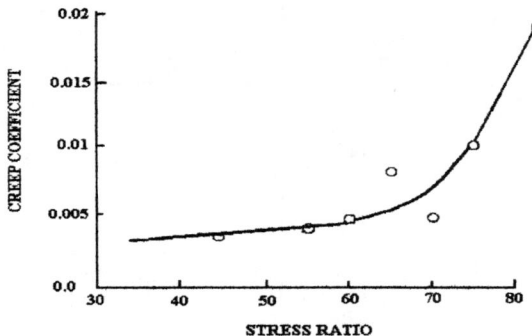

Figure 6 - Stress Controlled Tests

DESIGN

The embankment design was subdivided into two series of analyses, each representing the two major phases of the construction. The initial phase consisted of an evaluation of the stability of the embankment as fill was placed on the geotextile. This was accomplished using simplified wedge-type analyses to evaluate the fill placement. These analyses established the allowable lift thickness and setbacks. In addition, these analyses established the controlling tensile strength of the geotextile seams. The resulting filling criteria limited lifts to one meter and restricted side slopes to no steeper than 1 vertical to 20 horizontal.

Subsequent phases of the construction were evaluated using STABL analyses with progressive strength gains due to consolidation. These analyses established the required "waiting" periods for the embankment construction.

In addition to these semi-conventional analyses, finite difference analyses (FDM) were performed to simulate the subsoil behavior. Although the gross stability analyses indicated satisfactory factors of safety, there was a particular concern that localized overstressing of the subsoils could produce progressive failure and result in undesirable lateral deformations. Figure 7 presented the results of the FDM model for the phase one (initial) filling condition.

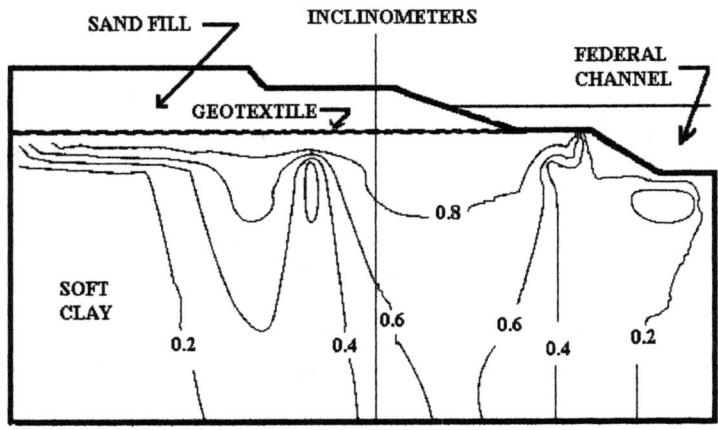

Figure 7 - Stress Ratio Contours $\{(\sigma_1-\sigma_3)/2Su\}$ (FLAC$^{©}$)

The stress levels at this stage of construction locally exceeded 65% of the clays ultimate shear strength. This suggested progressive creep movement. The FDM model was used to predict both the lateral displacements and creep movements at various locations within and beyond the embankment. Figure 8 presents the results for the crest of the embankment where stresses levels were the maximum.

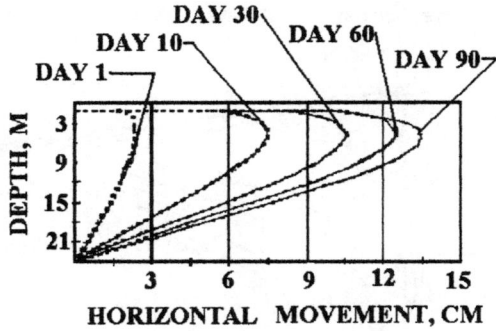

Figure 8 - Predicted Horizontal Creep (FLAC©)

FIELD PERFORMANCE

The performance of the embankment and subsoil were monitored throughout the fill placement. The primary concern was excessive horizontal movements of the subsoil. Lateral movements were measured initially using simple range markers. The range markers consisted of poles set in the soft bay mud and fitted with prismatic survey reflectors. The range markers indicated only modest horizontal displacements, on the order of 0.1 to 0.2 meters, during the initial hydraulic filling. This compared well with historic movements associated with end-dump filling that typically resulted in movements in the range of 2 to 4 meters.

Slope inclinometers, spider magnets and piezometers were installed through the embankment, once the fill reached above sea level,. The inclinometers were used to measure lateral movements and rates of movement. Figure 9 presents the typical lateral movement that occurred outboard of the new perimeter dike. The rates of movement were monitored to ensure that they stabilized prior to the placement of additional fill. A maximum rate criterion was established based on measurements taken during previous fill projects where it was observed that movement rates in excess of 0.75 cm/day indicated the onset of significant subsoil displacement. For the overrun project a movement rate of 0.25 cm/day was established as a criterion to halt fill operations until the soil movement rate stabilized. Figure 10 presents typical lateral movement rate data with the limit criterion shown.

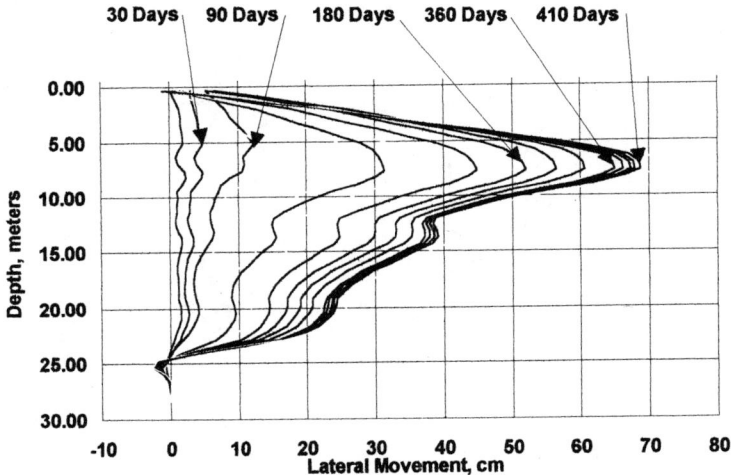

Figure 9 - Typical Perimeter Inclinometer

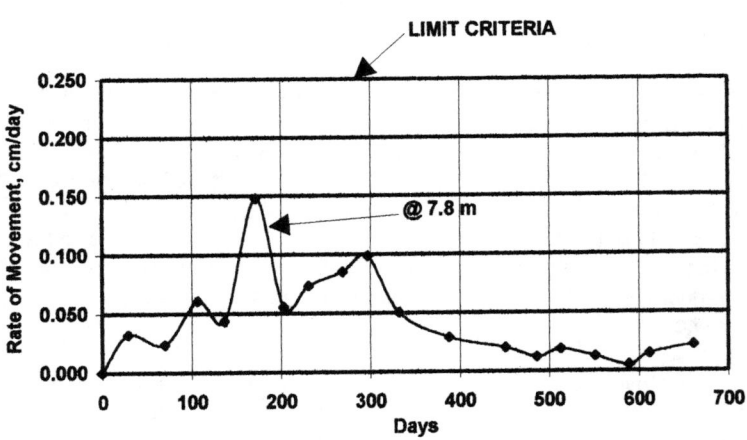

Figure 10 - Lateral Movement Rates

Spider magnets were installed to measure settlement and the results were used to confirm the degree of consolidation through the clay profile. Piezometer arrays were installed at three test locations. The tight drain spacing (1.2 meters) and the probable variation of their vertical alignment made it impractical to rely on individual piezometers. Multiple piezometers were installed a various depths to obtain a statistical measure of the pore pressure dissipation. Figures 11 presents data from one of the test areas. As has been noted on similar projects (REF. 4), the pore pressure dissipation lags the comparable settlement.

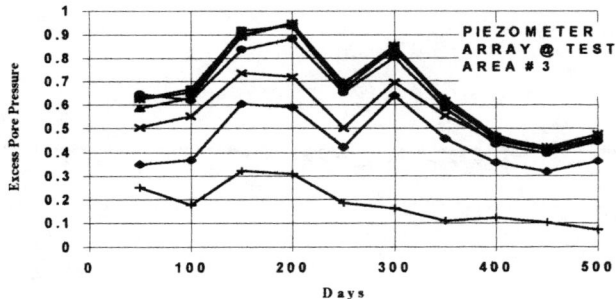

Figure 11 - Pore Pressure Dissipation

Numerous settlement plates were installed on the embankment. Figure 12 presents settlement plate data measured at a typical location along the perimeter dike.

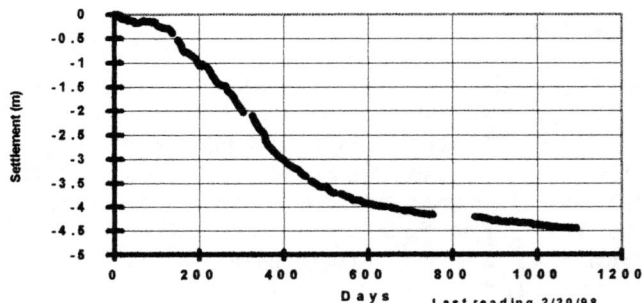

Figure 12 - Dike Settlement

Settlement plates proved to be quite useful for assessing the stability of the embankment. This was accomplished by calculating the theoretical consolidation (Ref. 2 and 5) for both vertical strip drain and horizontal boundary drainage. Strain isochromes were developed from these theories and converted into stress isochromes. Then, a relationship was developed relating surface settlement to strength gain (Figure 13). These settlement relationships were used in conjunction with the horizontal strain rate data to establish the rate at which embankment filling and dike construction could progress. These results were generally confirmed by strain measurements obtained from spider magnets.

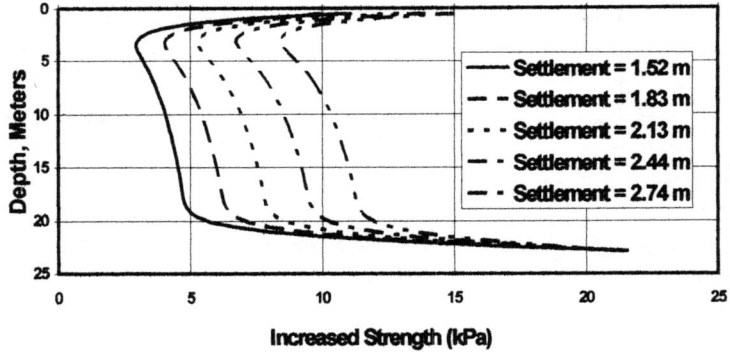

Figure 13 - Strength Gain Profiles

CONSTRUCTION

Two, five hectare sheets of geotextile were used for reinforcement. These were assembled on and then placed from barges. The tidal variation at the airport is typically two meters. This permitted barges with sufficient draft at high water to come within 15 meters of the shoreline and facilitated the placement of the geotextile by allowing the leading edge of the geotextile sheets to be pulled to shore and secured. Then, the barges were slowly pulled from shore with the geotextile unfurling into the water and progressively sinking onto the bay bottom. The unfurling process took approximately 90 minutes per sheet. The geotextile sheets were secured on the bay bottom with sandbags

A trailing suction hopper dredge was used to mine and transport sand from New York Harbor's Lower Bay to the site and then pump the sand onto the geotextile. The dredge was able to complete a full cycle of mining, transporting (45 KM)

The dredge was able to complete a full cycle of mining, transporting (45 KM) and pump-out in approximately 10 hours. A total of 400,000 cubic meters of sand were placed using this method.

Approximately two million linear feet of strip drains were installed on this project. These drains were installed at night over a period of approximately four months using three rigs. The staged filling process was accomplished over a one and one-half year period. The overrun dikes have been constructed with approximately one meter of freeboard to allow for future settlement.

CONCLUSIONS

The embankment construction was successfully completed on schedule without excessive subsoil movement or interference with the adjacent shipping channel. Subsoil performance was monitored with varying degrees of success using settlement plates, piezometers and spider magnets. Data obtained from the various instruments were used to assess specific aspects of the subsoil performance and, taken as a whole, were used to make construction staging decisions. The spider magnet and piezometer data required additional study and judgment to evaluate the soil behavior and were not as useful during construction as the slope inclinometers and settlement plates. However, these data hold promise of providing a better understanding of the clay's behavior.

REFERENCES

1. LaGuardia Airport - Geotechnical Evaluation of the East End Storm Protection System, June 1985, Engineering Department, The Port Authority of New York and New Jersey.

2. Davis, E.H. and Raymond G.P., 1965, A Non-Linear Theory of Consolidation, *Geotechnique*, London

3. Barron, R.A., 1948, "Consolidation of Fine Grained Soils by Drain Wells", Transactions, ASCE, Vol 113.

4. Kapp, M.S., York, D.L., Aronowitz, A. and Sitomer, H., 1966, " Construction on Marshland Deposits: Treatment and Results, HIGHWAY RESEARCH BOARD Number 133.

ACKNOWLEDGMENTS

The Port Authority would like to thank Jaw-Nan Wang and Naresh Samtani of Parsons Brinkerhoff Quade and Douglas, Inc. for performing the FLAC© analyses associated with this project; Dennis Cavaliere and Thomas Spero of the Port Authority's Materials Division for instrumentation installation and Andrew Leung from TAMS Consultants Inc. for collecting and reducing the instrumentation data.

Subject Index

Page number refers to the first page of paper

Bearing capacity, 59
Boring, 190
Boston, 161

Case reports, 177
Cements, 27
Centrifuge model, 239
Chemical reactions, 135
Chemicals, 285
Clays, 72, 161, 298, 326
Clogging, 298
Columns, 59, 111, 122
Composite structures, 27
Consolidation, 161, 326
Core walls, 41

Database management systems, 259
Deep foundations, 190
Deep soil mixing, 1, 27, 41, 59, 72, 84, 96, 111, 122, 135
Design, 259
Dewatering, 149
Drilling, 72
Dynamic response, 239

Earthquake engineering, 259
Earthquake loads, 239
Electrokinetics, 285
Embankments, 326
Equipment, 1, 111
Excavation, 84, 149, 202, 214, 226

Foundation settlement, 202
Freeze-thaw cycle, 161
Freezing, 149
Frozen soils, 149, 161

Geotextiles, 177, 298, 326
Granular media, 226
Gravity walls, 27

Installation, 190
Instrumentation, 298
Ion exchange, 285

Jet grouting, 72, 111

Kaolin, 285

Materials, 1
Micro piles, 239, 313
Monitoring, 226

Organic matter, 135

Phosphate deposits, 298
Pile foundations, 72
Pile groups, 239
Piles, 59, 190

Rehabilitation, 202
Reinforcement, 177, 326
Residual soils, 214
Retaining walls, 41, 84, 226
Retention, 41
Retrofitting, 313

Seismic design, 313
Seismic response, 259
Settlement analysis, 177
Shear strength, 161, 285
Shoring, 202
Slope stability, 177
Soft soils, 96, 122

Soil cement, 41, 59, 72, 96, 111, 135
Soil improvement, 1, 96, 111, 122, 149, 259, 326
Soil nailing, 214, 226, 259
Soil stabilization, 96, 122, 135, 149
Soil treatment, 1
Soil-pile interaction, 239
Stability, 84
Stabilization, 27, 285
Standard penetration tests, 190
Strength, 59, 135
Subways, 214

Supports, 214
Surcharge, 298

Temporary structures, 27
Tunnel construction, 122

Underpinning, 202

Vertical drains, 326
Viaducts, 313

Wick drains, 298
Wooden piles, 177

Author Index
Page number refers to the first page of paper

Aronowitz, Arnold, 326
Arrellaga, J. A., 72

Bahner, E. W., 27, 41
Bell, Roy A., 59
Benslimane, Aomar, 239, 259
Bruce, Donald A., 1
Bruce, Mary Ellen C., 1

Daly, Jim, 214
Das, Prabir K., P.E., 122
DiMillio, Albert F., 1
Donohoe, John F., 149
Drabkin, Serguey, 239

Edil, Tuncer B., 135

Ferris, Jeanine, 214

Gale, Robert J., 285
Greene, Christopher, 161
Gularte, Francis B., 313
Gularte, Greg G., 313

Hampton, Melanie B., 135
Hanna, Sherif, 239, 259
Heung, Wing, 177, 298

Isenhower, W. M., 72

Johnson, J. O., 72
Juran, Ilan, 239, 259

Khalil, Mike, 214
Kuo, Ching L., 177, 298

Law, Stephen, 326
Levy, Ofer, 259
Li, Jianchao, 190

Macari, Emir Jose, 84
Maishman, Derek, P.E., 149
Maswoswe, Justice J. G., P.E., 122
Mitchell, J. K., 27
Moriwaki, Y., 27

Naguib, A. M., 41
Nicholson, P. J., 27

Ozkan, Senda, 285

Qubain, Bashar S., 190, 202

Rhodes, Mark, 214
Ro, Kwang, 226
Roberts, John, 177, 298

Sandiford, Raymond E., 326
Schmall, Paul C., P.E., 149
Seals, Roger K., 285
Seksinsky, Eric J., 202
Shao, Yong, 84
Soliman, Nassef, 226
Swan, Christopher, 161

Taki, Osamu, 59
Taylor, Gary E., 313
Tejidor, Francisco J., 177

Wang, S. T., 72

Yagihashi, Jack N., 96, 111
Yang, David S., 96, 111

Yin, Edward Y. P., P.E., 122
Yoshizawa, Steve S., 96, 111

Zhang, Chunming, 84